Troubleshooting
Electronic Equipment

Author's Profile

 Mr. R. S. Khandpur is currently Director General, Pushpa Gujral Science City, Kapurthala, Punjab.

Prior to this, he was Director General, Centre for Electronics Design and Technology of India (CEDTI), an autonomous Scientific Society of the Ministry of Communication and Information Technology, Government of India. He was the founder Director of CEDTI, Mohali, which is the first ISO-9002 certified organization of the Ministry of Information Technology.

Mr. Khandpur is the recipient of the 1989 Independence Day Award by National Research and Development Corporation and IETE (Institution of Electronics and Telecommunication Engineers) for outstanding contribution towards the development of the electronics industry. He is Member, Board of Governors, Punjab Technical University; Director, Board of Directors, Electronics Corporation of Punjab; AICTE Distinguished Visiting Professor and Member; Vision Group on IT set up by the Punjab Government.

He has served as a scientist for 24 years in CSIO, Chandigarh, a constituent laboratory of Council of Scientific and Industrial Research (CSIR), Head of Medical Instruments Division (1975-1989) and Head of Electronics Division (1986-1989). He was the Project Coordinator for India's first Medical Linear Accelerator Machine for cancer Treatment, installed at PGI, Chandigarh in 1989.

Mr. Khandpur is a Member IEEE (Institution of Electronics and Electrical Engineers), USA; fellow of IETE (Institution of Electronics and Telecommunication Engineers) and Member, Society for Engineering in Medicine and Biology, USA.

He has over 37 years of experience in R&D, technology development, technology transfer, education and training, consultancy and management at national and international levels.

Mr. Khandpur holds 6 patents of innovative designs, has authored 7 books and published over 60 research and review papers

Contents

Preface

It is a great pleasure for me to present my book *Troubleshooting Electronic Equipment*.

The electronic industry has undergone a revolution in recent years. Technology has altered the way electronic products are designed, fabricated, and maintained. Accordingly, the personnel and the service bench need to keep pace with these developments. This has created the need for a book that keeps in mind the technology trends and the growing importance of maintenance engineering.

Traditionally maintenance engineering has not been a preferred career of most of the bright students. They find design and marketing jobs more challenging. It is also generally believed that the former is intellectually more challenging while the latter requires personality related skills in the competitive market. However, with the liberalization of economy and globalization of activities in the country, the field of maintenance engineering is expanding quite fast. It has been realized that at present, more than ever before, the reliability of a system can have a direct and immediate effect on the profitability of an operation or the efficiency of a plant, hospital, or a service organization. Thus, maintenance engineering today offers equally good challenging opportunities to the service engineers as other preferred professional fields.

Technology is an ever moving target. Continuing developments in hardware and software provide new features and increased performance of industrial products. Such developments, however, place increased demands on the engineering/maintenance departments of a facility. It is well understood now that the days of troubleshooting a piece of equipment only with an oscilloscope and multimeter and a vague idea about the hardware are over. Today, unless you have a detailed service and maintenance manual and the right type of test equipment, you may be far from any success in troubleshooting and repairing of equipment. The test equipments of the 1980s are now relics of the past.

Nevertheless, most of the equipment problems can still be located with a digital multimeter and oscilloscope if provided ample time and effort. As it costs money, time, and effort few managers are ready to accept it. The proper test equipment therefore, becomes a necessity if one has to work with current technology equipment.

Two fundamental changes have taken place in recent years. One is that today's equipment contain tiny components whose leads, terminals, and the tracks on the printed circuit boards are sometimes not visible to the naked eye. These components are jammed on a small surface of the circuit board area. This tight component packaging makes repair of the printed wiring boards difficult.

Secondly, signal-flow in modern digital circuits is not sequential. So the circuit path cannot be conveniently followed. The complex digital circuitry is so interconnected with various circuit blocks that the interrelated circuitry makes the repair up to the component level virtually impossible. Equipment today is just too complex—electrically and mechanically. The sophistication of hardware has ushered in a new era in equipment maintenance—i.e., repair by replacement. Maintenance professionals today must, therefore, take a systems approach to troubleshooting hardware.

"Repair by replacement" does not imply that you need less competent people. As technology drives equipment design forward, maintenance difficulties will continue to increase. Such problems can be handled by better technician training. It is well-established that a comprehensive training program can prevent equipment failure which impacts productivity, worker morale, and financial returns.

The goal of maintenance engineering should be to ensure top quality performance from each piece of hardware. On their own these objectives do not just get achieved. They are achieved by establishing and practicing a good maintenance management system. It is with this view that a separate chapter on `Maintenance Management' has been included.

The miniaturization of electronic equipment like cell phones, camcorders, and digital diaries, etc., has taken place primarily with the introduction and advancements in surface mount technology. Repair and reworking on SMD based printed circuit boards, have been included in the book.

Preventive maintenance is an essential component of a good maintenance management system and a requirement for reliability. Therefore, it needs to be given its due importance. Every piece of equipment requires a specific type of preventive maintenance action plan and therefore, it is advised to follow the manufacturer's recommendations too. Fundamentals of preventive maintenance have been explained in one chapter along with suggested preventive maintenance steps for a personal computer.

It is hoped that the book will enable the service engineers to do their job more efficiently and that students will be well served by the practice-oriented approach of the book and develop interest in pursuing maintenance engineering as a productive career.

R S. KHANDPUR

Troubleshooting
Electronic Equipment

1

Reliability Aspects of Electronic Equipment

1.1 ELECTRONICS TODAY

Electronics today permeates all walks of life. Few areas of human endeavour have remained unaffected by electronics and many are even dominated by the seemingly mysterious domain of this technology. We have all shared the excitement of outer-space exploration and the moon landings of the astronauts through the neat magical display of pictures on colour television. Today we have the opportunity to work with electronic computers which are silently operating at fantastic speeds and solving complex problems. Modern medicine would be seriously handicapped without the availability of electronic instruments and aids as electronic equipment is extensively employed for diagnostic, monitoring and therapeutic applications. Recent years have seen the emergence of several new non-invasive medical imaging systems like ultrasonic scanners, computerized axial tomography and nuclear magnetic resonance systems which are basically dependent upon the ability of modern computers to handle enormous amount of data needed for picture reconstruction.

Whether it is modern warfare or a research laboratory, agriculture or meteorology, the process control industry or telecommunications, the all–pervading spirit underneath all these operations is electronics. It is difficult to imagine our current lifestyle without the use of electronics.

The field of electronics has shown rapid advances during the last decade, particularly in terms of enormous developments in micro-electronics. Until 1948, the electron tube (Fig. 1.1) was the key component in almost every type of electronic equipment. However, this situation changed suddenly with the invention of the transistor (Fig. 1.2) by a team of three scientists led by William Shockley. Transistors have displaced a majority of electron tubes because a transistor is smaller and more efficient than a tube. The chief limitation of a transistor, however, has been its inability to handle large amounts of electrical power so far, particularly at high frequency, wherein vacuum tubes are still being used.

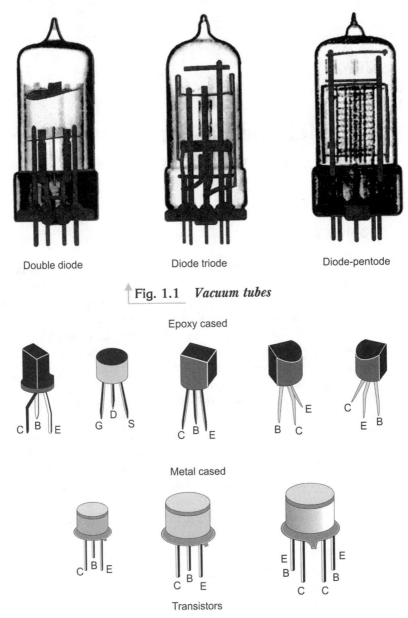

Double diode   Diode triode   Diode-pentode

Fig. 1.1   *Vacuum tubes*

Epoxy cased

Metal cased

Transistors

Fig. 1.2   *Transistors*

The integrated circuits technology evolved in the late 1950s. It facilitated the incorporation of a large number of electronic components on a single chip (Fig. 1.3). The evolution of the integration process offered enormous potential in matters of their utility and the subsequent developments proceeded at a rather spectacular rate. The number of transistors which could be made on a given area of silicon chip roughly doubled each year and in the meantime, semiconductor manufacturers were able to perfect techniques of producing high density chips with high yield. Integrated circuits have now been developed with 170 million transistors in a chip, which measures just 400 mm^2.

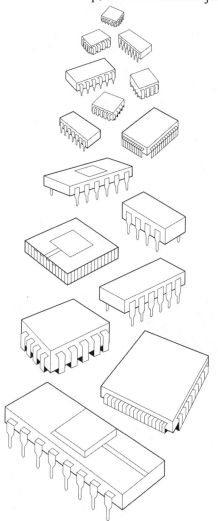

Fig. 1.3 *Typical integrated circuits*

The most exciting development in electronics took place in 1971 when Intel, a semiconductor device manufacturing company of the US released the first microprocessor—a

4-bit device. This was followed by 8-bit versions from several semiconductor manufacturers and today, we have microprocessors in 16 bit, 32 bit and 64-bit configurations. The microprocessor forms the central processing unit (CPU) of the present-day personal computers (PCs). Not only this, we now have single chip computers, which contain both processor and memory devices that are adequate for a host of applications in the same integrated circuit chip. Microprocessors have evolved at a dramatic pace in terms of numbers, technology, power, functionality and applications.

As a logical consequence of the developments in semiconductor technology, modern equipment tends to include some of the above referred components and devices. The handling of some of these devices needs special skills and precautions; while the troubleshooting techniques of this equipment are heavily based on a good knowledge of the circuits and systems and the use of modern sophisticated test and measuring instruments. However, the repair, servicing and maintenance of modern equipment continues to pose a problem for the engineering and technical staff. The managers face a difficult task in ensuring the smooth running of equipment in professional situations, necessitating the development of a sound maintenance policy for electronic equipment.

1.2   RELIABILITY ASPECTS OF ELECTRONIC EQUIPMENT

Reliability is defined as 'the ability of an item to perform a required function (without failure) under stated conditions for a specified period of time'. For example, if we say that the reliability of a microcomputer is 98% over a 200-hour period, at an ambient temperature of 25°C, this means that the probability of its satisfactory operation, without any failure is 98% during a 200-hour period.

For completely stating the reliability of an item, which may be a component, a piece of equipment or a system, it is necessary to specify the entire environment, in terms of the temperature and its variation, the climatic conditions such as humidity, dusty atmosphere and icy formation, and the mechanical conditions such as the amplitude and frequency of vibration and electrical and electro-magnetic conditions. The specified level of reliability cannot be expected if the item operates beyond the stated environmental conditions.

The reliability requirement for an equipment depends mostly on its intended application. For example, the reliability of communication equipment carried by satellites must be extremely high for it is generally not practical to carry out repairs in this case. On the other hand, there are situations wherein faults can be repaired, yet a very high degree of reliability is still demanded. Examples of such items are aircraft electronics, medical electronics, life support systems and process control instrumentation in an industry where the loss in production through a systems failure could be very high.

The development of miniature solid state electronic components, particularly LSI (large scale integration) and VLSI (very large scale integration) integrated circuits, has greatly improved the reliability of systems in which they are used. This is due to the complete isolation of the circuits from environmental stresses such as humidity, temperature, shock and the absence of soldered or plug inter-connections, most of which are made inside the encapsulation. This trend is also reflected in the development of highly reliable discrete components, mechanical parts and printed circuit boards. In spite of all these developments, equipment failures do take place and the items stop performing their required functions.

A piece of equipment or a system may break down due to a faulty component. Extensive research has been conducted to study component failure mechanism and failure phenomena. Experience and the vast body of data gathered by researchers and practitioners have shown that component failure rates follow one of the two patterns shown in Figs 1.4 and 1.5.

The traditional bathtub curve (Fig. 1.4) indicates component life in three stages. During the first stage, the failure rate begins at a high level and decreases rapidly with time. This stage is known as the infant mortality period (IFR) and it has a decreasing failure rate (DFR). The infant mortality period is followed by a steady-state failure rate period, which is usually long. This period is known as the random failure period and covers the useful life of a piece of equipment. This is one of the most interesting parts of the curve and is called the normal operating life. Failure rate data over this period facilitates the prediction of reliability by means of the probability theory. Finally, the curve ends in the third stage, beyond the useful life period, wherein a gradual increase in the failure rate is observed. This is a period of ageing and wear-out characterized by an increasing failure rate. This situation is commonly seen in the case of thermionic valves and mechanical devices, which follow the traditional bathtub curve.

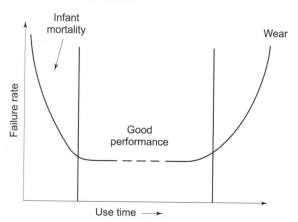

Fig. 1.4 *Traditional bathtub reliability curve*

In many systems, the components liable to suffer from this type of failure can be identified and replaced before reaching the final stage wherein failure can be anticipated. By doing so, the working life of the equipment can be prolonged much beyond the expected period without replacement. Obviously, the wear-out phase is largely irrelevant to solid state equipment.

A modification of the traditional bathtub curve, which is more relevant to modern microelectronic devices, is the generalized bathtub curve (Fig. 1.5). This curve consists of four stages, with the last two being similar to the last two stages in the traditional bathtub curve. The first stage has an IFR, indicating failure rate in this first stage which peaks quickly and is followed by a period of DFR. The infant mortality period includes all failures prior to the steady state, regardless of whether we have a traditional bathtub or generalized bathtub curve. Equipment based on solid state devices generally shows a somewhat longer period of infant mortality. Therefore, in the case of high reliability requirements of equipment, the equipment is operated for a minimum period before it is delivered to the customer. This is to ensure that most of the early failures are corrected before the equipment goes into service.

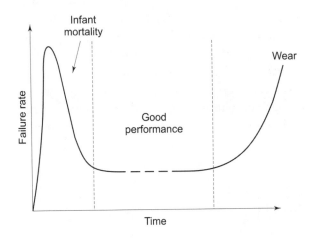

Fig. 1.5 *Generalized reliability curve*

Most of the infant mortality can be associated with poor quality control during the manufacture of the equipment assembly, while most of the mid-life mortality can be attributed to component manufacture.

Electronic components, most notably silicon integrated circuits, tend to either last a relatively long time or fail very early in their operation. One way to screen out defective components in order to improve systems reliability before customer delivery, is to burn-in the components under electrical and thermal conditions that approximate the real life of the component/system in a compressed time frame. The accelerated ageing tests are performed to predict, over a short-term assessment, the onset of the wear-out period under the conditions of service of the component or the assembly.

Usually, the conditions used to burn-in components are chosen arbitrarily. For example, 168 hours (one week) at an elevated temperature of 125 degree centigrade has been the most popular burn-in period for electronic components. This burn-in time can be equivalent to one year of operation when done at a normal operating temperature. Burn-in techniques eliminate infant mortality and thus fewer defective components are delivered, with the consequent reduction in failure rates and production costs. It also cuts down expenses on costly field repairs and helps in building the product image and the manufacturer's reputation. In addition, burn-in is absolutely necessary for devices in critical applications and for non-repairable items.

Equipment reliability is seen to improve considerably when it is operated under certain favourable conditions such as when the components are stressed well below the maximum ratings, operated at well-cooled environments and are subjected to minimum vibration and shock.

1.3 RELIABILITY PREDICTIONS

In the modern competitive market, the consumer has come to expect that producers would develop dependable products and if the product does fail, the consumer also expects that the product would be restored to an operational condition in the shortest possible time. The consumer's ability to depend upon an item is defined by its 'reliability' characteristics while the ability to restore the item to an operable condition is defined by its 'maintainability' characteristics. Although reliability and maintainability engineering activities have traditionally been focused at the front-end stages of a product's lifecycle, these activities can also be applied to products and systems already in use to potentially avoid or mitigate the re-investment option of costly major capital equipment.

In order to predict the usable service life of an electronic equipment assembly and the meantime between its functional failure, it is necessary to have, for each type of component in an electronic assembly, a measure of both (i) the mean random failure rate, and the (ii) mean wear-out failure rate. From these, the probability period within which each component and therefore, the whole assembly will survive for a given time can be calculated. This assumes that inspection and burn-in have enabled the satisfactory replacement of all the infant mortalities.

The occurrence of random failures results in a constant failure rate, which gives rise to the predominantly flat bottom of the bathtub curve. Since the random failure regime is independent of the time, predictive statistical analysis can be carried out by testing a sufficiently large number of components for failure within the random failure regime. Thus, if N components that survive infant mortality are tested for T hours and produce F failures, then the "mean time to fail" (MTTF) per component would be:

$$\text{MTTF} = NT/F \text{ hours}$$

The mean failure rate for a component (K_c) is the inverse of MTTF:

$$K_c = F/NT \text{ per hour}$$

If an assembly or a system consists of N components, the system failure rate (K_s) is given by

$$K_s = N K_c$$

This means that the mean time for the failure of the system is only 1/N of the mean time for the failure of each component.

The probability of occurrence of a failure f, is equal to the total number of defects occurring within the batch over their lifetime, divided by the total number of components at risk. For one assembled board with N components and joints, the average number of defects occurring during its lifetime is fN. Thus, in case of a board containing a large number of components and joints, whose individual reliability over its lifetime is known, it is possible to predict the probability that N defects will occur on a board over a specified life-time and the percentage of assemblies that may experience no faults, one fault or two faults, and so on.

Whereas while the random failures are not dependent upon the passage of time, it is clear from the bath tub curve that the failure rate from wear-out is highly dependent on time. Therefore, the most reliable method of measuring the wear-out failure function is to put the components on test until they fail. For most components, the wear-out failure should be in excess of 10 years extending up to 20 years. Obviously, this approach to reliability testing is neither practical nor viable. In practice, a representative sample of a component batch is subjected to an accelerated ageing process in which the time scale is compressed in a controlled manner. Here, the mechanisms that contribute to the failures are accelerated. This acceleration is achieved by increasing the temperature, humidity, the electrical parameters or a combination of all these parameters.

1.3.1 Failure Rate

Reliability predictions on any equipment or system would require information and data on the failure rate of each type of component that is used to make up the system. It has earlier been explained that the failure rate of components follows the bathtub curve. The failure rate is defined as the number of failures per number of component hours over the useful life of the component.

The failure rate of a component can be calculated by operating large numbers of the component for a known period and noting the number of failures that take place during that period. For example, suppose 1000 transistors are put on test, out of which 25 fail over a 1000-hour period, then by definition, the failure rate is given by the following equation:

$$\text{Failure rate} = 25/1000$$

$$= 0.025 \text{ per 1000 hours}$$

$$= 0.025/1000 \text{ per hour}$$

$$= 2.5 \times 10^{-5} \text{ per hour}$$

Often the failure rate is expressed as a percentage, in which case

$$\text{Failure rate} = 25/1000 \times 100\% \text{ per 1000 hours}$$

$$= 2.5\% \text{ per 1000 hours}$$

1.3.2 MTTF

From the failure rate data, it is possible to calculate the 'mean time to fail' (MTTF). If one transistor is used in the system:

$$\text{MTTF} = 1/2.5 \times 10^{-5}$$

$$= 40,000 \text{ hours} = 1666 \text{ days}$$

In cases wherein a system uses a large number of transistors, the chances of its failure are quite large. MTTF may also be calculated on the basis of the results of the life testing of components. For example, if 5 transistors are tested until failure, and the time to failure were 400, 500, 750, 300 and 600 hours, the total test time would be 2550 hours, and the MTTF is indicated in the following equation:

$$\text{MTTF} = 2550/5 = 510 \text{ hours}$$

It may be noted that MTTF is normally applied to items which cannot be repaired and are to be treated as 'throw-away' items. Examples of such items are resistors, capacitors, thermionic valves, diodes and transistors, etc.

1.3.3 MTBF

A prediction of the average time that an equipment/system will run before failing is provided by the term 'mean time between failures' (MTBF). The MTBF of a system is measured by testing it for a period T, during which M faults may occur. Each time the fault is repaired, the equipment is again put back on test. After excluding the repair time from the total test time, the observed MTBF is given in the following equation:

$$MTBF = T/N$$

This observed value is not necessarily the actual MTBF, since the equipment is usually observed for only a short period as compared to its total life. The observed value may suffer from a random sampling error, and calculations from the test data must thus allow for this error.

For many electronic systems, the failure rate is approximately constant during most of their working lives. In such a case, the MTBF is given in the following equation:

$$MTBF = 1/\text{failure rate}$$

To exemplify the failures in transistors given above, suppose a system contains 100 transistors. Since each transistor has a chance of failure of 2.5×10^{-5} in every hour, the chance of the system failing is 2.5×10^{-3} ($2.5 \times 10^{-5} \times 100$) in every hour. Therefore:

$$MTBF = 1/2.5 \times 10^{-3} = 400 \text{ hours}$$

The concept of MTBF is applicable to any system which can be repaired through the replacement of a faulty component. Obviously, an equipment with higher MTBF will be more reliable than others and will be usually preferred. MTBF, thus provides the most convenient index of the reliability of systems.

The MTBF of a system is estimated by first determining the failure rates of each component and then summing them all up to obtain the system failure rate. For a simple circuit using four components, the failure rate is given in the equation:

$$\text{Failure rate (circuit)} = FR_1 + FR_2 + FR_3 + FR_4$$

$$\text{MTBF (circuit)} = 1/\text{failure rate (circuit)}$$

The failure rates of components are usually available from the manufacturers of components. However, these failure rates depend upon the manufacturing methods of the components and the environment in which they are used. The typical failure rates of commonly used electronic components are given in Table 1.1.

Table 1.1 ■ *Typical Failure Rates for Common Components*

Component	Type	Failure rate $(10^{-6}/hour)$
Resistors	Carbon composition	0.05
	Carbon film	0.2
	Metal film	0.03
	Wire wound	0.1
	Variable resistors	3
Capacitors	Paper	1
	Polyester	0.1
	Ceramic	0.1
	Tantalum (solid)	0.5
	Electrolytic (Al foil)	1.5
Semiconductors	Signal diodes	0.05
(Silicon)	Rectifiers	0.5
	Transistors (1W)	0.08
	Power transistors (1W)	0.8
	Linear IC (Plastic DIL)	0.3
	Digital IC (Plastic DIL)	0.2
Switches	Per contact	0.1
Lamps	LED	0.1
	Filament type	5
Valves	Thermionic	5
Wound components	Power transformer (each winding)	0.4
	RF Coils	0.8
Connections	Soldered joint	0.01
	Plug and socket	0.05
	Crimped	0.02

1.3.4 Availability

Most of us are concerned with the time during which a system remains operative, i.e. the up-time of the system (u). If the system remains down for a period D, the availability of the system can be defined as:

$$\text{Availability} = u/u + D$$

and

$$\text{Unavailability} = D/u + D$$

System availability is the function of a large number of logistic parameters including:

- Operational and storage environment (temperature, humidity, vibrations, etc.);
- Maintenance procedures (periodic, preventive, etc.);
- System configuration and topology; and
- Availability of spare parts.

1.3.5 Maintainability

Maintainability is the measure of the ability of an item to be retained in or restored to specific conditions when maintenance is performed by personnel with specified skill levels, using prescribed procedures and resources at each prescribed level of maintenance and repair.

Alternately, maintainability is the ability of an item, under stated conditions of use, to be retained in, or restored to, a state in which it can perform its required functions, when maintenance is performed under stated conditions, using prescribed procedures and resources.

The focus of 'maintainability engineering' is on minimizing the time that a product spends in an inoperable state. The difference between maintainability and maintenance is that maintainability is a design characteristic embedded in the product under development while maintenance is the set of operative tasks which would occur once a product is in operational use.

1.3.6 MTTR

In practice, a system may have excellent reliability (with a very low chance of failure during operation), but when a failure occurs, and the down-time or the repair time is too large, the system will have poor availability. Therefore, the Mean Time to Repair (MTTR) is an important consideration for selecting a system as in some cases, even a few hours taken for repairs can prove to be very costly. By definition, MTTR is the average time required to bring a system from a failed state to an operational state. MTTR is strictly a design-dependent activity. It further assumes that the maintenance personnel and spares are available on hand or it does not include the logistics delay time. MTTR includes only the time taken to diagnose, locate and repair the fault. The term MTTR is used interchangeably with the Mean Corrective Maintenance Time (MCMT).

MTTR can be used in a reliability prediction in order to calculate the availability of a product or system. As *availability* is the probability that an item is in an operable state at any time, it is based on a combination of MTBF and MTTR. It can be expressed through the following equation:

$$\text{Availability} = \text{MTBF}/\text{MTBF} + \text{MTTR}$$

MTTR depends upon several factors, some of which need to be carefully considered during the design process. The layout of the various printed circuit boards and various other assemblies must be such that they can be easily approached or dismantled. Some of the factors which directly affect MTTR and reduce this time can be identified as follows:

(a) Comprehensive service manual with detailed circuit and component layout diagrams, parts list, dismantling procedure, test points along with voltage and waveform information, fault location trees, etc.;

(b) Correct constructional features of the equipment, such as assemblies which can be easily dismantled, plug-in circuit boards, hinged panels and assemblies, clearly labelled components and circuit cards, test points and internal test signal services;

(c) Availability of suitable test and measuring equipment and relevant tools;

(d) Easy availability of parts and components;

(e) Suitably qualified and experienced technical staff; and

(f) Location of equipment relative to the location of repairing and servicing facilities, availability of transport, etc.

Experience shows that the failure rate is generally greater during the initial period, which decreases as the initial faults or teething troubles are sorted out. Also, the repair rate generally improves after the early stages of operation, as the maintenance staff becomes conversant with the equipment. For both these reasons, the availability is likely to be less in the initial stages than in the later steady state period.

MTTRS: Mean Time to Restore System (MTTRS) is the average time it takes to restore a system from a failed state to an operable state, including the logistics delay time.

$$\text{MTTRS} = \text{Logistics delay time} + \text{MTTR}$$

The logistics delay time includes all the time required to obtain spares and personnel to start the repair.

1.4   ACCELERATED ASSESSMENT OF RELIABILITY

The mean lifetime of a type of a component can be assessed by accelerating the mechanisms of ageing and failure under controlled conditions of testing. The three parameters which are generally used to compress the time scale of the ageing mechanism that lead to failure are temperature, humidity and electrical over-stress.

1.4.1   Thermal Acceleration

The properties and behaviour of electronic components and assemblies degrade faster as the temperature is increased. It has been mathematically proved that the life of electronic components is reduced to half for every 6°C rise in temperature around room temperature and for every 10°C rise in temperature around 150°C. This leads to the following important conclusions:

- A serious deterioration in component lifetime takes place if proper care is not taken in thermal management of the assembly, and

- Very large acceleration factors can be achieved with a relatively small increase in the test temperature over service temperature.

The desired life of electronic components is generally about 20 years. In order to make reliability measurements over practically short periods, say about 1000 hours, a time compression factor of around 200 would be required. This factor can be achieved by increasing a service temperature of 70°C to a test temperature of 140°C. This means the lifetime at the higher temperature is predicted to be 1/200th of that at the lower temperature. Thus the reliability after 20 years can be predicted from the cumulative failures at just one temperature, viz. 140°C.

1.4.2 Electrical Acceleration

The ageing time scale for reliability testing can also be accelerated by electrically over-stressing the component or assembly. This can be achieved by increasing either the current or the voltage in a controlled manner to values in excess of their service operation levels. Current-induced failures are usually associated with an electro-migration phenomenon which symbolises the movement of metal atoms within the device architecture that is induced at high current densities. By studying the relationship between the mean lifetime and electro-migration failure with the current density, it is possible to have some prediction of the device lifetime.

The acceleration of device failure due to the use of high voltage is not generally studied because the functional dependence of the device on the voltage is quite complex and is not easy to translate the same to realistic conditions.

1.4.3 Damp Heat Acceleration

Apart from the effect of temperature, detrimental effects can also be seen on the active interfaces of semiconductor devices arising from the ingress of moisture, which can cause junction leakage and gain degradation. Also, the penetration of moisture into the chips may form corrosion cells at the chip surface. The accelerated ingress of moisture to an active semiconductor device is therefore used as a means of predicting long-term behaviour from short term tests.

It has been empirically found that over almost the entire range of humidity, the semiconductor lifetime varies inversely as the exponential of the square of the relative humidity. Since the diffusion of the moisture through a plastic encapsulation of the devices is greatly affected by the temperature, it is customary to accelerate the ageing effect by combining high humidity with a high temperature. Since they use both heat and moisture, the acceleration factors for ageing can be very large, but it should be ensured that they remain within the validity of the tests. It is normal practice to employ worst case values to set the acceleration factor for taking care of all the conceivable failure mechanisms. A commonly used damp heat ageing test condition is 85°C, 85 per cent relative humidity and the calculated test times required to simulate service conditions. Figure 1.6 shows the relationship between the service temperature and the duration of the test in hours at different levels of relative humidity.

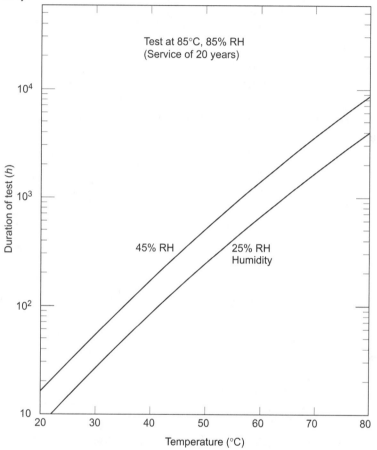

Fig. 1.6   *The calculated test for accelerated ageing at 85°, 85% relative humidity to simulate long-term service conditions. The service conditions illustrated are in the temperature range 20–80°C and either 25% or 45% RH*

It is also an accepted practice to include vibration along with temperature while specifying the reliability of a system. While in actual use, a system will, however, experience a wide range of environmental stresses. As a result, the actual or 'field' reliability may be considerably lower or higher than that experienced in testing or established by calculations. In real life, it may often occur that a calculated 10,000-hour MTBF is found to be a field MTBF of 1000 hours or vice versa.

1.5   PRACTICAL RELIABILITY CONSIDERATIONS

The reliability of an electronic assembly is a function of the intrinsic reliability of its individual components and the reliability of the inter-connections. Therefore, individual and unattached components are tested both in real time and in accelerated ageing tests. The reliability is expressed in terms of a typical failure rate per hour of service operation. The accelerated ageing tests that are commonly employed are:

- A dry heat over stress at, say, 125°C;
- A damp heat over stress at 85°C per cent relative humidity (RH); and
- A temperature cycling between –65°C and +150°C or –55°C to +125°C.

The results of these tests help us to determine the acceleration factors and compute the failure rates under normal service conditions.

Component reliability is often expressed in units of 'failure in time standard' or FITS. It is defined as one failure per 10^9 component hours. For chip resistors and capacitors, the failure rate is about 1 FITS, whereas for SOTs (small outline transistors) and SOICs (small outline integrated circuits), it is 10 FITS. For high reliability equipment, failure rates of less than about 50 FITS are recommended whereas for consumer electronic products, the failure rate could be relaxed to 100–500 FITS.

The reliability of inter-connections depends upon the assembly process and the inter-connection circuits. The primary long term failure mode of inter-connections is due to fatigue of the solder joints, brought about by thermally induced or mechanically induced cyclic strains.

The purchasers of components or assembly manufacturers carry out acceptance tests on the incoming components and statistically ensure that each batch meets specified acceptance quality levels (AQLs). These levels are specified by the component manufacturers which are realistic and practical. The manufacturer of components attempt to remove all the components that will fail in infancy. The actual defect levels are therefore considerably lower than the quoted AQLs. The defect level includes inoperative components, components outside the tolerance limits (electrical and physical) and those with illegible markings. The component manufacturers would normally aim for zero defects at delivery, and may establish an AQL as low as 5 in a million to sustain customer confidence.

It may, however, be noted that general confusion concerning the proliferation of reliability standards and the use of many methods for calculating reliability figures leads to a somewhat chaotic situation wherein various vendors provide different reliability figures, some of which are even artificially treated to fit customer requirements.

2

Fundamental Troubleshooting Procedures

2.1   MAKING OF AN ELECTRONIC EQUIPMENT

2.1.1   Electronic Circuits

Modern electronic circuits make use of both active and passive components. Active components may be transistors and integrated circuits, both linear as well as digital. There are three major techniques for physically inter-connecting these components. All three techniques have at least one point in common. They all start with the components inserted into a non-conductive board. The active components, particularly the integrated circuits (ICs), may be inserted directly into the board. Alternately, the IC may be inserted in a socket that is mounted into the board. The three techniques for inter-connecting the components are as follows :

(a) *Solder:* The oldest method used for inter-connecting the electronic components together is with solder and wire. This is a very slow method and is very cumbersome if a large number of devices are to be connected. The method is not amenable to the production of complex circuits, even though it is by no means an unsatisfactory method.

(b) *Wire-wrap:* The technique of wire-wrapping consists of tightly winding a small gauge wire around a wire-wrap metal post or terminal. There are special wire-wrap metal post sockets for ICs that have longer posts for wire-wrapping the wire. Also, special tools are needed for wrapping and un-wrapping the wire.

(c) *Printed Circuit Board (PCB):* A printed circuit board (Fig. 2.1) provides for inter-connections between points printed in metal on the non-conducting board. Printed circuit boards are generally made for completely checked out and working boards and it is difficult to make wiring changes on the PCB. Most of the equipment encountered in practice make use of printed circuit boards.

Fig. 2.1   ***Printed circuit board showing various components***

2.1.2   Inside of an Electronic Equipment

An electronic equipment is a combination of electrical and electronic components that are connected to produce a certain desired function. In the era of vacuum tubes and even later, electronic equipments were constructed by hand wiring and by point-to-point soldering. The wires were stripped of their insulation, tinned and soldered. Each discrete component was installed by hand, electrically and mechanically. The equipment was obviously large, sometimes awkward and bulky. It was difficult to meet the demand for its use in aircraft, the health sector and home emergency uses, necessitating the development of smaller and more compact electronic equipment.

A natural evolution took place in several areas. Smaller components were developed and modular design became popular basically to decrease the time between unit failure and repair due to easy replaceability. The advent of miniaturization in electronic equipment design gave birth to a new technique in inter-component wiring and assembly, commonly known as the *printed circuit*. The printed circuit board (PCB) is developed by a series of photographic and chemical procedures. This process involves achieving a conductive pattern, formed on one or both sides of one or more insulating laminates. The term *printed wiring* refers only to the conductive pattern that is formed on the laminate (base material) to provide a point-to-point connection.

An electronic equipment may be made up of just one circuit board or of many circuit boards. They are mounted inside a wooden or metallic cabinet with some arrangement for inter-connecting the circuit boards which is usually done by making use of edge connectors (Fig. 2.2).

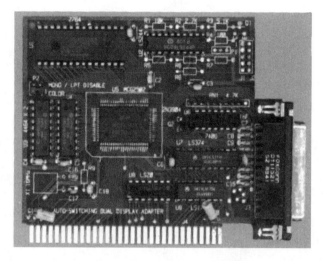

↑ Fig. 2.2 *Use of an edge connector for inter-connection of various circuit boards*

The purpose of the edge connectors is to bring signals and power to and from the circuit board without having to connect a wire to the circuit board itself. This facilitates easy removal and installation of circuit boards in an equipment.

At the same time, the placement of PCBs closely in a card rack poses its own problems. It is difficult to put a test probe on the circuit board for making any type of measurement. In order to solve this problem, special circuit boards called *extender cards* can be inserted into the card rack, and the circuit board is extended into the extender card. In fact, the extender card is just a wiring extension to make the circuit board accessible for testing. Figure 2.3 shows the use of an extender board for reaching a card for troubleshooting.

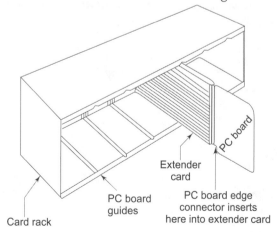

↑ Fig. 2.3 *Use of an extender card for reaching a card for troubleshooting*

2.1.3   Types of Printed Circuit Boards

Printed wiring boards may be classified according to their various attributes. An important classification is according to the number of planes or layers of wiring which constitute the total wiring assembly or structure.

Single-sided Printed Boards: Single-sided means that wiring is available only on one side of the insulating substrate. The side which contains the circuit pattern is called the 'solder side' whereas the other side is called the 'component side'. This type of boards are used mostly in case of simple circuitry and when the manufacturing costs are to be kept at a minimum. Figjure 2.4 shows the arrangement of a single-sided board.

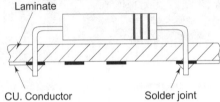

Fig. 2.4   ***Single-sided printed circuit board***

Double-sided Printed Boards: Double-sided boards have a wiring pattern present on either side of the insulating material, i.e. the circuit pattern is available both on the component side and the solder side. Obviously, the component density and the conductor lines are higher than the single-sided boards.

The two types of commonly available double-sided boards include those:
- With plated through-hole connection (PTH); and
- Without plated through-hole connection

Figure 2.5 shows the constructional details of the two types of double-sided boards. A double-sided PTH board has circuitry on both sides of an insulating substrate and is con-

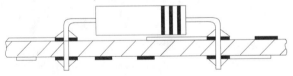

(a) No plated-through holes

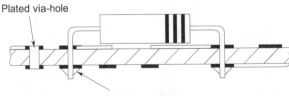

(b) Plated-through holes

Fig. 2.5   ***Double-sided printed circuit board***
(a) non-PTH (printed through-hole) type
(b) PTH type double-sided printed circuit board

nected by metallizing the wall of a hole in substrate that intersects the circuitry on both sides. This technology, which is the basis for most of the printed circuits produced, is becoming popular wherever the circuit complexity and density are high.

In the case of a double-sided non-PTH board, through contacts are made by soldering the component leads on both sides of the board.

Multi-layer Boards: The advent of modern VLSI (very large scale integrated) devices and other multi-pin configuration devices has tremendously increased the packaging density and consequently the concentration of inter-connecting lines. Such type of inter-connection cannot be achieved satisfactorily in single-sided or double-sided boards, necessitating the extension of the two-plane approach to the multi-layer circuit board.

The multi-layer board makes use of two planes of inter-connections, with three or more circuit layers, with some boards having as many as 16 layers. In these boards, the electrical circuit is completed by means of inter-connecting the different layers by plated through holes, placed transverse to the board, at appropriate places. Figure 2.6 shows the details of a multi-layer board.

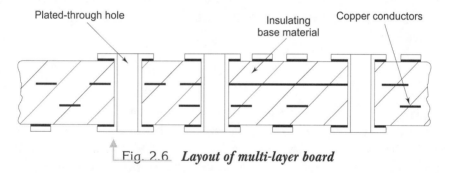

Fig. 2.6 *Layout of multi-layer board*

2.2 READING DRAWINGS AND DIAGRAMS

2.2.1 Block Diagram

All equipments can be considered as systems comprising a set of interacting elements responding to inputs in order to produce outputs. It is possible that a system may be too complex to be analysed in detail. It is therefore, necessary to divide it into sub-systems and then to integrate them. Each sub-system would then represent a functional block, and the combination of all the blocks would constitute the functional 'block diagram' of the equipment. A block will only be a 'black box' with certain inputs and outputs, but performing a definite function. Figure 2.7 shows a typical block diagram representation of a simple recorder.

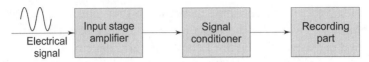

Fig. 2.7   *Concept of a block diagram. It shows various sub-systems in an equipment. For example, in a simple recorder, the electrical signal from any source is applied to an input stage amplifier. This is followed by a signal conditioner which makes the signal suitable for the writing or recording part. The signal conditioner may be again an amplifier, a filter or signal attenuator depending upon the requirement of the writing part*

The blocks are inter-connected by lines that represent some variable in the systems and indicate the signal flow path. In troubleshooting, once the fault is isolated to a particular block, the circuit diagram of that block can be studied in detail to localize the fault. The block diagram approach helps in understanding various functions and thereby helps in troubleshooting by replacing or repairing only the defective functional block.

The MSI and LSI devices such as microprocessors, counters, etc., are represented as individual blocks. These blocks are labelled with pin numbers, signals and associated inter-connecting wires.

2.2.2   Circuit Diagram

A circuit diagram is a graphical representation of inter-connections of various components constituting the equipment. Usually, every assembly in an equipment is assigned an assembly number which appears on the circuit board and on the diagram. The circuit diagram shows various components by means of symbols which are so arranged that they clearly show the working of the circuit. The component symbols are usually governed by various standards, which, unfortunately, vary widely. Therefore, one must find out which standard has been followed before attempting to read a circuit diagram. The latter is also called a schematic diagram.

Electronic components shown on the circuit diagrams are generally in the following units unless mentioned otherwise:

$$\text{Capacitors} = \text{Values one or greater are in the picofarads (pF)}$$

$$= \text{Values less than one are in microfarads (}\mu\text{F)}$$

$$\text{Resistors} = \text{Ohms (}\Omega\text{)}$$

Figure 2.8(a) and (b) are based on symbols recommended by the American National Standard Institute: ANSI Standard U 32.2–1970.

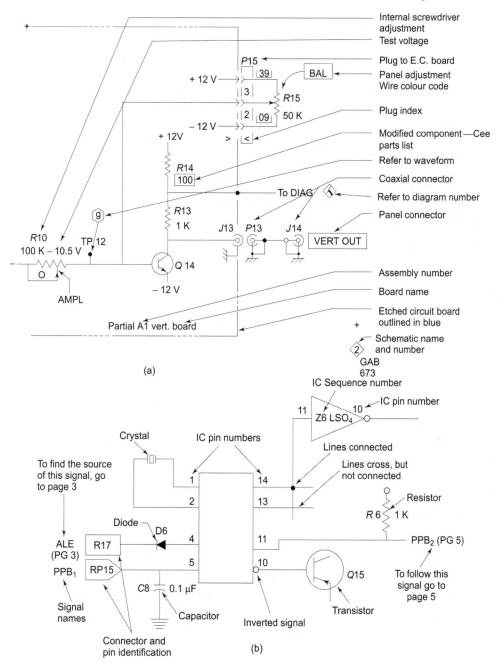

Fig. 2.8 *(a) Schematic circuit symbols*
(b) Special symbols used on circuit diagrams as per
American National Standard Institute

Guidelines have been developed over the years for drawing schematic diagrams. The main features of these guidelines are:

- The signal flow moves from left to right across the page with inputs on the left and outputs on the right.
- Electronic potentials (voltages) should increase as you move from the bottom to the top of a page. For example, in Fig. 2.9, +5V supply is shown upwards while the ground pin is downwards.
- Use the 'unit number' convention for assigning a unique package identification $U1$ with its internal gates identified by letter suffixes; $U1A$, $U1B$ etc. Only one of the common gates need show the power connections (pin 7 and pin 14). Power connections are often omitted, but it is better to include them as a reminder as well as to make one's schematic diagram complete.

Always begin troubleshooting work with a complete and readable engineering schematic that accurately reflects what has actually been built.

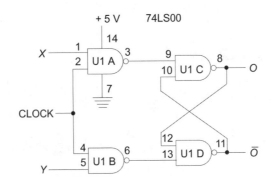

Fig. 2.9 *Example of an engineering schematic*

2.2.3 Wiring Diagram

Wiring diagrams show how components are connected on a circuit board or any other particular construction format. Figure 2.10 shows an example of a simple wiring diagram as is actively wired on a board. Barring for the inter-connections, no other information is conveyed except what you might conclude from the signal names. This makes troubleshooting difficult or sometimes, impossible from the wiring diagram alone.

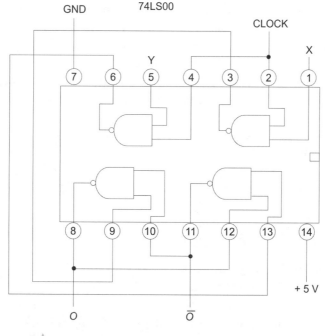

Fig. 2.10 *Example of a wiring diagram*

2.3 EQUIPMENT FAILURES

The failure of an equipment refers to its inability to perform its required function, i.e. when its characteristics change to such a degree that it cannot perform to its specified level of performance.

Failures may be partial failures resulting from deviations in characteristics or parameters beyond the specified limits but not such as to cause a complete breakdown of the required function. If the characteristics deviate beyond the specified limits such as to cause a complete breakdown of the required function, it is called complete failure.

Quite often, failures can be anticipated by prior examination (gradual failure); while in some cases, it could be sudden failures that cannot be predicted on routing examinations.

A sudden and complete change in the performance of an equipment is called a catastrophic failure. This may often be caused by an open circuit or a short circuit failure which is irreversible. An example of this kind of failure is a short-circuit in a capacitor or an open circuit in a wire-wound potentiometer. The only remedy for this type of failure is the replacement of the faulty component or its removal or other types of fault-creating situations. Failure may also occur gradually and in a partial manner. This type of failure is referred to as degradation failure, which may impair the system without completely preventing its operation. Such types of failures are encountered mostly with analog systems, particularly if some noise or disturbance is present, but is less likely in digital systems. With the introduction of integrated circuits for complex functions, the possibility of degradation faults has been considerably reduced. However, they are still prominent in equipment which operates under adverse environmental conditions.

2.4   CAUSES OF EQUIPMENT FAILURES

Equipment failures take place due to various reasons. Essentially, they can be classified as follows:

2.4.1 Poor Design

(a) Improper choice of components;
(b) Inadequate information on the stress and failure analysis of components;
(c) Poor mechanical layout of components, assemblies and panels;
(d) Excessive development of heat inside the equipment and lack of cooling, etc; and
(e) Insufficient prototype testing for meeting both reliability and performance specifications.

2.4.2   Production Deficiencies

(a) Lack of inward inspection of goods, sampling tests and inspection;
(b) Unsuitable storage methods and unduly long storage period of components;
(c) Poor motivation, lack of skill and sense of involvement among staff members;
(d) Absence of training programmes for workers for employing correct and most effective production techniques;
(e) Use of sub-standard manufacturing equipment and tools;
(f) Lack of proper working environment, ill-ventilated, poorly illuminated and dusty assembly shops;
(g) Insufficient testing and inspection of finished products; and
(h) Negligence in performing environmental tests like temperature cycling and operation of equipment at elevated temperatures for specified periods.

2.4.3   Careless Storage and Transport

(a) Unduly long storage of equipment before its dispatch to the customer;

(b) Improper packaging which may fail to protect the equipment from corrosion and mechanical damage; and

(c) Excessive vibrations and mechanical shocks during transportation of equipment form the manufacturer to the user.

2.4.4 Inappropriate Conditions during Working Life

(a) Hostile working environmental conditions like lack of air-conditioned room and dust-free areas;

(b) Carelessness in handling and operation and usage of equipment without following manufacturer's instructions, cautions and warnings;

(c) Poor operatability of equipment with the possibility of a high degree of a operator error;

(d) Poor maintenance policy, wrong selection of equipment and provision of inadequate after-sales service by manufacturers;

(e) Fluctuations in the mains voltage (a common problem in developing countries);

(f) Running the equipment beyond its prescribed lifetime;

(g) Aging of the equipment; and

(h) Lack of preventive maintenance.

One or more of the above factors may lead to either a sudden breakdown of the equipment or slow degradation in its performance. This failure may manifest itself through a deterioration in the stability, reproducibility, accuracy and overall performance of the equipment.

2.5 NATURE OF FAULTS

The nature of faults occurring in different types of electronic equipment may vary from simple mechanical faults to complex faults in electronic circuitry. However, experience shows that about 30% of the faults in an instrument are minor in nature, such as a blown-up fuse, broken or shorted power cord, sticky panel meters, loose sockets or defective connectors, improper ground leads causing spurious artifacts in recording instruments, unfavourable environments causing equipment to get unduly heated up leading to insta-bility and drift, and the presence of another disturbing equipment in the vicinity of the equipment under use. Most of these cannot be considered as faults in the system, but for a non-technical equipment user, they could be the cause of severe anxiety.

About 20% of the faults are of a very common nature. These are mostly mechanical faults and are also conspicuous. Such faults include loose mechanical fixtures, knobs, handles, power sockets and stylus of recorders, mounting of printed circuit boards in connectors and loose valve, transistor and IC base contacts. Sometimes, these faults may be due to drained out dry batteries, uncharged Ni–Cd cells, burnt-out pilot lamps, defective sensors, etc.

About 30% to 40% of the equipment develops faults which are neither simple nor common in nature. These faults are specific to the equipment concerned and include burnt transformers, erratic performance due to instability and failure of one or more stages in the equipment due to voltage fluctuations or carelessness of the user.

Another 10% of the faults which are observed in the equipment are chronic in nature which become a headache for both the service engineer and the user. These faults can be termed as 'repetitive faults' and occur due to either poor designing, or the use of substandard components in its manufacture, or poor quality control or the prolonged use of equipment beyond its expected working life.

One of the most common problems in electronic equipment is related to the power supply consisting of tripped over current protection devices or damage due to over-heating. Although the power supply circuitry is usually less complex than the circuitry that is being powered, it is nevertheless more prone to failure as it generally handles more power than any other part of the system.

Active components tend to fail with greater regularity than passive devices, due to their greater complexity. Semiconductor devices are notoriously prone to failure due to electrical transient (voltage/current surge) overloading and thermal (heat) overloading.

Passive components are the most rugged of all. However, the following failure probabilities have been generally noticed :

- Capacitors (shorted), especially electrolytic capacitors, and thin dielectric layers that may be punctured by over-voltage transients;
- Diodes open (rectifying diodes) or shorted (zener diodes);
- Inductor and transformer windings open or shorted to conductive core, failures related to over-heating, insulation breakdown; and
- Resistors open, almost never shorted, which is usually caused due to over-current heating or physical damage due to vibration or impact.

2.6 MAINTENANCE TERMINOLOGY

Troubleshooting

This is the process of isolating and correcting the problem in a malfunctioning equipment so that it returns to its expected performance level.

Breakdown Maintenance (Repair)

This refers to troubleshooting to isolate the cause of device malfunction and replacement or adjustment of components or sub-systems in order to restore normal function, performance, safety and reliability.

Preventive Maintenance

The use of periodic procedures to minimise the risk of failure and to ensure continued proper operation of an equipment is called preventive maintenance.

Inspection

It is a procedure used to check the physical integrity of a device and to ensure that it meets the appropriate performance and safety requirements as laid down by regulatory, licensing and accrediting agencies. Inspection consists of the following:
- Visual examination, verification of functional operation and measurements to determine safety (electrical, mechanical, chemical, thermal and pneumatic, etc.); and
- Testing of important performance characteristics to verify that the device can fulfil its intended purpose and operate according to manufacturer's specifications.

Acceptance Test

This is a detailed procedure used to verify the performance and safety of an equipment/device before use, either at the time of the initial receipt (incoming inspection) or following major repairs.

Calibration

Calibration is a process used to determine the accuracy of a device by using test equipment of verified and appropriate accuracy and adjusting the device to meet recommended accuracy requirements.

Overhaul

It involves the replacement of worn parts and their upgradation by carrying out calibration according to the manufacturer's recommendations. It may be required at fixed intervals or it could be determined from an elapsed time meter.

2.7 GETTING INSIDE ELECTRONIC EQUIPMENT

In the troubleshooting process, it is often necessary to open or dis-assemble the equipment. After repairs, it has to be re-assembled again. The process of dis-assembling and re-assembling requires great care, particularly when access to inside of the equipment apparently looks difficult. The following guidelines will assist in dis-assembling and re-assembling the equipment.

2.7.1 Dis-assembly

For dis-assembly, it is better to consult the service manual or documents on specific equipment provided by the manufacturer. However, sometimes this information is not available in the product literature. In fact, some manufacturers seem to take pride in being very mysterious about instructions, as to how to open their equipment. Therefore, opening the equipment non-destructively may be the most difficult and challenging part of many repairs. A variety of techniques are used to secure the covers on electronic equipment, including:

Screws: Somewhat antiquated technique screws are usually of the Phillips type. Sometimes, there are even embossed arrows on the case indicating which screws need to be removed for getting inside the equipment. In addition to the obvious screw holes, there may be some that are only accessible when a battery or cassette compartment is opened or a trim panel is popped off.

Hidden Screws: These will require the peeling off of a decorative label or prying up a plug. Sometimes, the rubber feet can be pried out thus revealing screw holes. For a stick-on label, rubbing your finger over it may permit you to locate a hidden screw hole.

Snaps: Look around the seam between the two halves. You may see points wherein gently or forcibly pressing with a screw driver will help unlock the covers. Sometimes, popping the cover at one location with a knife or screw driver may reveal the locations of other snaps.

Fused Casings: Fused casings, particularly those made with plastics, and the disposable type are commonly used with AC wall adapters. Some devices are even totally potted in epoxy as throughways. Some of these can be cut open with a hacksaw blade, and after repairs, re-assembled with plastic electronic tape.

Plastic Catches: LCD display housings are usually secured by plastic catches built into the case. They still may have a couple of screws that are positioned in the most innovative places such as under stitches. Have patience in locating the catches and screws.

It must be remembered that it is not advisable to force open any equipment unless you are sure that there is no alternative. Once you have understood the method of fastening, the covers will come apart easily. If they get hung up, there may be an undetected screw or snap still in place.

2.7.2 Re-assembly

When re-assembling equipment after repairs, all parts should go together without being forced. You must remember that all the parts you are re-assembling were dis-assembled by you. If you can't get them back together, there must be a reason, which you should find out.

Also, make sure to route cables and other wiring such that they will not get pinched or snagged and possibly broken, or have their insulation nicked or pierced, and that they will not get caught or entangled in moving parts. Replace any cable ties that were cut or removed during dis-assembly and add additional ones of your own, if needed.

2.8 TROUBLESHOOTING PROCESS

The process of troubleshooting comprises the following steps:

(a) Fault Establishment

Before any other action is taken, it is important to establish the presence of a fault in an equipment. In some cases, a system may be reported faulty, but it may be a case of faulty operation or a system failure may be reported with either very little or misleading information. It is essential that a functional test be made for checking the system's actual performance against its specification, and that all fault symptoms be noted.

It is also important to check with the equipment operator the history of the equipment and the repair and servicing work carried out earlier by any other person. Sometimes, incomplete work or mishandling of the equipment by an inexperienced worker could prove disastrous and defy all efforts to repair the equipment.

(b) Fault Location

The fault location procedure will comprise a study of the literature relevant to servicing, maintenance and repairs, and pinpointing the cause of the fault, first to a sub-system and finally to a single component within the sub-system. The fault location methods are discussed later in this chapter.

(c) Fault Correction

Fault correction consists of replacing or repairing the faulty component. This must always be followed by a thorough functional check of the whole system.

2.8.1   Fault Location Procedure

For fault location, it is advisable to follow systematic and logical approach as it is often easy to reach the trouble-spot by proceeding step by step. However, the degree of success in locating a particular fault will depend upon the technician's knowledge of the equipment and his ability to troubleshoot. It must be clearly understood that there is no substitute for familiarity with the equipment when it comes to troubleshooting a system. For example, if you are not familiar with a microcomputer to the extent that you can analyse its operation, interpret its indications, read its printouts, and analyse its program, you will certainly have a tough time in isolating a failure within that equipment.

Troubleshooting procedure must always start with a preliminary analysis of the trouble symptoms, from which various possibilities of malfunction are deduced. These are analysed in order of probability and various quick checks are usually made to eliminate or verify mutual deductions. Figure 2.11 gives a summary of the suggested generalised troubleshooting procedure.

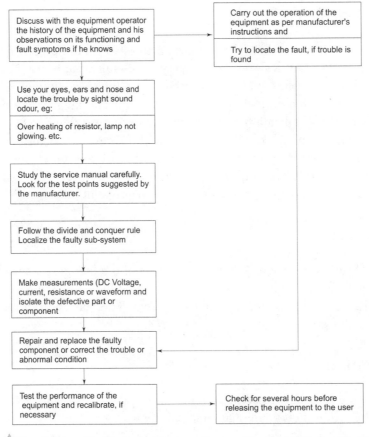

Fig. 2.11   *Fault diagnosis tree—generalized troubleshooting procedure*

The following questions should be asked and facts ascertained before proceeding with troubleshooting:

(i) Has this ever happened before ? If the device has been historically known to fail in a certain particular way, check for this first.

(ii) If a system has been having problems immediately after some kind of maintenance or other change, the problems could be linked to those changes.

(iii) If a system is not producing the desired end-result, look for what it is doing correctly. In other words, identify where the problem is not present, and focus your efforts elsewhere.

(iv) On the basis of your knowledge of how a system works, think of what kind of failures would cause this problem or phenomenon to occur.

Effective troubleshooting requires a blend of both art and science. There is always a safe way and a technically accurate way to do something, but the precise choice of techniques and strategies used to fix the problem is largely decided by the technician's own experience and background.

2.9 FAULT-FINDING AIDS

In order to achieve rapid fault localization and its subsequent repair, the technician will obviously look for certain aids to supplement his technical skills. The most commonly needed aids are:

(a) Service and maintenance manuals and instruction manuals;

(b) Test and measuring instruments; and

(c) Special tools (instruments, mechanical tools).

2.9.1 Service and Maintenance Manuals and Instruction Manuals

Reputed manufacturers usually supply a service and maintenance manual along with the equipment sold. If properly prepared, the manual provides information on:

(a) List of test instruments and special tools;

(b) Method of dismantling;

(c) Safety procedures to be observed while dismantling and while carrying out tests;

(d) Test points with DC voltages and operating waveforms;

(e) Fault location tree or tables showing typical symptoms for various fault conditions, together with the most probable causes and suggested course of action;

(f) Block diagram description;

(g) Circuit diagrams;

(h) Spare parts list, with component values, rating and tolerances;

(i) Printed circuit board layouts for various components identified by means of callouts; and

(j) Mechanical layout, line diagrams, photographs and exploded views of various mechanical parts.

After checking the obvious things, if you do not find a clue about the fault, it is time to consult the service manual and to try to understand, firstly, the result that you should be getting and secondly, the result that you are getting. By doing this, either by measurement or observation, the technician performing the fault finding is guided to the faulty component.

Sometimes, the service manuals are not available. No doubt, based on the experience of other similar systems, the technician can attempt to repair the fault. However, if the system is unknown, proceeding with tests without knowing exactly how the system performs can often lead to incorrect conclusions and in some cases, in causing extra faults. Therefore, it is wise to procure the service manual, as far as possible, before trying to locate a serious fault in the system. However, if you are not able to get the schematic diagram, in some cases, some reverse engineering will be necessary in drawing the circuit diagram. The time will be well-spent as you would have learnt something in the process that can be applied to other equipment problems. Your success in finding and repairing the fault, will be much more likely when you understand how a device works.

The instruction manual or user manual or operator's manual provides necessary information for operating the equipment, start-up and shut down instructions, general design concepts, specifications and the installation procedure. Operator level maintenance instructions including preventive maintenance manuals are useful for preliminary diagnostics establishing preventive maintenance schedules.

2.9.2 Test and Measuring Instruments

The task of troubleshooting is essentially carried out by using numerous test instruments in various areas of electronic servicing. Leaving aside any special instrument that may be required for working on digital and communication equipment, a majority of the system faults can be located and rectified by using the following three instruments:

(a) Multimeter — Analog
              — Digital

(b) Oscilloscope

(c) Signal generator or pulse generator or function generator.

The type of measurements usually carried out in electronic circuits are basically DC and AC voltage measurements. Besides these, a variety of measurements in terms of amplitude, frequency and phase, and a detailed waveform analysis are needed to be performed on complex circuit functions in order to aid the troubleshooting procedure.

2.9.3 Special Tools

Special hand tools and chemicals are required to efficiently carry out troubleshooting and servicing routines. These are described in Chapter 4.

2.10 TROUBLESHOOTING TECHNIQUES

2.10.1 Preliminary Observations

When an equipment is received for repairs, it is essential to carry out some preliminary checks on it before the actual work is started. These checks are necessary for your safety and often help in quickly approaching the trouble spot. The checks are as follows:

(a) Carefully examine the equipment on all sides for pertinent information given by the manufacturer on the panels. Safety precautions are generally printed on the panels with the following nomenclature:

    (i) Caution—indicates a personal injury hazard not immediately accessible as one reads the marking, or a hazard to property including the equipment itself.

    (ii) Danger—indicates a personal injury hazard immediately accessible as one reads the marking. Danger sign (4) will be normally marked at places where high voltages exists on the equipment.

    (iii) ⏚ Protective ground (earth terminal).

    (iv) ⚠ Refer to manual.

(b) Be sure about the power supply requirements of the equipment. Some equipments are battery-operated while others may be operating on the mains power supply. Identify the mains power supply voltage, i.e. whether it is 110 volts, 60 Hz or 220 volts, 50 Hz and the maximum voltage that can be applied from the power source. Determine the supply conductors (if the plug is not present or if it is suspected to be wired wrongly) and the ground lead. A protective ground connection by way of the grounding conductor on the power cord is essential for safe operation.

(c) Almost all modern electronic equipment which is mains-operated, is grounded through the grounding conductor of the power cord. In order to avoid electrical shock, the power cord must be plugged into a properly wired receptacle before connecting to the product input or output terminals.

(d) Before switching on the equipment for preliminary examination, study the 'Service Manual' thoroughly and look for the following terms in the manual:

    (i) Caution statements specify conditions or practices that could result in damage to the equipment or other property.

    (ii) Warning statements specify conditions or practices that could result in personal injury or loss of life.

(iii) ⚠ indicates where applicable cautionary or other information is to be found.

(e) The service manuals usually provide information for removing the equipment panels and to gain access to various circuit boards and individual components. Such instructions must be followed to the letter, otherwise by opening the wrong screws, the internal assemblies sometimes get dislodged resulting in damage to the fragile parts.

(f) Disconnect power to the equipment before attempting to remove the cabinet panels in order to avoid electric shock hazard.

(g) In some equipment, dangerous potentials exist at several points throughout the equipment. When the equipment is operated with the covers removed, do not touch exposed connections or components. Some transistors have voltages present on their cases. Always disconnect power before cleaning the equipment or replacing parts.

(h) Do not try to break open potted assemblies or components about which you have limited information. You would end up in further spoiling the equipment in case something gets damaged during this process.

(i) As far as possible, do not perform internal service or adjustment with dangerous voltages unless another person capable of rendering first aid resuscitation is present.

(j) The painted metal panels of equipment are generally vulnerable to scratching. Avoid rough handling of the panels.

(k) When the equipment cover is removed from the chassis of the equipment, bare metal edges are exposed which can scratch the work surface. Be careful that the sharp edges and corners of the chassis do not scratch the work surface area.

(l) The CMOS logic family of integrated circuits used in some equipment can be damaged by an uncontrolled static electricity discharge. Before handling any of the circuit boards, firmly grasp the equipment chassis to eliminate any static charge difference between your body and the equipment.

(m) Handle all circuit boards by the edges. CMOS circuits operate with currents in the nanoampere range and leakage paths caused by skin oils, dirt, dust, etc. can cause inaccurate circuit performance in some equipment.

(n) After removing the cover of the equipment, inspect all exposed screws for tightness. Check that all printed circuit boards are firmly seated in their connectors or are in position. Check the conditions of all external cables, especially for splits or cracks and signs of twisting. If any serious damage is evident, the cable should be replaced immediately.

To summarise, an effective system of troubleshooting should be quite logical and it is useful to remember the following three points:

1. *Know your equipment*: Without familiarity with the equipment to the extent that one can analyse its operation, interpret its indicators and read and decipher printouts, one will have difficulty in isolating problems.

2. ***Think before you act***: Do not straight away start replacing parts, disassembling various parts, etc. without thinking and analysing for possible causes of trouble. Unorganized way of troubleshooting leads to more trouble and can take an unusually long time.

3. ***Establish a general troubleshooting procedure***: Depending upon your knowledge of equipment, the availability of proper tools, test equipments, spare parts and time, a general troubleshooting procedure needs to be developed. It could be component level repair or board level maintenance. In the latter case, the defective boards are then repaired at a later time or sent back to the manufacturer for repairs.

2.10.2   Troubleshooting Methods

The various troubleshooting techniques given below are representative of those used in troubleshooting in a majority of electronic systems. The type of system you are handling, along with your personal abilities, will decide the troubleshooting technique or techniques you adopt.

Functional Area Approach: An electronic system may comprise several functional parts which may include power supplies, amplification, generation, transmission, conversion, logical manipulation, and storage of signal voltages. Obviously, when the system fails to give the expected performance, the trouble could lie in any of the functional areas. This implies that it is essential to troubleshoot the system in an effort to isolate the fault to the failing functional area and then to the failing component within the functional area. The logical approach of isolating a fault is through a process of elimination of the functional areas that are performing properly. Once a failure is isolated to a specific functional area, further analysis of the circuitry within this area is carried out to isolate the malfunction to the faulty component. This functional area approach is also called the block-diagram approach to troubleshooting. The service manuals normally carry a block diagram description of the equipment which assists in a proper understanding of the performance of the electronic system. It is often more useful than the full circuit diagram, particularly when a system is handled for the first time.

Split-Half Method: In this technique of troubleshooting, once the absence of an output has been established, the circuit is split into half and the output is checked at the half-way point. It will help to isolate the failing circuit in the first or second half. After determining the faulty half, the circuit is split into half for further isolation of the failure. The process is continued until the failure is isolated to one function or component. The half-split method is extremely useful when the system is made up of a large number of blocks in series. A typical example is that of the division of frequency of an oscillator by decade counters for generating various timing pulses. Figure 2.12 shows one such scheme.

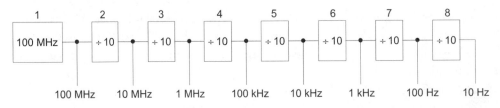

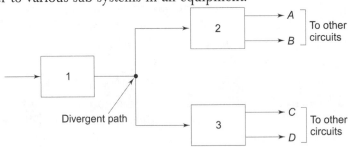

Fig. 2.12   *Split-half method for isolating trouble in a circuit*

Suppose that in this circuit, the expected output (100 Hz pulses) is not present. Let us suppose that the fault is in block 4. The sequence of carrying out tests would be as follows:

(a) Split the circuit into half (mentally) and measure the output of block 4. The output will not be present. So the fault lies anywhere in the first half (blocks 1 to 4).

(b) Split the circuit further into half and measure the output of block 2. The output will be found to be correct as 10 MHz. The fault thus lies in blocks 3 and 4.

(c) Measure the output of block 3. It will be found to be correct as 1 MHz. Therefore, the fault lies in block 4.

The example given here and the number of tests indicated may be an over-simplification. In practice, more measurements would be necessary to localise a fault, in case there are multiple faults in the system.

Unfortunately, most electronic systems do not involve only series of connected blocks. They may have feedback loops or parallel branches in a part of the circuit. Such situations will make the task of localizing faults complicated, but the half-split rule still remains applicable. The following situations may be carefully watched and fault analysis systematically carried out:

Divergent Paths: In divergent paths (Fig. 2.13), the output from one block feeds two or more blocks. The most common example is that of the power supply circuit which supplies DC power to various sub-systems in an equipment.

Fig. 2.13   *Troubleshooting circuits with divergent paths*

In such a scheme, it is best to start by checking the common feedpoint. Alternatively, if one output (say at A or B) is normal, check after the divergence point. Conversely, if one output is abnormal, check before the common point.

Convergent Paths: In convergent paths, two or more input lines feed a circuit block (Fig. 2.14). To check the performance of such a scheme, all inputs at the point of convergence must be checked one by one. If all are found to be correct, the fault lies beyond the convergent point. However, if any of the inputs is incorrect (at C, D) then the fault lies in that particular input circuit. For example, if C and D are correct, and there is no output at E, the fault lies in Unit 3. But if the input at C is incorrect, the fault lies in Block 1 or before that.

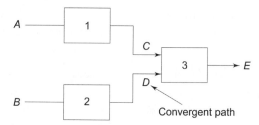

Fig. 2.14 *Isolating faults in convergent paths*

Feedback Paths: The presence of a feedback loop in a circuit presents one of the more difficult problems in fault location. The feedback loop normally corrects the output of some block with the input of an earlier block via some network, known as the feedback network (Fig. 2.15). Since the circuit behaves as a closed loop, any fault within the loop will appear as if the outputs of all blocks within the system are at fault.

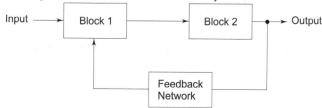

Fig. 2.15 *Feedback network circuit fault location*

Feedback paths are provided basically for the following functions:

(a) *Sustaining circuit function:* In this case, the feedback is totally essential for an output to exist. For example, an oscillator circuit.

(b) *Modification of circuit function:* The feedback loop is provided to modify the characteristics of a system, for example, an automatic gain control circuit in a superhet radio receiver.

Before attempting repairs on a faulty system having a feedback loop, it will be essential to understand the type of feedback used and its purpose. Having done so, proceed as follows to locate the fault:

With *modifying* feedback, it may be possible to break the feedback loop and convert the system to a straight linear data flow. Each block can then be tested separately without the fault signal to be fed around the loop. In some cases, instead of completely breaking the feedback loop, it may be preferable to modify the feedback at or near the point where the feedback path rejoins the main forward path. If the output appears normal, check the feedback circuitry. If not, check the forward path.

In case of the *sustaining* type, feedback is disconnected from the input and a suitable test signal is injected to check the performance of various circuit blocks. Due to a wide variety of feedback circuits, it is difficult to suggest a rule of thumb for fault location in this type of feedback.

Switching Paths: In a system which has switchable parts (Fig. 2.16) and if the circuit function is found to be faulty in one position of the switch, throw the switch to another position. If the trouble persists, check after the switch in the common circuitry. If the trouble disappears with this action, check that the circuitry is switched out.

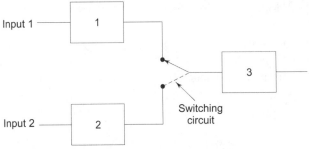

Fig. 2.16   *Fault location in switching circuits*

2.10.3   Systematic Troubleshooting Checks

The types of trouble in a non-functional equipment could often be simple but are in many situations, quite complicated. Therefore, the checks should be arranged in such a manner that simple trouble possibilities are eliminated before proceeding with extensive trouble-shooting. The simple and preliminary checks help in ensuring proper connection, operation and calibration. If the trouble is not located by these checks, the more elaborate tests aid in locating the defective component. When a defective component is located, it may be serviced or replaced in a systematic manner. The following steps are usually helpful in troubleshooting:

(a) *Check Control Settings:* Incorrect control settings can indicate a trouble that does not exist, particularly in equipment that have a large number of controls on the front and sometimes on the back panel. In case of any doubt about the correct function or operation of any control, the operating manual should be consulted.

(b) *Check Associated Equipment:* Before proceeding with troubleshooting of an equipment (particularly test and measuring instrument), check that the equipment used with this is operating correctly. For example, if the working of an oscilloscope is in doubt, check that the signal is properly connected and that the inter-connecting cables are not defective. Also, check the power source.

(c) *Visual Check:* Visual checks help in locating many troubles and therefore, the position of the equipment in which the trouble is located should be thoroughly inspected. Visual checks can aid in quickly detecting unsoldered connections, broken wires, damaged circuit boards and damaged components, etc.

(d) *Calibration:* Some troubles are caused due to misadjustment of certain pre-set or calibration controls. Therefore, check the calibration of the equipment or the affected circuit. The calibration procedure as suggested by the manufacturer may be carried out to correct the trouble.

(e) *Isolate the Troubling Circuit:* To isolate the trouble to a particular circuit, the trouble symptoms should be thoroughly noted and analysed. The symptoms often identify the circuit in which the trouble is located. For example, poor focus control in an oscilloscope indicates that the cathode ray tube circuit including high-voltage supplies is possibly faulty.

   The incorrect operation of all circuits or major parts of an electronic equipment often shows trouble in the power supplies. The power supplies must operate within close tolerances, which can be judged by measuring voltages between power supply test points and the ground. If the voltages are outside the tolerances, the supply may be misadjusted or operating incorrectly. If incorrect operation of the power supplies is diagnosed, the equipment may be connected to a variable auto-transformer. Check for correct regulation with a DC voltmeter and correct ripple with a test oscilloscope. These checks should be carried out while varying the auto-transformer throughout the regulating range of the instrument.

(f) *Measurements:* Often the defective component can be located by checking for the correct voltage or waveform in the circuit. Usually, information on typical voltages and waveform are given for various points on the circuits in the service manual of the equipment.

   It may, however, be kept in mind that voltages and waveforms are usually not absolute and may vary slightly between equipments. Therefore, these measurements and observations should be used as a guide for diagnosing a fault, rather than making efforts to adjust the circuit components to achieve the desired voltage levels and waveform patterns. Nevertheless, any gross variation should be carefully looked into and rectified for correct operation of the equipment.

(g) *Individual Components:* When an individual component is suspected, it must be properly tested. Components which are soldered in place are best checked by first disconnecting one end. This isolates the measurements from the effects of the surrounding circuitry.

While detailed procedures for different types of components will be given in the following chapters, it may be remembered, at this stage, that power must be turned off before removing or replacing semiconductor devices.

Some useful tips for troubleshooting are given below:

1. Whenever possible, try to substitute a working unit, particularly when the equipment is of modular construction. This will help to narrow down the problem to a single unit which should be the first priority.

2. Many problems have simple solutions. It should not be immediately assumed that the problem is always complex requiring detailed measurements. For a VCR, it may just be a bad belt, for a CD player a dirty lens or the need for lubrication. The majority of problems being mechanical, they can be dealt with by using nothing more than a good set of precision hand tools, some alcohol, de-greaser, light oil and grease coupled with your experience and power of observation..

3. Confirm the problem before venturing into repairs. Try to get as much information as possible about the problem from the user of the equipment. For example, 'did the problem come and go before finally staying bad for good' can be a useful hint for troubleshooting.

Rules of Thumb in Troubleshooting

1. Problems that result in a totally dead unit or affect multiple functions are generally related to the malfunctioning of the power supply. These are usually easy to diagnose and rectify.

2. Problems that are intermittent or erratic—that come and go suddenly—are almost always due to bad connections. These could be due to cold solder joints or internal or external connectors that need to be cleaned and re-seated.

3. Problems that change gradually, usually they decrease or disappear as the equipment warms up, are often due to dried up electrolytic capacitors. While they occasionally leak making diagnosis easy, in most cases, there are no obvious signs of failure. In that case, they need to be tested with an ESR meter.

4. Catastrophic failures often result in burnt, scorched, cracked, exploded or melted components. They can be easily identified by using senses of sight and smell for the preliminary search for such evidence.

5. Listen carefully for signs of arcing or corona, which could be snapping or sizzling sounds. They should be dealt with immediately since they can lead to more serious and expensive consequences. However, there are components like the flyback

transformer, yoke or other magnetic parts, which may also emit a buzz constantly or intermittently. The two types of sounds should not be confused with each other.

6. A number of organisations have compiled data base covering thousands of common problems with some of the electronic equipment like TVs, VCRs, computer monitors, etc. They provide information based on actual repair experiences and case histories. Most of the companies charge for the information but a few are accessible via the Internet. In that case, you can greatly simplify your troubleshooting or at least confirm a diagnosis before ordering the parts.

2.11 APPROACHING COMPONENTS FOR TESTS

Most manufacturers provide test points at convenient locations on the circuit board. These points are defined by specific DC and AC voltages, along with the waveform pattern. Figure 2.17 shows a test point as it would appear on a circuit board. This is usually a vertically mounted pin to which a test prod can be attached.

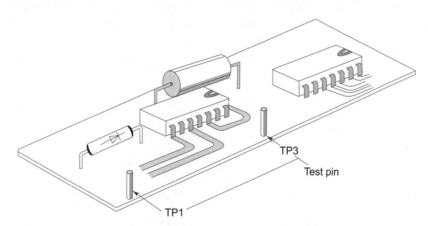

Fig. 2.17 *Typical test point indication in a printed circuit board*

If specific test points are not provided, measurements can be made at various points on the circuit by approaching various components. In that case, proceed as follows:
 (a) For transistors, make test prod connection to the legs under the case.
 (b) To read the signal on a circuit board trace, locate a component that is connected to the trace (Fig. 2.18). Clip your test lead onto the leg of the component that is connected to the trace.

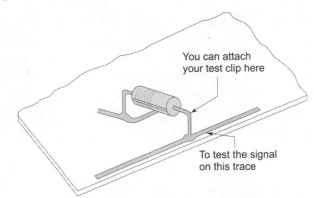

Fig. 2.18   *Taking measurements from the circuit trace by connecting a test prod on the component*

(c) Connections to the ICs can be made more conveniently by using an IC test clip (Fig. 2.19). Be careful not to touch more than one conductor at a time, otherwise you can easily create a short circuit. Since digital circuits are usually densely packed on a board, make use of only as narrow a prod as possible.

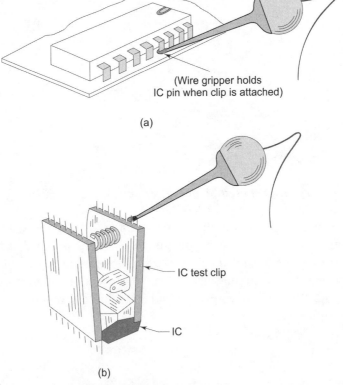

Fig. 2.19   *Use of test clip for taking measurement on IC pins*
(a) Test clip on the IC directly (b) Test clip on the IC connector

(d) Flexible flat wires with connectors often offer a good place to take readings. The connector pins themselves are usually well-protected, but you can take readings at the conductors behind the connectors (Fig. 2.20).

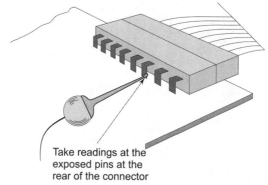

Take readings at the exposed pins at the rear of the connector

Fig. 2.20 **Taking test readings from a connector**

(e) Reading from an insulated wire can be taken by piercing the insulation as shown in Fig. 2.21. Carefully push a needle through the insulation and attach the test prod to it. After finishing the test, cover the pierced part of the wire with insulated tape.

Needle or pin

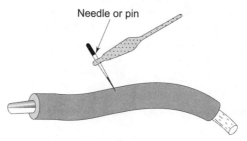

Fig. 2.21 **Method of taking measurements from insulated wire**

2.12 GROUNDING SYSTEMS IN ELECTRONIC EQUIPMENT

For making any electrical measurement, two test points are needed—the test point itself and the ground. In the case of electronic equipment, the term 'ground' may have different meanings in different situations as explained below:

(a) In the case of old electronic equipment, components were usually mounted on a metal frame or chassis. This frame was usually grounded through the power cord with the third prong of the AC plug.

'Chassis ground', in the use of modern equipment, generally refers to the grounding traces on the circuit boards, ground leads on motors and solenoids and the case of the equipment. Figure 2.22 shows a typical chassis ground.

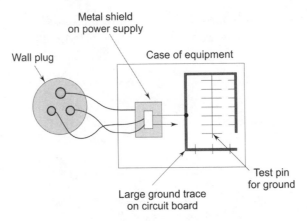

Fig. 2.22   ***Chassis ground network in an electronic equipment***

Chassis ground traces on the circuit board are usually wide and run around the outer edge of the board, from where they are carried to all parts of the circuit. There is generally a test pin mounted on the circuit board which is dedicated to this ground.

(b) In most equipment, the digital circuits (TTL) are provided with 'logic ground' which is isolated from the chassis ground. Such an arrangement is shown in Fig. 2.23. This system of grounding is adopted to prevent transients and electrical noise from contaminating the power to the TTL ICs. In this system, the logic ground will be at a slightly different potential than the chassis ground.

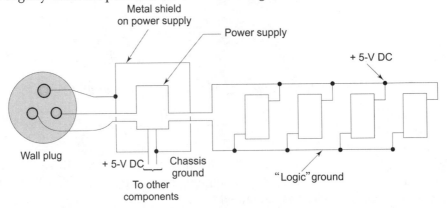

Fig. 2.23   ***Chassis ground and logic ground in a digital electronic equipment***

(c) A third type of ground exists in a special situation in which a 'signal' ground is also provided. The signals are referenced to this independent ground. An example of this ground is found in the link between a computer and a disk drive.

In view of the presence of different type of grounds in the equipment, always try to use a ground connection as close as possible to the component that is to be tested. It will remove most of the confusion between the three types of ground.

2.13 TEMPERATURE-SENSITIVE INTERMITTENT PROBLEMS

You cannot find troubles that are not there. The trouble that goes away when you start looking for it is the most exclusive kind. Often that condition is a temperature-sensitive one and you can make it occur by raising the temperature of one of the boards. The heat from the bench lamp or hand-held hair dryer is usually enough to do the job. But take care, you can sometimes quickly overheat the board that way.

When you observe a trouble to come and go by applying and removing heat to a whole area, you can usually finish the job quite quickly by combining the heating act with a cooling act. Use a pressurised can of circuit coolant and spray different components while the board is hot. When the defective component is cooled, the trouble will come on immediately. When the others are cooled, it makes no difference. The blast of coolant should be brief and closely confined to the component suspected of being faulty. It may be noted that sometimes you will think you have located the trouble spot when you are actually only close to it.

The faulty ICs often feel warm or hot. You can use your finger to make a quick 'touch test' to find the faulty IC. It is better to use the back of your finger to make the test as shown in Fig. 2.24. This part of the finger is actually more sensitive to heat than the finger tip. Professional service engineers sometimes use digital thermometers to measure the temperature of a suspect component.

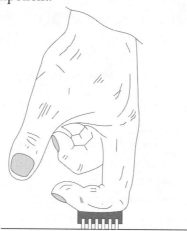

Fig. 2.24 *Feeling of temperature of a component with the back of finger*

It may be remembered that some components are hot even when they are operating normally. If a transistor has a large heat sink, you can expect it to be hot. Take care that you do not touch any part for temperature feeling which is connected to the mains supply.

2.14 CORRECTIVE ACTION

After establishing the fault and localising the defective component, special techniques are required to carry out the repairs. The steps involved are given below:

2.14.1 Arranging Replacement Parts

Replacement for all electrical and mechanical parts for most of the equipment can be obtained from the manufacturer of the equipment. At the same time, many of the standard electronic components can be obtained from vendors, which can save considerable time in procurement. Before purchasing or requesting the manufacturer for replacement parts, the parts list should be carefully studied for the value, tolerance, rating and description of the component required.

When selecting equivalent replacement parts, it may be remembered that the physical size and shape of a component can sometimes affect its performance in the equipment, particularly at high frequencies. It is therefore imperative that all replacement parts should be direct replacements unless it is established that a different component will not adversely affect equipment performance.

In addition to the standard electronic components, the equipment may contain some special parts. Such components are manufactured or selected by the equipment manufacturers to meet specific performance requirements. The mechanical parts are mostly especially made and must therefore be directly obtained from the equipment manufacturer.

While ordering replacement parts from the equipment suppliers, the following information is generally needed by them for making early supplies:
  (a) Name of the equipment;
  (b) Equipment model or type;
  (c) Equipment serial number;
  (d) A description of the part needed; i.e. name of the part and circuit number; and
  (e) Part number of the part specified by the manufacturer in the service manual.

Changes in equipment are sometimes made to accommodate improved components, as they become available by the manufacturers. Information about such modifications is usually issued by them as 'addenda' to the service manual. When any discrepancy is noticed about a component mentioned in the circuit diagram and the one actually present in the equipment, do look for the modifications sheet. In case of doubt, the manufacturer can be contacted for clarification.

2.14.2 Component Replacement

Before attempting the replacement of components, the equipment must be disconnected from the power source. The service manuals usually contain exploded-view drawings associated with the mechanical parts and they may be helpful in the removal or disassembly of individual components of sub-assemblies.

Unnecessary component substitution should be avoided as far as possible, otherwise it can result in damage to the circuit board and/or adjacent components.

Sometimes, it may be found that a circuit board is damaged beyond repair; the entire assembly including all soldered-on components should be replaced. Spare boards can be normally ordered from the equipment manufacturer.

Semiconductor devices should not be replaced unless found to be actually defective. Unnecessary replacement of semiconductors may affect the performance or calibration of the equipment. If removed from the sockets during routine maintenance, they should always be returned to their respective sockets.

Replacement semiconductor devices should be of the original type or a direct replacement. Some plastic case transistors have lead configurations which do not agree with standard casing as used for metal-cased transistors.

An extracting tool should be used to remove the integrated circuits (dual-in-line package, 14- and 16-pin types) to prevent damage to the pins. If an extracting tool is not available when removing one of these integrated circuits, pull slowly and evenly on both ends of the device. Try to avoid having one end of the integrated circuit disengage from the socket before the other, as this may damage the pins.

The power transistors are usually mounted on the heat radiator. After replacing a power transistor, check that the collector is not shorted to the ground before applying power.

The switches used in the equipment, if found to be defective, are not usually repairable and should be replaced as a unit.

Almost all electronic equipments make use of a power transformer, unless the equipment is battery-operated. If a fault is encountered in the power transformer, it must be replaced with a direct replacement transformer. When removing the transformer, tag the leads with the corresponding terminal numbers to aid in the new transformer. After the transformer has been replaced, check the performance of the complete equipment.

2.14.3 Performance Check

After any electrical component has been replaced, the performance or calibration of that particular circuit should be thoroughly checked. Since power supplies are used to operate all the circuits, the entire equipment must be checked to assess if work has been done in this section or if the power transformer has been replaced. In order to avoid unnecessary adjustment of other parts of the circuitry, adjust only if the tolerance given in each 'CHECK' part is not met.

In the service manuals, the following terms are often used as instructions in relation to the performance check:

(a) *Check:* Indicates that the instruction accomplishes a performance requirement check. If the parameter checked does not meet the indicated limits, an adjustment or repair is normally required.

(b) *Adjust:* Describes which adjustment to make and the desired result. Adjustments should not be made unless a previous 'Check' instruction indicates that an adjustment is necessary.

(c) *Interaction:* Points out that the adjustment described in the preceding instruction interacts with other circuit adjustments. The nature of the interaction is generally indicated and reference is made to the procedures affected.

2.14.4 Replacement of Circuit Boards

Usually, the service/instruction manual accompanying the equipment contains sufficient information to guide an experienced and skilful electronic technician in fault analysis and repair of some circuits in the equipment. If a malfunction is located to one board (or more), that is not readily repairable, it should be returned to the manufacturer for repair. Many manufacturers recommend that for economical and prompt replacement of any circuit board, the exchange board should be ordered. Usually, its price is considerably less than that of a new board.

For requesting an exchange board, the manufacturer usually expects the following information:

(a) *Equipment Description:* Name of the equipment, catalogue number, and serial number. This information is generally available on the front and rear panels.

(b) *Part Number of the Board:* This information is normally given on the parts list supplied in the manual. The number printed on the printed circuit board is not the part number.

(c) *Purchase Reference:* This helps in finding out if the unit is under or out of warranty and what type of billing is to be made.

2.15 SITUATIONS WHEN REPAIRS SHOULD NOT BE ATTEMPTED

It is quite difficult to suggest an approach to decide when the equipment is worth repairing because proper evaluation in terms of monetary, sentimental or other values plays a significant role. However, the following specific situations lead to the inevitable conclusion that the equipment is not worth attempting to repair, on your own or possibly at all:

  (i) Serious damage due to water, fire or smoke.
 (ii) Serious physical damage, especially to the mechanical parts, unless they are completely replaced.
(iii) Equipment wherein prior attempts at repair may have resulted in an undetermined number of new unidentified problems.
 (iv) Equipment with known design or manufacturing problems as it may not have a known solution.
  (v) When it is difficult to obtain service information for equipment like cellular phones, pagers, cordless phones, PC mother boards, disk drives, etc.; normally, you may not have the documentation, test equipment, rework equipment or access to spare parts. You may be lucky if the problem is an obvious broken connector or broken trace on the printed circuit board or possibly a dead power supply.
 (vi) If you really do not know about the equipment, leave it to be checked by a trained professional. Not only could it be dangerous in some cases, but such attempts will also cause additional, possibly fatal damage to the circuitry. If you can't justify a professional repair, leave it aside for sometime until you have gained more experience and can deal with the equipment safely.
(vii) Do not attempt to repair a piece of equipment for which you are not equipped in terms of test equipment and tools. Attempting to remove a part from a multi-layer printed circuit board without proper de-soldering equipment will make an unsalvageable mess.

2.16 GENERAL GUIDELINES

- Check your test equipment periodically.
- Never trust that brand new components will always be good.
- Do not assume that there could be only one failure which accounts for the problem. Sometimes there can be two or more component failures which contribute to a single noticed problem.
- If several events have occurred at nearly the same time, it does not necessarily mean that one event caused the other. They may all be consequences of a common cause or they may be totally unrelated. Do not mistake causality for co-incidence.

- After a long attempt at troubleshooting a difficult problem, you may get worn out and overlook crucial clues. Let someone else look at it for a while or let them help you. Take a break. You will be amazed at what a difference this has made.

The steps outlined in this chapter, if followed systematically, will help a technician to undertake troubleshooting work on electronic equipment in a systematic way and enable him to complete the repair and servicing jobs in an efficient manner.

3

Electronic Test Equipment

3.1 MULTIMETERS

A multimeter is simply the most common and most useful tool of the electronics trade. This instrument facilitates the measurement of DC voltage, AC voltage, DC current and resistance values. The multimeter can be particularly useful for the following tests:

(a) Measurement of in-circuit resistance, i.e., checking for dry joints, taking resistance readings around transistors and diodes;

(b) Measurement of DC voltages around suspect transistors, to determine whether the transistor is conducting or non-conducting; and

(c) Measurement of supply voltages on all ICs.

It may be appreciated that for all such measurements, a very high degree of accuracy is not required. A commonly available analog multimeter with an accuracy of ±1% is adequate for a majority of requirements in which merely the presence of a value near the one specified is required rather than a measured value that is exactly as expected. Such requirements are conveniently met by an analog multimeter in preference to a digital reading type instrument since an analog indication of approximate voltage level can be more quickly observed on this instrument. A 20,000 ohms/volt multimeter can be used to check the voltages in the equipment if allowances are made for the circuit loading of the multimeter at high-impedance points.

A digital multimeter is preferred when a high accuracy is required, specially when very small changes in a level need to be detected. The digital multimeter also has a high input impedance, typically 10 MΩ, so that its loading effect is negligible in the circuit. A digital multimeter with 0 to 500 V range and Ω meter, 0 to 2 MΩ and an accuracy of +1% is adequate for most of the work.

3.1.1   Analog Multimeter

The analog multimeter or VOM (volt ohm milliammeter) is the most widely used electrical instrument in design, development, repair, service and maintenance laboratories. It is usually a moving coil meter, which, by switching and the selection of probe jacks, can become a DC voltmeter, an AC voltmeter and DC milliammeter or an ohm meter. Sometimes, AC current measuring facility is also present and occasionally, it is not included.

Figure 3.1 shows the principle of the moving coil meter used as a test instrument. A coil of fine wire wound on a rectangular aluminium frame is mounted in the air space between the poles of a permanent horse-shoe magnet. Hardened-steel pivots attached to the coil frame fit into the jewelled bearings so that the coil rotates with a minimum of friction. An indicating pointer is attached to the coil assembly, and springs attached to the frame return the needle (coil) to a fixed reference point.

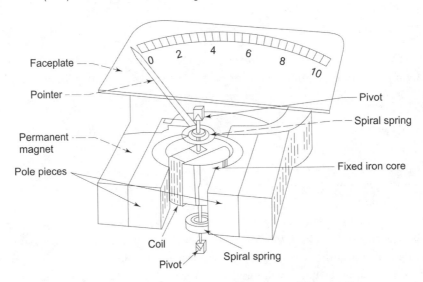

Fig. 3.1   *Principle of the moving coil meter*

When electric current flows through the coil, a magnetic field is developed that interacts with the magnetic field of the permanent magnet to force the coil to rotate as in an electric motor. The direction of rotation depends on the direction of electron flow in the coil. The magnitude of the pointer deflection is proportional to the current. In a typical multimeter, there will be different scales for different ranges. Usually, the full-scale deflection is about a 90 degree arc. Figure 3.2 shows a typical scale of a multimeter.

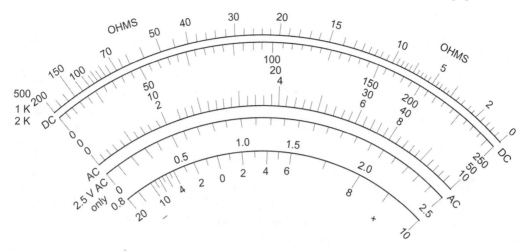

Fig. 3.2 *Typical scale of an analog multimeter*

The meter movement has three characteristics which determine its effectiveness for a given measurement. These characteristics are sensitivity (specified as milliamperes full-scale or microamperes per centimetre or microamperes per degree), internal resistance (the DC resistance of the coil in ohms) and accuracy (normally 0.5 to 3% of full-scale deflection for general purpose metres and 0.1 to 1% for laboratory types). A multimeter has a lower sensitivity rating on its AC voltage function than on its DC voltage function. For example, a 20 Kohm/V instrument may have a sensitivity of only 5 Kohms/V on its AC voltage function.

Using the Multimeter

Measurement of Current: The moving coil meter is basically sensitive to current and is, therefore, an ammeter. However, it is often necessary to increase the range of a meter in order to measure large currents. This is done by shunting part of the current around the coil so that only a fraction of the total current passes through the coil. When different ranges are switched for current measurement, different shunt resistances get placed across the coil.

For *direct current* measurement, the meter is placed in series in the circuit of interest. Therefore, the circuit must be broken to connect the ammeter. The ammeter thus becomes a part of the circuit. Since the ammeter has some internal resistance, its insertion into the circuit may decrease the current in the measured branch. Usually, this resistance is small and can be ignored.

The moving coil meter depends only on the unidirectional current through its coil.

For *alternating current* measurement, rectifier type meters are used which will respond to the average value of the rectified alternating current. If the meter is calibrated in amperes rms (root-mean square), it is assumed that sine waves are being measured and the factor 0.707/0.637 (rms/average) is included in the dial calibrations. If some waveform other than a sine wave is being measured, so that the rms/average factor is different, the meter will not read the true rms value of the current. The frequency characteristics of the rectifier impose an upper limit on the frequency of which alternating current can be measured. At frequencies above 1KHz, it is important to consider the possibility of error from this source.

Measurement of Voltage: The moving coil meter has a constant resistance, so that the current through the meter is proportional to the voltage across it. In this way, the current meter can be used to measure voltage. The full-scale deflection sensitivity in volts is the full-scale deflection current times the resistance of the meter. To extend the voltage range of the meter, it is necessary to only add resistance in series with the meter circuit.

The potential difference between any two points in a circuit is measured by connecting the two voltmeter leads to these points. Thus, the voltmeter is connected across or in parallel with the circuit whose potential is to be measured. The application of the voltmeter to the test points may change the current magnitudes in the circuit enough to cause a considerable change in the potential being measured. To avoid this kind of error, a voltmeter of sufficiently high internal resistance must be used to make voltage measurements.

The ohms per volt rating of a meter is constant. It will be higher if a moving coil meter of greater current sensitivity is used. The current sensitivity of the meter with a 20,000 Ω/volt rating is 50 microamperes full-scale deflection. A 20 kohm/V meter will have a resistance of 30 kΩ on the 1.5 V scale and 600 kΩ on the 30 V scale. The resistance of the meter movement would be about 300 Ω and is so small that it can be neglected in calculations.

In order to measure AC voltage, rectification is required. As in the AC current meters, AC voltmeters respond to the average value of the rectified voltage but are calibrated in volts rms for a sine wave. If a non-sinusoidal voltage is measured, a factor other than 0.707/0.637 will have to be used to determine the rms value of the voltage. Analog multimeters operate accurately over a limited frequency range. A typical 20 Kohms/V meter loses its accuracy at 20 KHz. At higher frequencies, the meter indicates more than the actual value. Digital multimeters are preferred if the measuring frequency range is higher.

For making measurements of extra high tension (EHT) voltages, in access of 1000 V, the multimeter is supplied with special adaptors called EHT probes. Great care should be taken while taking measurement at these high voltages because they can be lethal. Test probes must be well insulated to prevent accidental shorting.

Most diodes do not work at very low current values. Therefore, very sensitive meters are not used as AC voltmeters, and the ohms per volt rating of AC voltmeters is usually much lower than is typical for a DC voltmeter. The diodes also impose a frequency limitation, and care should be taken to ensure that accuracy is maintained when an AC voltmeter is used to measure voltages of frequencies over 1 kHz.

In the case of AC measurements, the meter movement responds to the average value of the rectified current and therefore, there could be inaccuracy in measurement because of different wave-shapes. If the applied waveform is non-sinusoidal, such as a square and triangular waveform, all rectified type AC volt meters are subject to indication errors. Therefore, it is advisable to consult the manufacturer's chart for the appropriate factors to be taken into consideration to get the correct value.

Measurement of Resistance: The moving coil meter can be used to measure unknown resistance by using a circuit configuration as shown in Fig. 3.3. With the test prods short-circuited $(R_u = 0)$, the ohms adjust control is turned so that the current through the total circuit resistance $(R_m + R_f + R_e)$ deflects the meter exactly full scale. Now, by connecting the test prods across the unknown resistance R_u, the current is decreased to a value I_2. If the battery voltage is 1.5 V and a 1 mA meter movement is used,

$$R_u = \left(\frac{I_1}{I_2} - 1 \right) 1500$$

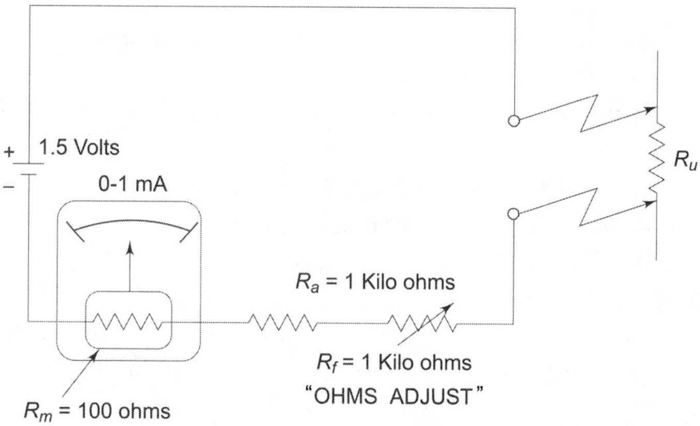

Fig. 3.3 *Resistance measurement circuit in an analogue multimeter*

When $I_2 = (1/2)\, I_1$ (mid-scale deflection), the unknown resistance $R_u = 1500$ ohm, and with $I_2 = (1/3)\, I_1$, $R_u = 3000$ ohm, etc. It is apparent that the scale is non-linear. The value of R_u becomes much higher than 1500 ohm become crowded on the 'infinite ohms' end of the scale, and the values much lower become indistinguishable from zero.

The resistance of devices that might be damaged by moderate currents cannot be measured with an ordinary ohmmeter. Such devices include meter movements and some fuses, lights, relays, tube filaments, some diodes, etc. When the danger of damage exists, some other means must be devised to make the resistance measurement.

An ohmmeter is never used while the circuit is in operation and thus there is no circuit distortion introduced by the measurement. For resistances that depend on circuit conditions, the only solution is to establish normal operating conditions, measure the voltage across the resistance, measure the current through it and calculate the resistance value.

Many meters incorporate a buzzer for use in continuity measurements. Before using the multimeter, it is essential to make a check of its zero setting and the condition of the batteries. To do so, the range switch is set to OFF and the two test probes are shorted together. If the meter pointer does not read exactly zero, turn the screw on the meter movement slowly until the proper zero reading is obtained. Similarly, to check the condition of internal batteries, the range switch is set to $R \times 1$ and the two test probes are again shorted together. Try to adjust the zero ohm adjuster control so that the pointer reaches zero ohm at the right end of the ohms scale. If the pointer cannot be brought to the zero mark, the batteries need replacement.

Precautions in the Use of a Multimeter

(a) The precautions, limitations and errors mentioned for each measurement circuit apply to the multimeter as well.

(b) The batteries in the multimeter should be checked frequently for correct operation in the resistance ranges.

(c) The instrument should be stored away from corrosive atmospheres for protecting the switch contacts from environmental conditions.

(d) The function switch of the multimeter should always be kept on a high DC volts scale to avoid:
   (i) draining batteries by accidental short-circuiting of leads during storage; and
   (ii) burning out the rectifier by accidentally connecting to the DC voltage when on the AC scale.

(e) Never ground yourself when making electrical measurement. Keep your body isolated from the ground by using dry clothing, rubber shoes, rubber mats or any other suitable insulating material.

(f) Some meters provide a fuse in the resistance mode for protection from over-voltage, while others include thermistor protection. It is unwise to substitute these devices with other kinds of resistance for fusible components when these devices fail due to overload. This can be dangerous and should be avoided.

3.1.2   Digital Multimeter

Digital multimeters (Fig. 3.4) are characterized by high input impedance and better accuracy and resolution. They usually have auto-ranging, auto-polarity and auto-zero facilities, which means that the user need only set the function switch and get the reading.

Fig. 3.4   *Digital multimeter (courtesy: M/s Fluke, USA)*

The digital multimeter converts an input analog signal into its digital equivalent and displays it. The analog signal input might be a DC voltage, an AC voltage, a resistance or an AC or DC current. Figure 3.5 shows a block diagram of a typical digital multimeter (DMM).

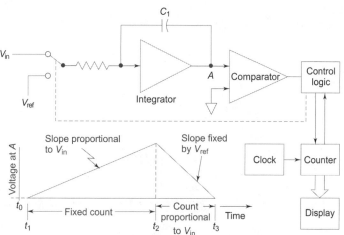

Fig. 3.5   *Block diagram of a digital multimeter using analogue-to-digital converter by the dual-slope technique*

The heart of the DMM is the analog-to-digital (A/D) converter. There are several different principles on which A/D converters can be designed. However, the most popular and widely used method is the dual-slope technique. Referring to the diagram, at time t_1, the unknown input voltage V_{in} is applied to the integrator. Capacitor C_1 then charges at a rate proportional to input voltage V_{in}.

The counter starts totalizing clock pulses at time t_1 and when a pre-determined number of clock pulses has been counted, the control logic switches the integrator input to V_{ref}, a known voltage with a polarity opposite to that of V_{in}. This is at time t_2. Capacitor C_1 now discharges at a rate determined by V_{ref}.

The counter is reset at time t_2 and again it counts clock pulses, continuing to do so until the comparator indicates that the integrator output has returned to the starting level, stopping the count. This is at time t_3.

The count retained in the counter is proportional to the input voltage. This is because the time taken for capacitor C_1 to discharge is proportional to the charge acquired, which, in turn, is proportional to the input voltage. The number in the counter is then displayed to give the measurement reading.

The front-end of the DMM contains circuit blocks which govern its basic characteristics such as its number of digits, ranges, sensitivity, etc. Since the input to the A-D converter must be DC voltage, all other quantities should be reduced to DC. Therefore, for measurement of AC voltage, it is first rectified to derive a DC voltage proportional to the average value of the AC waveform calibrated to the rms value for a sine wave, and the resulting DC voltage is applied to the integrator.

Resistances are measured by supplying a 1-volt reference signal to the input amplifier through a range resistor and configuring the amplifier to place the unknown as the feedback resistor, as shown in Fig. 3.6. The amplifier output is proportional to the ratio of the unknown resistance to the range resistor.

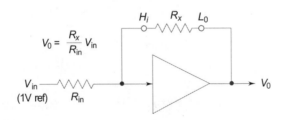

Fig. 3.6 *Input configuration for resistance measurement*

Most DMMs are similar in terms of voltage, current and resistance measuring capabilities. These three electrical properties are part of every circuit, whether it is for a medical electronic equipment or computer or a motor. However, they differ in these areas chiefly in terms of accuracy, selection of ranges, AC bandwidths, etc. With the help of optional accessories, a digital multimeter can also be used to measure temperature, frequency, duty cycle, capacitance and other parameters.

In general, the basic DMM comes with a pair of general purpose test leads. Most manufacturers, however, offer a variety of optional accessories for their instruments, specially among these being battery eliminators and special-purpose probes. The probes either extend existing ranges or the utility of the DMM. Standard available accessories are high-voltage probes that permit measurements up to and beyond 30 kV, temperature measuring attachments and current measuring clamp-ons.

The choice of accessories depends on the specific applications for which it is intended to be used. For example, for handling appliances repair, a relatively high current clamp-on facility is required. Most DMMs can measure up to 2 A while some appliances often draw 10 A or more. Temperature probes are required for the heating and air conditioning equipment and for monitoring the operating temperatures of power transistors. TV technicians may select a probe which extends a 15 Meg ohm input meter to 10,000 V DC operation for checking horizontal output stages. Its high input impedance also reduces loading in oscillator, power and IF stages. Although LCD display type DMMs appear to be the most popular, of late, a field technician might find an LED display more useful when working in a subdued lighting area.

DMMs can be portable, bench type or both. Portable instruments are preferable for troubleshooting work. They are available in a variety of sizes, the most common being the hand-held type. Bench models usually provide more facilities than portable instruments.

Since the DMM becomes a part of the circuit when it is being used, its impedance is likely to affect the current flow through the circuit under test. Obviously, the meter should have a very high input impedance so that it causes negligible loading effect. This is an important factor to be considered when choosing a DMM. Most DMMs have an impedance of the order of 10 M Ω to 10 G-Ω.

Most DMMs feature auto-range facility in which the appropriate scale offering the best resolution for set functions is automatically selected by the meter. The resolution, which is defined as the smallest detectable change in the reading of the multimeter, varies from meter to meter. The resolution is described in terms of digits and counts. For instance, a 3½ digit meter will display 3 full digits ranging from 0–9 and 1 (if the reading has a fractional portion of one half) or a blank if the reading has no fractional portion. When the resolution is described in terms of counts, a 3½ digit meter will display up to 1999 counts of resolution. The auto-polarity feature in the meters indicates negative reading with a minus sign; so even if you connect the test leads in reverse order, the meter is not damaged.

The accuracy of the meter indicates how closely the measurement is displayed by the meter to the actual value of the signal. Usually, accuracy is expressed as a percentage of the reading. For example, an accuracy of 0.5% of the reading means that for a particular displayed reading of 100 V, the actual value of the voltage could be 99.5 V or 100.5 V. However, in a digital multimeter, digits are sometimes added to the basic accuracy values which indicates how many counts the digit to the extreme right of the display might vary. For example, $\pm(0.5\% + 1)$ means that for a display reading of 100.0 V, the actual voltage will be anywhere between 99.4 V and 100.6 V.

Most DMMs can accurately measure AC voltage with frequencies ranging from 50 Hz to 500 MHz. However, its accuracy specification for AC voltages and currents should clearly indicate the frequency range along with its related accuracy.

Low priced DMMs usually offer 3½ digits display which is adequate for most applications. The meter reads 1.999 mV on its most sensitive range. The half digit is, of course, the 1 in a display that goes to 1.999. It is turned on by the overflow of the last full digit. Do not be confused by the fact that the range selector switch is marked 2,20,200. The maximum reading, range-wise, would be 1.9, 19.9, 199.9. DMMs operate usually on dry cells or rechargeable Ni–Cd cells. Some DMMs come with a PC connection which allows you to download data to the PC for recording and analysis. The meter can be interfaced with the PC using the appropriate software.

3.1.3   Graphical Multimeters

Graphical multimeters have the ability to test electrical/electronic components through pattern recognition. This is done in an un-powered circuit. The instrument supplies a sine wave voltage to the component under test and plots the voltage vs. current relationship on the LCD display. The component test allows you to select one of the test frequencies: 2 Hz, 20 Hz, 200 Hz, 2 kHz and 18.75 kHz. This broad range of test frequencies offers the user the ability to test circuit with a wide range of capacitive characteristics. With multiple test frequencies, it is thus possible to troubleshoot many types of analog and digital circuits.

The characteristics of some of the basic components displayed on a graphical multimeter are shown in Fig. 3.7. Testing a component in-circuit will create troubleshooting patterns that are combinations of the basic pattern representing the characteristics of individual components.

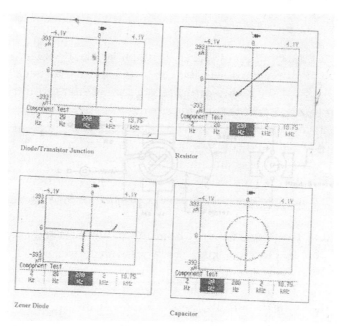

Fig. 3.7 **Characteristics of some of the basic components as displayed on graphical multimeter (a) diode/transistor (b) Zener diode (c) Resistor (d) Capacitor**

While it is possible to recognize individual component patterns, it will be advantageous to have a good circuit board available for comparison purposes. This will allow the most efficient use of the component test mode where there are multiple components influencing the pattern. In situations, where it is not possible to have a good board available, you can store component patterns for later recall and comparison.

A typical example of a graphical multimeter is the 860 Series available from M/s Fluke, USA. These instruments can store three test patterns in the waveform storage memory. Once stored, a pattern can be recalled to the screen for comparison. The FlukeView 860 software allows to upload component test patterns to a personal computer for long-term storage and circuit documentation. The computer makes it possible to make a large number of comparisons.

Although a graphical multimeter shortens the troubleshooting time, it has its own limitations. For example, the three-volt peak test voltage will not forward bias multiple p-n junctions in series. The display will remain a flat horizontal line indicating an open circuit. While this is not a severe limitation in most troubleshooting situations, it can present a serious problem when testing high power devices.

3.1.4 Megger

In servicing electronic equipment, there is often a need for the measurement of insulation resistance between the winding of a transformer, between the chassis and input terminals, between the phase and the ground, etc. The ordinary ohmmeter or VOM meter measures high resistance by applying only 9V or 15V across the resistance which is inadequate for insulation tests. Therefore, for these measurements, a special portable ohmmeter called 'Megger' is used.

The Megger has a small permanent magnet DC generator capable of developing 250V, 500V and 1000V. The generator is hand-driven, through gearing and a centrifugally controlled clutch, which slips at a pre-determined speed so that a steady voltage can be obtained.

The meter used in the Megger differs from the standard D'Arsonaval movement, in that it has two windings. One winding is in series with resistor R_2 (Fig. 3.8) across the output of the generator and is wound in such a way as to move the pointer towards the high resistance end of the scale when the generator is in operation. The other winding and resistance R_1 are series connected between the negative pole of the generator and the line terminal. This winding is wound so that when the current flows through it from the generator, it tends to move the pointer towards the zero-end of the scale. When an extremely high resistance appears across the terminals such as in the case of an open or near open circuit, the pointer reads infinity. On the other hand, when a relatively lower value of resistance appears across the test points, such as occurs when the cable insulation is set, current through the series winding causes the pointer to move towards zero resistance (short-circuit). However, the pointer stops at a point on the scale determined by the current through the series resistor which is governed by the value of the resistance being measured.

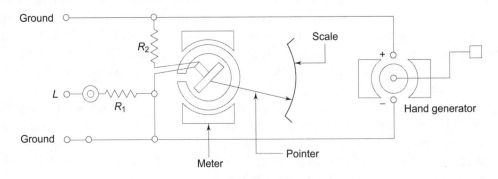

Fig. 3.8   *Circuit diagram of a megger*

In the modern Megger, the hand generator is replaced by a battery operated DC to DC converter to generate test voltage. Also, the analog display is replaced by digital display. For field use, the generator-based Megger is still preferred.

3.2 THE OSCILLOSCOPE

The oscilloscope (Fig. 3.9) is probably the most versatile, informative and useful electronic test instrument. It gives a visual indication of what a circuit is doing and often it can show what is going wrong quicker than any other instrument. Besides, some faults are often virtually impossible to pinpoint without using an oscilloscope. Pulse and digital circuits are extremely difficult to troubleshoot without this instrument because steady voltage readings are meaningless in pulse circuits as all one can check is whether an IC has a voltage supply or not.

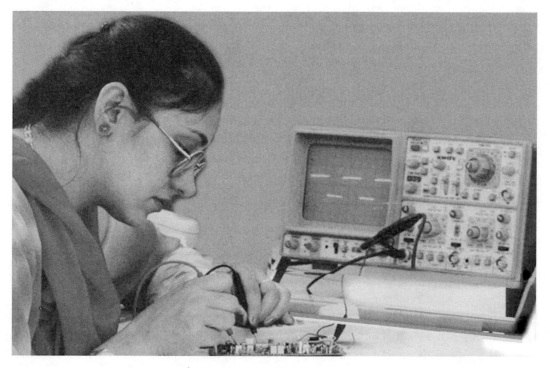

Fig. 3.9 *Oscilloscope in use*

A cathode ray oscilloscope is the most versatile of the test instruments. It can be used for wave-form analysis, signal frequency measurement, peak-to-peak voltage measurement and the most important for signal tracing. It is usually necessary for fault-finding on digital circuits, when correct operation depends not on voltage levels but on the presence (or absence) of a fixed level pulse which is too fast to register on a multimeter. A transistor curve tracer facility with an oscilloscope helps to test the semiconductor devices used in the equipment.

Broadly speaking, there are two types of oscilloscopes in the market today: Analog and Digital. Traditionally, oscilloscopes have been analog instruments, but as digital electronics has become cheaper and more powerful, digital scopes have increased in popularity. In fact, digital scopes now out-sell analog scopes as they offer many advantages to the users at an affordable cost.

In analog scopes, an electron beam sweeps across a phosphorescent screen, lighting up the screen wherever the beam hits. Circuits in the scope deflect the beam horizontally and vertically, thereby displaying a signal continuously. Digital scopes, which are often called digital storage oscilloscopes or DSOs, work very differently from analog scopes. A digital scope measures the voltage of the input signal at discrete time intervals. Using an analog-to-digital converter, the DSO converts a waveform into a series of numbers, which it stores in a table in its memory. The scope then uses the table of numbers to create the waveform display since a true display would contain only a series of dots.

The sensitivity offered by an oscilloscope is usually high, typically 10 mV/div, and in some cases 2 mV/division. Its impedance is generally greater than 1 Mohm. An oscilloscope with frequency response of DC to 15 MHz (preferably DC to 50 MHz) and a deflection factor of 5-volts/division is necessary for most of the troubleshooting requirements. A 10 X probe is generally used to reduce circuit loading.

Just to refresh the memory, the heart of an oscilloscope is the cathode ray tube (CRT). The working of a CRT depends upon the generation of electrons by a heated cathode, focusing it to a thin beam and making it travel towards a positively charged anode. The electrons strike on a glass screen, coated with phosphor which gives off light, making a spot of light on the screen.

The brightness of the spot can be controlled and so also its position. The spot can be deflected (guided) to any part of the screen by applying a varying electric field to the deflection plates—altogether four of them arranged in pairs called the X-plates and the Y-plates (Fig. 3.10). The Y-plates deflect the spot vertically, up or down, while the X-plates move it from side to side. Unlike the needle of a meter, a beam of electrons has practically no mass, so it can be moved around to trace out complicated patterns at very high speeds.

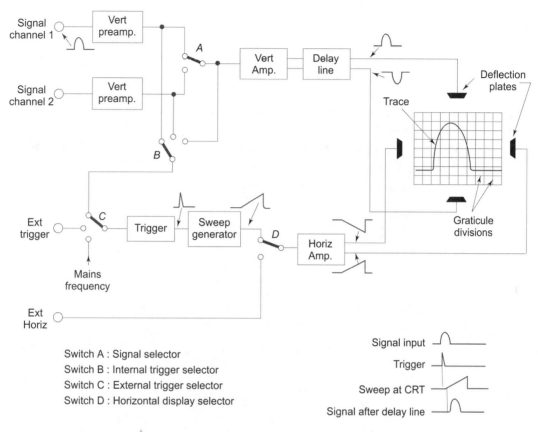

Fig. 3.10   *Block diagram of an oscilloscope*

Every oscilloscope has a built-in oscillator, the time base or horizontal sweep circuit. This circuitry generates a voltage waveform with a saw-tooth shape and feeds it to the X-plates. This results in moving the spot on the screen at a steady speed, from left to right. The speed can be controlled and measured conveniently and its value can be read in time per centimetre (time/cm, sometimes time/div.) control on the front panel of the oscilloscope.

During the time the spot is moving across the screen, a voltage fed to the Y-plates will make the spot move vertically showing the wave shape of the voltage in the light which is being fed to the Y-plates. An amplifier (Y-amplifier) is provided in the circuit whose gain control is calibrated in volts per centimetre of vertical movement so that the peak-to-peak voltage of the waveform can be measured. Thus, an oscilloscope can be routinely used to:

(a) Display the wave shape;

  (b) Measure its frequency; and

  (c) Measure the peak-to-peak amplitude.

3.2.1 Understanding an Oscilloscope

If you are not very conversant with the oscilloscope and how to use it, proceed as follows:

  (a) Carefully observe all the controls (Fig. 3.11) on the front panel. They may not be the same or designated was the same on all the instruments, but some of them have to be there somewhere and in some form. The essential controls are:

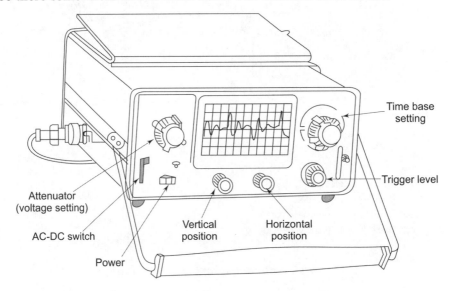

    Fig. 3.11   *Controls on an oscilloscope*

    (i) Intensity or Brilliance control;

   (ii) Focus control;

  (iii) X and Y position controls;

  (iv) Trigger, Sync or Level control, auto-mode; and

   (v) On/off control; it may be a separate control or combined with Brilliance/Intensity control.

  (b) Before switching on the instrument, make the following settings:

    (i) Intensity control fully anti-clockwise (off);

   (ii) Stability control to auto;

  (iii) Vertical and horizontal position controls to midway round;

  (iv) Volts/cm control to highest value of the range; and

    (v) Time/cm control to 1 ms/cm or its nearest value.

(c) Plug in the instrument to the mains supply and switch on the instrument. Wait for a minute or two so that the CRT heater warms up. Then gradually advance (clockwise) the setting of the brilliance control until you observe the horizontal line of the trace on the screen. Sometimes, the trace may not appear on the screen. In some oscilloscopes, a push button control 'TRACE LOCATE' is provided which, on pressing, produces a spot at the centre of the screen. When the switch is released, the spot slowly moves off to wherever it was before. This indicates that position controls are not properly set. If this control is not provided on the oscilloscope, proceed as follows:

    (i) Turn the Brilliance control right up to the fully clockwise position.

    (ii) Time/cm control to the slowest speed, but not the off position. A light spot should appear on the screen moving slowly from left to right.

        If still nothing is seen:

    (iii) Adjust the Trig/Level control in the clockwise direction. Observe if something seems to be happening.

    (iv) Operate the vertical position control until the trace appears. Some adjustment of the vertical gain and horizontal position control may be necessary.

        If these steps do not result in showing a trace on the screen, there is some problem with the instrument. Unplug the mains and check the fuses before attempting anything else.

        Assuming that the above-mentioned steps have produced the trace on the screen, the following additional steps will help in making various measurements:

    (v) The first step is to centre the trace with the help of horizontal and vertical position controls. The trace should start at the left hand side of the screen and lie along the centre-line. If there is a control labelled X Gain or TRACE EXPANSION, set it so that the trace is just enough to stretch across the screen, but no more.

    (vi) Reduce the Brilliance setting to a comfortable viewing level and adjust the focus control so that the line is as thin as possible. It is usually difficult to obtain a fine line if the Brilliance control is set too high.

Quite often the waveform does not appear to be stationary on the screen. If the whole wave is moving, adjust the control labelled Sync or Trig level. This control is used to start the time base at the same part of the waveform on each sweep, so that the sweep appears stationary (Fig. 3.12).

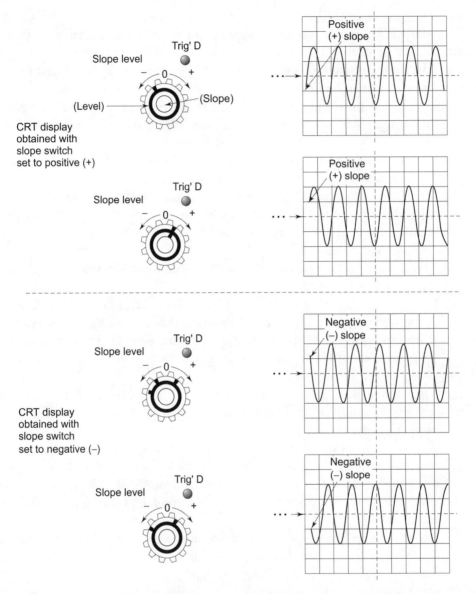

Fig. 3.12   *Effect of LEVEL control and SLOPE switch on CRT display*

If the trace is not still locked, check if there is a switch labelled TRIG INT-EXT (or SYNC INT or INT-EXT). In case it is present, place it on the INT (internal) setting. In this position, the time base is locked to the signal into the Y-input (TRIG, EXT or the X-input). Sometimes, on a few oscilloscopes, a FINE TIME/cm control may have to be adjusted to obtain a perfect lock.

It is usual for oscilloscopes to provide a choice of AC or DC coupling (Fig. 3.13) by means of an AC/DC switch at the Y-input. In the AC position, the signal on the Y-input is passed via a coupling capacitor and, therefore, any DC voltage also present in the signal is blocked. With the switch in the DC position, however, the Y-amplifier is completely DC-coupled from the input all the way to the CRT plates.

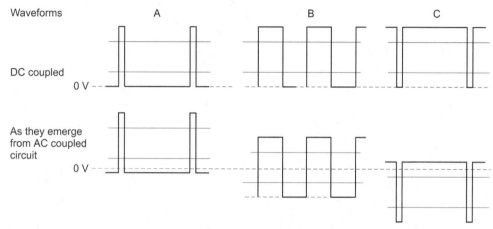

Fig. 3.13 *AC/DC coupling—the display of the same waveform on two different coupling positions*

When making oscilloscope measurements, a pair of probes is very valuable which facilitates making a contact on the point of measurement in a convenient manner. For inexpensive instruments, two lengths of ordinary insulated wire are sufficient, though a set of ordinary test leads, the black or earth lead having a crocodile clip on one end and the red or signal lead with a test prod at its end are preferred.

The ordinary test leads are adequate for the measurement of signal voltages of low frequencies. However, for high frequencies, it is essential to use a fully screened probe so as to avoid the possibility of signal degradation by way of signal amplitude attenuation and phase distortion occurring in a coaxial cable due to its large capacitance. The use of a compensated probe unit will, however, reduce these effects considerably.

3.2.2 Making Measurements with Oscilloscope

All oscilloscopes enable measurements to be made on the displayed waveform. The most common method is to have an engraved plastic sheet, called the 'graticule', which is fitted over the screen. The graticule is engraved with parallel lines, 1 cm apart, with small divisions on the centre lines to show 0.2 cm. Both horizontal and vertical lines are engraved, so that both time and voltage measurements are possible.

Amplitude (Voltage) Measurement: For voltage measurement, count the number of centimetres on the vertical scale from the negative peak to the positive peak and then multiply this number by the setting of the volts per centimetre switch. For example, if the volts/cm switch is set to 5 V/cm, and the waveform measures 4.8 cm from peak-to-peak, the waveform voltage is $4.8 \times 5 = 24.0$ V peak-to-peak (Fig. 3.14).

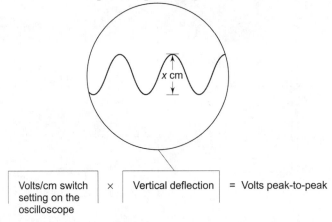

Fig. 3.14   *Voltage measurement with an oscilloscope*

Frequency Measurement: For frequency measurement, the method is to measure the time (period) of one complete cycle (Fig. 3.15) on the screen, i.e. the horizontal distance between two identical points on the neighbouring waves. This distance is then multiplied by the setting of the time/cm switch to calculate the period of one cycle. The reciprocal of this time (i.e., 1/time) is the frequency of the wave. For example, if the peaks of the waveform are 5 cm apart and the time/cm switch is set to 200 μs/cm, the time of one complete cycle is $5 \times 200 = 1000$ μs $= 1$ ms and the frequency is $1/1000$ μs $= 1$ kHz.

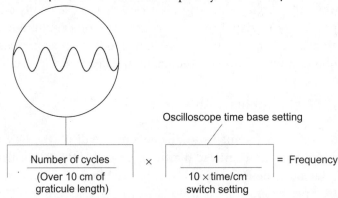

Fig. 3.15   *Frequency measurement with an oscilloscope*

An oscilloscope can be used to compare frequencies which are to be adjusted so as to be equal or in some simple relation to each other. The method is an old one called '*Lissajous Figures*' which are obtained by feeding two different signals into the scope at the same time, one into the vertical input and the other into the horizontal input. Under these circumstances, the internal time base is switched off, i.e. it is set to the external horizontal time base.

If the two signals are sine waves, and are synchronized, the pattern produced by this arrangement will be stationary. For equal frequency sine waves, the pattern can vary from a diagonal line to a circle due to phase difference between the waves—a difference of 90 degrees produces a circle whereas a 0 degree or 180 degree difference produces a straight line. If the frequencies are not identical, then the pattern will change and the number of complete cycles of change per second is equal to the difference in frequency of the two signals. Using this method, minute frequency differences between two sources can be measured with an accuracy of better than 0.01 Hz. It is an excellent way of testing the frequency stability of one crystal, in an oscillator, as compared to another.

Waveform Analysis: An oscilloscope is an excellent tool to see what is going on in a circuit and with experience, much can be gained from the correct interpretation of what is displayed. For example, if you are feeding in a pure sine wave signal into an amplifier and the oscilloscope displays a flat-topped waveform when connected at its output, it means that clipping is taking place in the amplifier as a result of over-driving one of its stages.

Similarly, when working with fast repetitive pulses (TTL or CMOS), it is often necessary to look at the leading or trailing edge. In order to facilitate this measurement, oscilloscopes incorporate a pulse delay (Fig. 3.16) facility. The delay is needed because a triggered time base cannot be started instantaneously. By the time a normal triggered time base has started, the pulse you want to see is just about finished so that all you ever see even with a fast time base, is the end of the pulse. To make use of the pulse delay facility, the input signal is also used to trigger a monostable circuit which produces a delayed pulse which, in turn, operates the EXT TRIG circuit of the oscilloscope. The delay and time base controls are adjusted until the edge of the pulse can be seen, which enables one to estimate the rise or fall of time.

3.2.3 Double Beam vs Dual Trace

A double beam oscilloscope is helpful in comparing waveform. In a double beam instrument, two traces appear on the screen, each using the same time base but with separate Y-input controls. The two traces are separate and different waveforms can be displayed. Double beam arrangements are very useful when looking at circuits which make use of pulse triggering and synchronization.

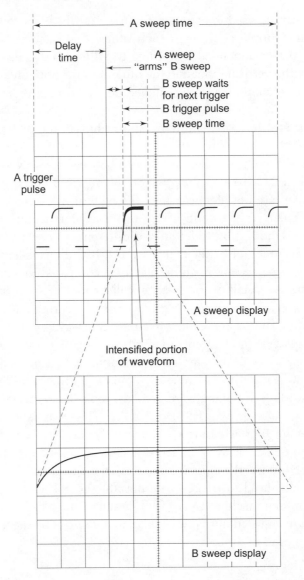

Fig. 3.16 ***Delay facility for viewing rising (and trailing) edge of a pulse***

Several different techniques are used to obtain displays. Some of these are:

(a) *Separate Guns:* In this, two separate electron guns, with the X-plates connected together but with separate Y-plates, are used.

(b) *Beam-Splitting:* In this, the beam from an electron gun is split into two after it has passed the X-plates but before reaching the Y-plates.

(c) *Beam Switching:* This makes use of DC coupling into the scope's Y-amplifier. At the start of sweep, one of the two input signals is applied to the Y-input of the amplifier, with some DC level determined by a Y-shift control. On the next sweep, the other signal is applied to the Y-input but at a different DC level, so that the traces are at a different vertical position.

When this is done at a fast rate, it appears as if two traces are present simultaneously. Obviously, the method cannot work at slow sweep rates wherein you see first one trace, then the other, but never both at once.

An alternative method is to use the 'chopping' technique in which input 1 is displayed for a short time, then the trace is shifted up (or down) so that input 2 can be displayed. Each trace would appear continuous if the beam and the inputs are switched at a frequency many times that of the time base sweep. At speeds approaching the chopping frequency, the trace, however, will appear as a dashed line.

It may be remembered that nothing quite beats the satisfaction of using an oscilloscope for yourself. The more practice you get in using an oscilloscope, the more useful the instrument would prove to be.

3.2.4 Precautions in the Use of an Oscilloscope

(a) Keep the beam intensity down to the minimum required for a particular setting. Take care to turn down the glare on slow sweep speeds.
(b) When using the oscilloscope in the external horizontal time base mode, avoid displaying a stationary bright dot for any length of time. This can result in burning the phosphor on the screen.
(c) While making measurements, it should be ensured that the time base and vertical amplifier controls are in their calibrated positions.
(d) Ensure that the vertical gain control is set above the voltage of the signal to be measured. If in any doubt, start with maximum attenuation (highest voltage setting, minimum sensitivity) and work down the range until the correct setting is reached.

3.3 DIGITAL OSCILLOSCOPES

In contrast to an analog oscilloscope, a digital oscilloscope uses an analog-to-digital converter (ADC) to convert the measured voltage into digital information. Basically, it acquires the waveform as a series of samples, and stores these samples in large enough numbers to describe a waveform. The digital oscilloscope then re-assembles the waveform for display on the screen.

Digital oscilloscopes can be classified into the following three categories:
(i) Digital Storage Oscilloscope;
(ii) Digital Phosphor Oscilloscope; and
(iii) Sampling Oscilloscope.

3.3.1   Digital Storage Oscilloscope (DSO)

Figure 3.17 shows a block diagram of a digital storage oscilloscope primarily consisting of the serial processing architecture. The input stage is a vertical amplifier wherein the vertical controls facilitate adjustment of the amplitude and position range. Next the analog-to-digital converter (ADC) in the horizontal system samples the signal at discrete points in time and converts the signal's voltage at these points into digital values called sample points. This process is referred to as digitizing a signal. The horizontal system's sample clock determines how often the ADC takes a sample. This rate is called the *sample rate* and is expressed in samples per second.

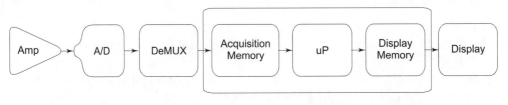

Fig. 3.17   *The architecture of a digital storage oscilloscope (DSO)–serial processing architecture*

The sample points from the ADC are stored in acquisition memory as waveform points. Several sample points may comprise one waveform point. Together, the waveform points comprise one waveform record. The trigger system determines the start and stop points of the record. The signal path further includes a microprocessor through which the measured signal passes on its way to the display. This microprocessor processes the signal, coordinates display activities, manages the front panel controls, etc. The signal then passes through the display memory and is displayed on the oscilloscope screen.

A DSO is an ideal instrument for low repetition rate or single-shot, high-speed, multi-channel design applications. Most of today's digital oscilloscopes provide a selection of automatic parametric measurements simplifying the measurement process.

3.2.2   Digital Phosphor Oscilloscope (DPO)

The digital phosphor oscilloscope employs a parallel processing architecture (Fig. 3.18) to capture, display and analyse signals in comparison to the DSO which uses a serial processing architecture. The input stage is similar to that of an analog oscilloscope, i.e. a vertical amplifier and the second stage is similar to that of DSO, i.e. an ADC. But the DPO differs in the subsequent stages from its predecessors. The DPO transfers the digitized waveform data into a digital phosphor database through a raster system. Every 1/30th of a second—about as fast as the human eye can perceive it—a snap shot of the signal image that is stored in the database is sent directly to the display system. This direct transformation of waveform data and direct copy to display memory from the database removes the data processing bottleneck due to the speed of the microprocessor, inherent in other architectures. The result is an enhanced real-time display of the waveform. The microprocessor in the DPO works in parallel with the integrated organisation system for measurement automation and instrument control so that it does not affect the oscilloscope's acquisition speed.

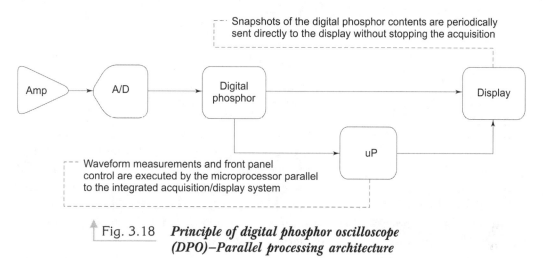

Fig. 3.18　*Principle of digital phosphor oscilloscope (DPO)–Parallel processing architecture*

Unlike an analog oscilloscope's reliance on the chemical phosphor, a DPO uses a purely electronic digital phosphor that is actually a continuous updated database. This database has a separate cell of information for every single pixel in the oscilloscope display. The DPOs are equally suitable for viewing high and low frequencies, repeated waveforms, transients and signal variation in real time.

3.3.3　Digital Sampling Oscilloscope

For measurement of the high frequency signal, it is possible that the oscilloscope may not be able to collect enough samples in one sweep. A digital sampling oscilloscope is an ideal tool for accurately capturing such signals whose frequency components are much higher than the oscilloscope's sample rate. Figure 3.19 shows the architecture of a typical digital sampling oscilloscope. Here the position of the attenuator/amplifier and the sampling bridge is reversed. The input signal is sampled before any attenuation or amplification is performed. A low bandwidth amplifier can then be employed after the sampling bridge because the signal has already been converted to a lower frequency by the sampling rate, resulting in a much higher bandwidth instrument. This arrangement facilitates achievement of a bandwidth and high speed timing that is ten times higher than other oscilloscopes for repetitive signals. Sampling oscilloscopes are available with bandwidth of up to 50 GHz.

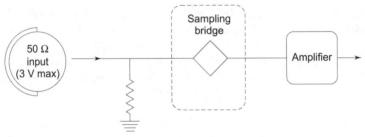

Fig. 3.19   *Principle of digital sampling oscilloscope*

The high bandwidth, however, limits the dynamic range of the sampling oscilloscope. Since there is no attenuator/amplifier in front of the sampling gate, there is no facility to scale the input. The sampling bridge must be able to handle the full dynamic range of the input at all levels and times. Therefore, the dynamic range of most sampling oscilloscopes is limited to about 1V peak-to-peak, whereas the digital storage and digital phosphor oscilloscope can handle 50–100V. Also the safe input voltage for a sampling oscilloscope is about 3V as compared to 500V available on other oscilloscopes.

3.3.4   Controls on Digital Oscilloscope

A basic oscilloscope consists of four different sections—the vertical system, horizontal system, trigger system and display system. It is necessary to understand the function of each of these systems to be able to effectively apply the oscilloscope to take specific measurements. The front panel of both digital and analog oscilloscopes is divided into three main sections: vertical, horizontal and trigger. When using the oscilloscope, you need to adjust these three basic settings to accommodate an incoming signal. Some oscilloscopes may have other sections depending upon the model and type—analog or digital.

Digital oscilloscopes have settings that let you control how the acquisition system processes a signal. It is necessary to understand the acquisition position on your digital oscilloscope from the operator's manual of the instrument. One of the greatest advantages of digital oscilloscopes is their ability to store waveforms for later viewing. For this, there are usually one or more buttons on the front panel that allow you to start the acquisition system so that you can analyse waveforms at your convenience. Using the power of their internal processors, digital oscilloscopes offer many advanced mathematical operations: multiplication, division, integration, Fast Fourier Transform and more.

Besides the basic controls on the oscilloscopes which are mostly common for both analog and digital oscilloscopes, some of the special facilities and controls on digital oscilloscopes include:

- Automatic parametric measurements;
- Keypads for mathematical operation or data entry;
- Measurement cursors;
- Printing capability; and
- Interfaces for connecting your oscilloscope to a computer.

If you can afford it, get a good digital storage scope. You can even get relatively inexpensive scope cards for PCs, but unless you are into PC-controlled instrumentation, a standalone scope is much more useful.

3.3.5 Oscilloscope Probes

An important part of the measurement system in an oscilloscope is the *probe*. Probes actually become part of the circuit introducing resistive, capacitive and inductive loading. For accurate results, the probe should be selected in such a manner that it has a minimal loading effect on the device under test.

For measuring general types of signals and voltage levels, *passive probes* can serve the purpose. Most passive probes have some attenuation factor such as 10X, 100X, and so on. The 10X attenuater probe reduces circuit loading and improves accuracy of measurement in comparison to a 1X probe but it reduces the signal's amplitude at the oscilloscope input by a factor of 10. Therefore, it becomes difficult to look at signals less than 10 mV peak-to-peak with the 10X attenuator probe.

The 10X attenuator probe works by balancing the probe's electrical properties against these on the oscilloscope. Before using a passive probe, you need to adjust this balance every time you set up the oscilloscope, otherwise a poorly adjusted probe may make the measurements less accurate. This adjustment is known as compensating the probe. Most oscilloscopes have a square wave reference signal available at a terminal on the front panel which can be used to compensate the probe. The following instructions are useful to compensate the probe:

- Attach the probe to a vertical channel;
- Connect the probe tip to the square wave reference signal on the front panel of the oscilloscope;

- Attach the ground clip of the probe to the ground terminal on the oscilloscope;
- View the square wave reference signal on the screen; and
- Make adjustments on the probe so that the corners of the square wave are perfectly square (Fig. 3.20).

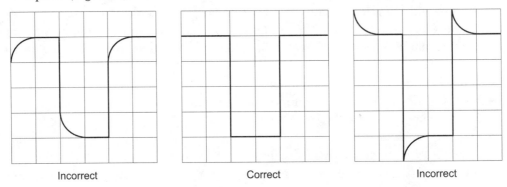

Incorrect   Correct   Incorrect

Fig. 3.20   ***Probe compensation method***

General purpose passive probes cannot accurately measure signals with extremely fast rise times. The measurement solution at high speeds includes high speed, high fidelity probing solutions to match the performance of the oscilloscope. *Active* and *differential probes*, which use specially developed integrated circuits, have been developed to preserve the signal during acquisition and transmission to the oscilloscope, ensuring signal integrity when measuring high speed and/or differential signals. In some oscilloscopes, *intelligent probe* interfaces are provided in which the act of connecting the probe to the instrument notifies the oscilloscope about the probe's attenuation factor, which, in turn, scales the displays so that the probe's attenuation is figured into the readout on the screen.

3.4   LOGIC ANALYSER

The measurement problems associated with digital systems are many and do not lead themselves to solutions by conventional instrumentation such as voltmeters and oscilloscopes. The multi-line, aperiodic digital signals that characterize digital equipment require test instruments that are fast and provide lucid analyses. The most popular test instrument for troubleshooting in the case of digital and microprocessor-based systems is the Logic Analyser. Its appearance is similar to an oscilloscope but it looks into the activity on the communicating links in a digital system known as *interface buses* or *data highways*. Since an interface bus is a parallel assembly of connections, data is transmitted as parallel bit streams. Thus, at discrete instants in time, data words are formed and this form of activity takes place in the data domain. The logic analyser is to the data domain what the oscilloscope is to the time domain and the spectrum analyser to the frequency domain. The data domain is characterized by state-space concepts, data formats, data flows and equipment architecture.

The logic analysers are basically of the following two types:

(a) *The logic timing analyser.* It samples the data according to a clock signal (at rates faster than the system rates) generated at regular intervals, stores them in the memory and then displays these stored data in a timing waveform. This method aims at measuring the time relation for each signal of the digital circuit and the presence of glitches, and is suitable for analysing problems arising in the computer hardware. This method is called 'asynchronous measurement'.

(b) *The logic state analyser:* It uses clock signals generated inside the measured equipment, samples the data and stores them in the memory only when they are synchronized with the measured equipment, providing either binary or hexadecimal state display for these data. This method can be used for the measurement of devices such as microprocessor-based equipment operating in synchronization with clock signals that become a standard for the total system. This method, known as 'synchronous measurement', is suitable for solving problems arising in connection with software such as execution state of the programme, etc.

Modern logic analysers incorporate both these functions in one unit. In the state analysis mode, captured data may be displayed as a list (a binary table) map or graph, while in the timing analysis mode, it is displayed as a multi-channel logic timing diagram. Once the desired display is obtained, a particular class of problems can be analysed.

3.4.1 Principle of Operation

Figure 3.21 shows a block diagram of a logic analyser. In order to examine bus activity, multi-line data probes are required. Data on each probe line is sampled under the control of a system clock which could be typically qualified to sample every processor instruction cycle once. The samples are stored in an internal memory with a capacity which could be anywhere from 16×16 bit words to 1000×20 bit words. A digital delay is introduced to establish a unique trigger word which facilitates movement of the display window downstream in discrete clock periods, and also a viewing of events leading to the trigger word. By this method, faults as also incorrect data, branching statements, etc. can be detected. In the modern logic analysers, the options available for setting up the equipment are displayed as a menu on the screen with a cursor to indicate the next input required. The keyboard is used to enter the information to give the required operating sequence. The typical options for trigger selection are: clock source, edge polarity, trigger word, clock cycles delay, trigger start or end, block pattern recognition, etc.

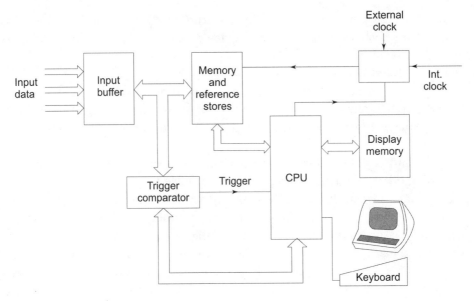

↑ Fig. 3.21   ***Block diagram of a logic analyser***

It is important to define the display in a relevant format. In a typical situation, 32 channels of information can be displayed in a standard digital format, i.e. the most significant bit on the left and the least significant bit on the right with the first word at the top and each successive word under the previous word. The data sequence table presents data into blocks of four for reading in hexadecimal code or blocks of four for reading in octal code or BCD. The possible data display modes on the screen are:

Table: This is a listing of the sampled data states in binary or to some other numerical base such as hexadecimal.

Timing: Data is displayed across the screen as several channels showing the HIGH–LOW activity.

Graph: The horizontal axis represents analyser store addresses and the vertical axis, the numerical value of the stored data.

Map: Each 16-bit sample is divided into its upper and lower 8-bit bytes. The values of these produce the vertical and horizontal deflection.

'Mapping' is a dynamic view of a system's operation wherein a pattern of dots and lines that are unique for each programme, is displayed. Each dot is a specific data word. Its location indicates the binary magnitude and its brightness indicates the relative frequency of occurrence. The map is termed as a 'personal fingerprint'. It is different for every programme.

When monitoring the data on a microprocessor bus with a logic analyser, it is possible to re-convert the binary data back into its mnemonic assembly language automatically.

The analyser has a personality module according to the microprocessor in the system under development or test and the table display can then be a list of assembly language statements which are interpreted for programme de-bugging.

The logic analysers are usually provided with a series of 'Personality Modules' to re-configure the equipment for a wide range of microprocessors. A general purpose 'Personality Module' is usually available for non-microprocessor devices for which a specific Personality Module is not available.

Conventional oscilloscopes can be used for logic analysis by using logic trigger generators as accessories. These units may be as simple as four-input AND gates for connection to the oscilloscope trigger input and would synchronize the oscilloscope with the occurrence of a particular parallel word. They may also include counters to delay triggering by a set number of events or shift-register comparators to allow bit-serial word recognition for triggering. A logic trigger generator is adequate when the cost is considered and an oscilloscope is readily available.

In recent years, so many new types of logic analysers adopting novel concepts have been introduced that potential buyers often get confused as to which model to choose. It should, however, not be too difficult to find the most appropriate model among the many different types available, if the basic performance characteristics and functions required for measurement are selected first and the related sub-functions picked out afterwards.

3.5   SIGNATURE ANALYSER

Signature analysis, to a common man, may mean studying the way people write their names, but in digital circuit testing, signature analysis is a specific troubleshooting technique based on coded representations of serial bit streams. Using a known input signal, a signature analysis system generates such a coded representation at each point on a known good printed circuit board. The signature at each point on a board under test should then be the same as the signature at the corresponding point on the known good board.

Signature analysis is a data compression technique. For example, if a piece of equipment is made to repetitively execute a certain sequence of instructions, its correct operation can be identified by monitoring the changing logic levels at each node in the circuit. However, it would produce massive information which would be ordinarily unmanageable in a test situation. This problem can be overcome by using the 'signature analysis' technique. In this technique, the data appearing at a given node is sampled for a known period, between the start and stop signals, by clocking it with a system clock into a feedback shift-register. The residue at the end of the sampling period is a characteristic (signature) of the activity at that node. Because signature-analysis data is clocked from the node into the shift register, time is effectively compressed. What the operator observes is the result of all the data bits passing the node in a specific period.

In Hewlett-Packard's Signature Analyser, a 16-bit shift register is used (Fig. 3.22) which has feedback taps at 7, 9, 12 and 16. The feedback bits are summed up (module-2) in a 5-input exclusive—OR gate to derive the shift-registers input signal. The fifth register input is the bit stream from the point under test.

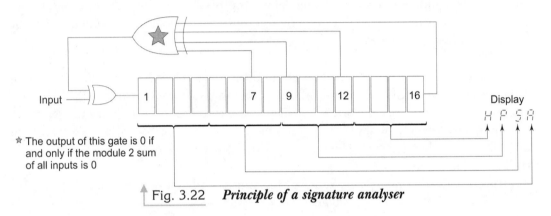

☆ The output of this gate is 0 if and only if the module 2 sum of all inputs is 0

Fig. 3.22   *Principle of a signature analyser*

Using the 16-bit shift register and arranging the feedback such that a maximal length sequence is produced will give 65,536 possible residual states. The parallel 16-bit output from the register is used to drive four hexadecimal displays and the resulting number is known as the '*signature*' of that node. Errors in the data stream will normally cause a different signature to be displayed.

In signature analysis, the input from the node under test is only one input to the test circuit. The shift-register is re-set to zero and the clock begins upon receipt of a 'start' signal. A clock signal provides a stroke edge for each good data bit and causes the register to sum and shift synchronously with the data in the bit stream. A 'stop' signal stops the measurement and displays the residual pattern (Fig. 3.23).

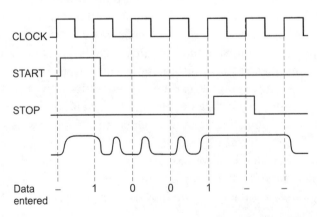

Positive edges selected for START, STOP, and CLOCK

Fig. 3.23   *Timing sequence in a signature analyser*

It is experienced that all single-bit errors (dropped or incorrect bit) will change the signature and the probability of multiple-bit errors being missed is less than 0.002%. This order of accuracy is far better than the performance of other techniques such as bit or transition counting.

Let us now look at the difference in a logic analyser and signature analyser. Both these instruments have a measurement window, i.e. the clock is enabled for a specified time. The logic analyser saves the system's state during the time the clock operates whereas the signature analyser clocks the data that occurs during the window into the shift register. The result in the shift register at the end of the measurement window is the signature of the node.

A logic analyser requires considerable skill on the part of the operator to view each time slot independently and to interpret traces correctly. In contrast, the signature needs no interpretation, it is either correct or incorrect. By studying signatures at various nodes, the source of the failure can be located.

Signature analysis is a very powerful service aid. The equipment required for it is also relatively inexpensive.

3.6 SIGNAL GENERATORS

A signal generator is a device that supplies a standard voltage of known amplitude, frequency and waveform for test and measuring purposes. Signal generators are classified according to the shape of the output waveform, e.g. sine wave oscillators, sine square generators, saw tooth generators, etc. They may also be classified according to the range of frequencies they generate. For example, there are audio generators, IF RF generators, VHF-UHF generators, microwave generators, etc. These are mostly sine wave generators and usually the output frequency from any one of these types is variable over a wide range. Also, each type usually includes several ranges or bands. Any one of these ranges can be selected by a switch on the front panel of the instrument.

Among the most common types of signal generators are function/arbitrary waveform generators which provide standard waveforms such as sine, square, pulse, exponential rise and fall, cardiac waveforms. These generators use direct digital synthesis techniques to create waveforms over a wide range of frequency characteristics from down to 1 μHz frequency resolution. The frequency range covered for standard waveform is from 1μHz to more than 50 MHz. A colour graphical LCD panel displays not only the frequency of the output signal but also the waveshape.

The application of the signal generators in troubleshooting electronic equipment are almost unlimited. Signal generators are used for signal tracing, testing and adjusting amplifier response, alignment of radio and television receivers, and testing digital circuits, particularly with pulse generators. They may also be used for the precise measurement of time and frequency.

The circuit description of almost all types of signal generators are generally described in standard texts on electronic measurement techniques. The pulse generator has, however, some special features and is therefore described in detail in the following section.

The need for a pulse generator arises because of the extensive use of many different varieties of digital logic in industrial, laboratory and consumer applications. It is usually difficult to service logic-based electronic systems using only a sine wave or a square wave generator. A sine wave or square wave generator cannot vary the pulse width of its output signal. It also cannot generate a wide enough range of frequencies. In addition, frequently the output impedance of a sine-wave generator or a square-wave generator is not low enough to drive logic circuits. On the other hand, a function generator with DC offset, variable symmetry control and a wide frequency range may fulfil the need for a pulse generator to a considerable extent.

3.6.1   Pulse Generators

The tremendous growth in the use of digital circuits and devices has made the pulse generator a valuable instrument for any service shop. Besides the special features, which will differ from one make to the other, this instrument provides pulses of variable repetition rate, delay, width and amplitude. These pulses are made use of in testing various digital and logic circuits.

Figure 3.24 shows a simplified block diagram of a pulse generator. The basic pulse repetition rate is controlled by the repetition rate generator. Rate generation can be controlled in the following two ways:

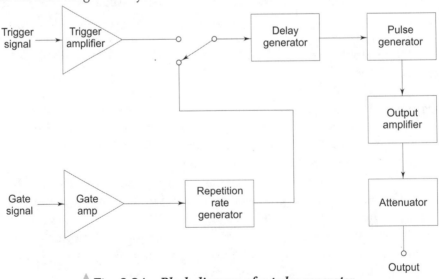

Fig. 3.24   *Block diagram of a pulse generator*

(a) *External Trigger:* The signals applied to the external trigger pass through triggering circuitry much like that found in an oscilloscope. Those circuits establish the triggering amplitude and polarity. The trigger amplifier may also have inputs like line frequency and a manual push button for one-shot operation.

(b) *Gated Mode:* In this mode, signals from the gating amplifier permit the repetition rate generator to output pulses only when a gating signal is applied, otherwise no pulses are generated. A burst of pulses can be produced in that mode.

Signals from the delay pulse repetition-rate generator or from the trigger circuits are applied to the delay generator. Once the delay generator has completed its cycle, the main pulse generator is triggered. Independently variable rise times and fall times can be generated in some pulse generators.

The output amplifier is a DC-coupled, variable-gain, wide band amplifier. In some pulse generators, where extremely close control of pulse characteristics is desired, two output amplifiers are used in the output stage. One of the amplifiers handles the positive pulses while the other handles the negative pulses.

Pulse amplitude is usually controlled by a variable-gain control in the output amplifier. Some instruments also include a step attenuator at the output. As the output amplifier is DC-coupled, a DC offset signal can be added to the amplifier, along with the pulses, so that the pulse base line can be changed with reference to 0V.

Pulse characteristics can be adjusted from the instrument by the following controls: Pulse repetition rate, pulse width, pulse delay output amplitude, pulse polarity, offset control and trigger modes.

The repetition rate is usually specified in terms of frequency but sometimes as a pulse period. The accuracy of the repetition rate is generally unreliable, and whenever an accurate pulse repetition rate is needed, it should be obtained either by triggering the pulse generator from a known frequency source, such as a signal generator or by measuring the repetition rate at the trigger output, using a digital frequency meter or oscilloscope.

The maximum amplitude is usually specified with the amplifier terminated into a load that is equal to its characteristic impedance, which is normally 50Ω. Typical maximum outputs range from 5 to 10V.

Some pulse generators are provided with a special switch setting that permits the pulse generator to be operated in a square-wave mode. When operated in this mode, the duty cycle is automatically maintained at 50%.

Some generators facilitate presentation of an internal counter by using thumbwheel switches. When the pulse-burst mode is used, each pulse is counted, and when the count equals the preset value in the counter, no more pulses are given out. This feature is useful for checking the accuracy of counters and similar instruments.

Complementing output is produced by some generators. It implies that a pulse generator that normally produces a positive pulse with a 25% duty cycle produces one with a 75% duty cycle.

A pulse generator is usually not calibrated very precisely. When using a pulse generator, especially over extended periods of time, it is necessary to re-set the operating parameters, many of which can vary over time or with changes in temperature and other factors. Adjustment of some controls can offset other parameters also. For example, any changes in the rise time and fall time of the output pulse can cause changes in the pulse width and amplitude.

The most frequent errors in using a pulse generator are human errors. For example, a common mistake is exceeding the allowable duty cycle. As explained earlier, a 70% duty cycle is common for most pulse generators. For example, a pulse repetition rate of 1 kHz cannot be established if the pulse width is chosen as 10 milliseconds instead of 10 microseconds. It is important to remember that the best way to get the most out of a pulse generator is to use it with a good oscilloscope.

3.7   UNIVERSAL BRIDGE

A universal bridge is used for measurements of passive components which include resistors, capacitors and inductors. Although inductor and capacitor defects usually appear as leakages, open circuit, shorts or partial short conditions, and a change in capacitor or inductor value is rare, a universal bridge is still considered an important test equipment in a servicing and maintenance laboratory.

A general technique for measuring R, L and C is by using some form of bridge circuit with the unknown element as one of its arms and balancing the bridge by varying a standard known element. The circuit of a universal bridge would therefore include an AC source excitation for energizing the bridge, a switching arrangement for selection of the appropriate range, and a detector and display device to indicate bridge balance.

Principle of Operation: The principle of a transformer ratio arm bridge is shown in Fig. 3.25. The secondary winding of voltage transformer T_s consists of a centre-tapped winding whose two halves feed the standard and unknown impedances Z_s and Z_u respectively, in phase opposition and whose centre tap is at the neutral potential.

Having passed through the impedances, the two opposing signals are applied to the primary of the current transformer T_d, whose secondary winding is connected to the detector circuit. Since $E_1 = E_2$ when $Z_u = Z_s$, the two halves balance, and there is no output from T_d to the detector.

If Z_u is not equal to Z_s, the bridge can be balanced by: (i) Changing tapping points on T_s secondary. This alters E_1 and E_2 and hence controls the current in Z_u and Z_s, (ii) Changing tapping points on T_d primary. This alters the turns ratio between the arms so that if the currents are differen t, the ampere-turns can be made equal.

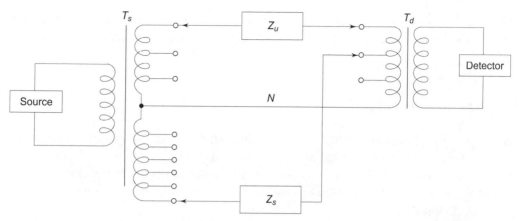

Fig. 3.25 **Principle of a transformer: ratio arm bridge for measurement of R, L and C**

In practice, the voltage transformer secondary winding is tapped in decades on one side (1–10–100–1000) and in tens on the other side. The current transformer primary also contains decade taps (Range selector switch). It is possible to cover the entire range of the instrument.

The bridge source is derived from a phase shift oscillator, providing for either 80 Hz or 1 kHz according to the setting of the function switch. Off-balance output from the bridge is amplified by a multi-stage amplifier. The last stage feeds the meter via a pair of diode detectors.

Phase detectors are used to generate DC voltage proportional to the real and imaginary parts (in-phase and quadrature components) of the signal at the inputs to the phase detectors. In order to do this, it is necessary to maintain the precise phase relationship between the reference signal and the detected signal.

Modern RLC bridges are automatic digital instruments. In these instruments, some form of digital-to-analog converter is used to adjust either the standard or the voltage across the standard to balance the bridge. The general purpose RLC meters that are presently commercially available provide for various test frequencies depending upon the application. Typically, the test frequencies are available from 40 Hz to 100 MHz. Impedance measuring instruments are, however, available with test frequencies of up to 3 GHz. Besides measuring RLC, these instruments also calculate the total impedance Z of the component, Q (quality factor) and the phase angle θ. The measurement range for different parameters is as follows:

Resistance (R)	1 mΩ to 100 MΩ
Capacitance (C)	1 pF to 1 F
Inductance (L)	10 nH to 100 kH

In addition, RLC meters provide for storage and retrieval facilities for measurement, verification of reliability of test connections and correction for test fixtures and cable errors.

3.8   POWER SUPPLIES

For the testing of circuit boards and individual components, regulated and stabilized power supply is an essential requirement in a repair and maintenance laboratory. Usually, different power supplies are required for different circuits but in general, the following voltage and current rating power supplies are essential for mixed signal applications.

For digital circuits	6V, 5Amp
For analog circuits	±25V, 1Amp

Depending upon the load conditions, a power supply may be a constant voltage power supply or a constant current supply, with the following characteristics:

- A constant voltage power supply stabilizes the output voltage with respect to the changes in load conditions. Thus, for a change in load resistance, the output voltage remains constant while the output current changes by whatever amount is necessary to accomplish this.
- A constant current power supply stabilizes the output current with respect to changes in the load impedance. Thus, for a change in load resistance, the output current remains constant while the output voltage changes by whatever amount is necessary to accomplish this.

Normally, power supplies have more than one output. A dual output range usually has one range with high voltage and low current, and another range with low voltage and high current. These supplies are protected from over-voltage and over-current situations.

3.9   FIBRE-OPTIC TEST EQUIPMENT

The important system parameters which are encountered in testing fibre-optic based systems are:

Operating wavelengths: The three major transmission windows are 850, 1300 and 1550 nm (nanometers).

Source type: This can be LED or laser. Most low speed local area networks (LANs) in the operating range of less than 100 MBs (Megabytes/sec) use LED sources. They are economical and useful for short haul applications.

The high speed systems operating at speeds greater than 100 MBs use laser sources to extend the signal over long distances.

Fibre type (single mode/multi-mode): A standard single mode fibre is 9/125 μm. However, other special single mode fibres also exist and should be properly identified.

Typical multi-mode fibre sizes include 50/125, 625/125, 100/140 and 200/230.

The following test equipment is generally adequate for testing fibre-optic based systems:

3.9.1 Optical Power Meter (OPM)

This instrument is used to measure optical power or optical power loss over a fibre-optic path. The measurement of optical power is the most fundamental measurement in the fibre-optic system. Therefore, OPM is the workhorse in fibre-optic laboratories, much like the digital multimeter in electronics laboratories. It facilitates the measurement of the absolute power being injected into or emerging from the network.

Figure 3.26 shows a block diagram of an optical power meter. This instrument contains an optical detector which is a solid state photo-diode. The photo-diode receives light coupled in from the network and converts it into an electrical signal. The electronics circuitry processes the detectors, output signal, applies a calibration factor based on the wavelength and displays the optical power on a digital readout in dBm units (absolute dB referenced to 1 mW, i.e. 0 dBm = 1 mW.

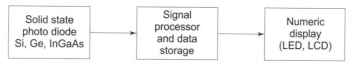

Fig. 3.26 *Block diagram of an optical power meter*

When an OPM is used together with a stabilized light source, the combination can measure link loss to verify continuity and help assess the quality of the transmission path through the optical fibre.

The most important selection criterion for an optical power meter is matching the appropriate optical detector type with the expected range of the operating wavelength. Table 3.1 shows the optimum detector choices for different ranges of operating wavelengths.

Table 3.1 ■ *Operating Wavelengths and Detectors*

Operating Wavelengths	*Optimum Detector Choice(s)*
850 nm only	Silicon (Si)
850 nm and 1300 nm only	Germanium (Ge) and InGaAs
1300 and 1550 only	InGaAs
850 nm, 1300 nm and 1550 nm	InGaAs

It is obvious that the detector InGaAs has superior spectral flatness across all three windows. Also, it has better accuracy as well as temperature stability and low noise characteristics.

3.9.2   Stabilized Light Sources

During the loss measurement process, a stabilized optical source provides light of known power and wavelength into the optical system. The light source is used along with a power meter to measure the optical loss of a fibre-optic system. If the system is already installed, the system transmitter can be used as a stabilized light source. Only when the system transmitter is inaccessible is a separate stabilized light source required. A power meter/ optical detector calibrated for use as the wavelength of the light source, receives light from the network and converts it into an electrical signal. In order to ensure accuracy in the loss measurement, the light source should preferably be of the same type (LED or laser) and wavelength, offers the same connector interface with magnitude of its output power sufficient to measure the worst case system loss.

Frequent measuring of the end-to-end loss of a system after installation is necessary to determine if the link loss, including connector, splice and fibre losses meet the design specification.

3.9.3   Optical Loss Test Sets

The combination of an optical power meter and stabilized light source constitute an optical loss test set, and are used to measure optical power loss over a fibre-optic path. These instruments can either be two individual component devices or a single integrated unit. Figure 3.27 is a typical system configuration for optical power and loss measurements.

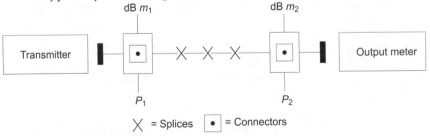

System loss (dB) = 10 log (P_1/P_2) = dB m_1 – dB m_2

Fig. 3.27   *System configuration for optical power and loss measurements*
P_1 = Power transmitted from the transmitter (in mWs)
P_2 = Power received by the receiver (in mWs)
dBm_1 = 10 log P_1 (in dBm)
dBm_2 = 10 log P_2 (in dBm)

When short LAN spans within walking and talking distance of the ends, the technicians at either end can use the economy of employing a stabilized light at one end, and the optical power meter at the other end. For long distance network, an integrated optical loss test set is preferred. Integrated systems are generally more expensive but usually offer higher levels of sophistication and functionality than their component counterparts.

3.9.4 Variable Optical Attenuators

These are essential for testing optical receivers to determine if the system will work over the specified range and linearity.

The performance reliability of a fibre-optic high-end system is usually denoted as the bit-error rate (BER) which characterizes the system performance under various conditions. BER is expressed as the number of failures per bit. A typical error rate of 10^{-9} means that one bit out of 100,000,000 may be wrong.

With a BER generator connected to the electric input of the optical transmitter, the attenuator is used to reduce power (measured with an optical power meter) and verify acceptable BER performance down to the minimum signal requirement of the optical receiver.

Attenuators are of three types: continuous, step or combinational step/continuous configuration wherein the stepping function momentarily blocks the optical signal. When BER measurements are being performed, both pure step and combinational models are undesirable due to the momentary blocking of the optical path between attenuation settings.

3.9.5 Optical Time Domain Reflectometers (OTDR)

Optical time domain reflectometers provide necessary information regarding the link under test. They are used to characterize fibre loss as a function of distance. Basically, the instrument is a one-dimensional, closed circuit optical radar, requiring the use of only one end of a fibre to take measurement. A high intensity, short duration light pulse is launched into the fibre while the high speed signal detector records the returned signal. The instrument provides a visual interpretation of the optical link. Splices, connectors and faults may be identified in both magnitude (dB loss) and distance away from the operator.

The industry divides the optical time domain reflectometers into two groups based on spatial resolution performance: high resolution and long haul. Based on an instrument's ability to distinguish between two connector events, a high resolution instrument can resolve two connectors separated by 10 mts. which will not be possible with long haul instruments whose dead zone may exceed 10 mts. In general, if the expected distance of the fibre segment (distance between two connectors) is expected to be short (less than 100 mts.), a short haul high resolution type instrument is selected. No single instrument can do all the things. There is a trade-off between distance and resolution.

The above set of instruments individually or in suitable combinations can help to make all necessary measurements on fibre-optic based systems and detect and locate faults. However, the optical systems will not perform if their interfaces are not clean. As a rule of thumb, the optical interfaces, particularly connectors, should be cleaned before any measurement is made. Even a 1–2 μm dust particle can cause havoc in the performance of a power-optic system.

The user should also consider the calibration requirements of different instruments. How often does the instrument have to be re-calibrated? Can the maintenance/re-calibration be performed by the user or must the equipment be returned to the manufacturer? What is the cost of a routine calibration and turn-around time? Therefore, before procurement of the fibre-optic test equipment, the user should closely evaluate the manufacturer and the performance of the latter's service operation.

4

Tools and Aids for
Servicing and Maintenance

4.1 HAND TOOLS

A repair and servicing technician is only as good as the tools with which he works. Tools are no substitute for knowledge, of course, but a lack of key items or the use of improper ones can put severe limitations on the technician's activities. The type and variety of tools and test equipment that an electronics technician possesses have a direct bearing on his efficiency and on the effectiveness with which he handles the job.

The human hand is a unique instrument with which we work, construct our civilisation and create our greatest works of art. In fact, it is with the hand that we have shaped the world we live in. But the hand has its limitations. To augment its range and capabilities in terms of grasping and delicate manipulations, many hand tools have been developed over the years. The serviceman's dedication, interest and application are all important, but if he does not possess the right tools to carry out his work, he might as well try windsurfing on a skateboard.

The tools which are generally used in equipment servicing practice are:
(a) Hand drill, power drill, drill bits, twist drill;
(b) Taps and dies, punches, glass cutter, cramps;
(c) Spanners (wrenches);
(d) Pliers-circlip pliers of various types, tweezer, wire strippers;
(e) File-round, half round, oval, knife square, three square, needle files;
(f) Hammer-clow, ball peen, cross peen;
(g) Screw drivers of various sizes;
(h) Bench grinder, oil stone;
(i) Bench vice;
(j) Brushes—soft and wire type;
(k) Saws—hacksaw, piercing saw;
(l) Crimping and anti-crimping tool;

(m) Instruments—phase testers, side caliper, micrometer caliper, test lamps, rules and tapes;

(n) Soldering irons, solder sucker, desoldering pump;

(o) Eye protectors; and

(p) Oil can, grease gun.

All these tools are important in an electronic workshop. However, in this book, only those tools which are more frequently used by instrument technicians shall be described. These are pliers, wrenches, screw drivers, soldering irons and desoldering pumps.

4.1.1 Pliers

One of the most commonly used tools in electronics shops are pliers. The frequent tasks for which pliers are used include holding wires in place during soldering, acting as a heat sink to protect a delicate component, bending component leads to fit mounting holes on a circuit board and pulling wires through a panel or chassis hole.

Obviously, one single design in pliers cannot meet all the demands. Therefore, there are many types of pliers. The important types are shown in Fig. 4.1. Some pliers also have cutting knives. A single pair of long nose pliers is adequate for most jobs, but having several on hand can simplify a task.

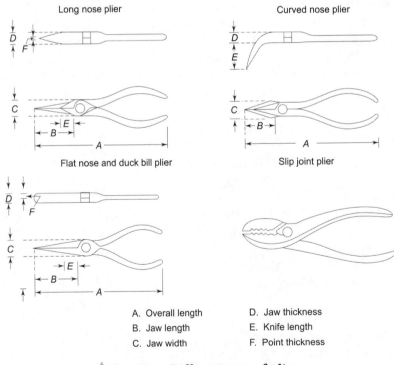

A. Overall length
B. Jaw length
C. Jaw width
D. Jaw thickness
E. Knife length
F. Point thickness

Fig. 4.1 *Different types of pliers*

Pliers have different shapes of handles, though the most common are those with curved handles. They are designed for maximum comfort and efficiency. Extended handles are provided on some long nose and Duck Bill pliers for longer reach and increased leverage.

Pliers do need maintenance. Dull cutting knife edges may be sharpened with small, medium grade honing stone. The knurled inside of the plier nose may be cleaned with a wire brush. An occasional drop of oil at the plier hinge will help preserve the plier and assure its easy operation. Any plier which is cracked, broken or has nicked cutting knives should be discarded and replaced.

While using pliers to cut a wire or metal piece, ensure the protection of eyes with goggles.

A good pair of pliers is invaluable as a precision extension of your fingers when holding and forming components for PCBs. When bending resistor or axial capacitor leads to the correct pitch for your design, the pliers will give you a professional finish and avoid stress to the lead/component joint. Usually, a pair of 'snipe nose' pliers will suit most applications. The fine tip will enable you to use it like a strong pair of tweezers and the serrated jaws will give you a good grip when holding and forming different wires and parts.

4.1.2   Cutters

Good cutters (Fig. 4.2) are an essential tool for component lead cutting and removing insulation prior to soldering or performing. Ideally, they should be of the side cutter type with insulated grips. They should be slimline and lightweight for precision work.

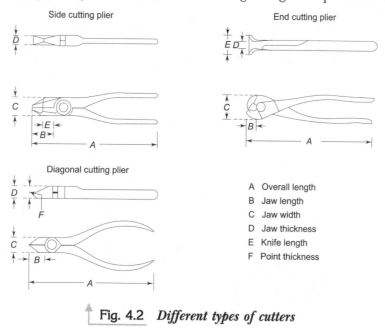

Fig. 4.2   *Different types of cutters*

Cutters up to 15 inches long are employed in electronics applications. The cutters should be made of high quality tool steel so that they will make a sharp, clean cut. The tips of the cutters should be tapered to allow the user to reach a particular wire in a crowded area. Cutter jaws should be very well aligned so that cutting edges meet squarely and allow little or no light to pass through when held together. Cutter action should be smooth and clean.

Cutters should be used only for cutting copper wire or leads and not for trimming PCBs or metal parts. The cutting blades are easily blunted if misused. A blunt pair of cutters is worse than useless as it will not crop leads cleanly or strip insulation from wire without snagging and breaking strands. Some cutters have a safety clip incorporated, which traps the cut-off lead and stops it from flying and ending up all over the room.

Sometimes, it is difficult to strip the wire with the cutters. In that case, one can use wire strippers and cutters, which are available with adjustable stops for different wire sizes. The stop ensures that the cutting action is limited to the thickness of the insulation and will prevent nicking the actual wire.

4.1.3 Wrenches (Spanners)

The directory meaning of the noun 'wrench' is 'violent twist' or 'oblique pull'. So, a hand tool made to grip and turn nuts and bolts is called a 'wrench'. Another name for the wrench is 'spanner' (Fig. 4.3).

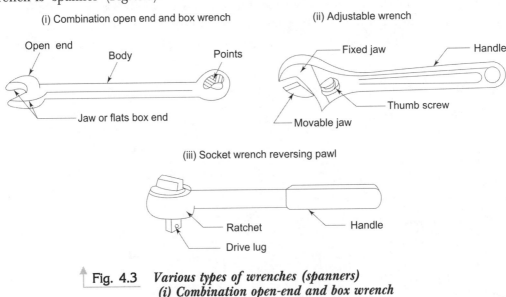

Fig. 4.3   *Various types of wrenches (spanners)*
(i) Combination open-end and box wrench
(ii) Adjustable wrench
(iii) Socket wrench

The various types of wrenches are: combination, open-end and box wrenches, adjustable wrenches, socket wrenches, combination wrench set, torque wrenches and pipe wrench.

Open-end wrenches are used primarily on the large hexagonal nuts that secure switches and controls to project panels and to operate chassis punches. The box wrench has similar applications but has the advantage of completely enclosing the nut, thus eliminating the danger of slippage that can spoil a finished front panel.

The plier wrench or vise-grips is an excellent tool for applying brute torque to the task of turning bolts, nuts and shafts. Vise-grips come in several sizes, but a 10 inch model is sufficient for most workbench applications. Since hexagonal keys (allen keys) do the same job as a wrench, they are called hexagonal wrenches. Allen wrenches (Fig. 4.4) are most useful for opening many control knobs. Adjustable wrenches are made 102 mm (4″) to 457mm (18″) in size with capacities of 13 mm (1/2″) to 52 mm (2-1/16″) and pipe wrenches of sizes from 20.3 cm (8″) to 91.4 cm (36″) with 1″ to 5″ capacities.

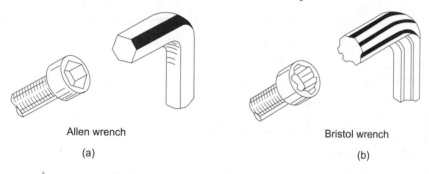

Allen wrench

(a)

Bristol wrench

(b)

Fig. 4.4 *(a) Allen spanners (wrench)*
(b) Bristol wrench

Attempts to repair box, open-end or combination wrenches/spanners are not recommended. Any of these wrenches with bent handles, spread, nicked or battered jaws or rounded or damaged box points, should be discarded and replaced. However, socket and adjustable wrenches can be repaired by the replacement of damaged parts. Periodic inspection, cleaning and light lubrication help to maintain these wrenches in good working condition.

4.1.4 Screw Drivers

A number of screw drivers in a variety of sizes and types are required in the equipment servicing practice. Figure 4.5 shows various parts of a screw driver. The handle is usually made of tough, transparent coloured plastic and shaped to provide a firm, comfortable grip. Smooth, semi-rounded heels fit the palm comfortably for the application of extra power if needed. The blade or shank is made of steel which is heat-treated and tempered to apply torque to the screw head. The blade and tip are chrome plated.

Screw drivers are classified in several ways, including:

(a) Size and shape of the shank; shanks are square or round;

(b) Tip shape-like Philips (+), blade type regular and square; and

(c) Type of application or use; heavy duty, light duty, electrician's, instrument, pocket, watch-maker's, etc.

Blade screw drivers come in a number of sizes. A minimum of three sizes are generally required for handling electronic equipment. These sizes are 1/8 inch, 3/16 to 1/4 inch and 5/16 inch or larger. If finances permit, a larger selection of blade screw drivers should be at hand.

Philips screws, i.e. those with start-shaped holes in their heads as opposed to straight slots, are often needed. Philips screw drivers (Fig. 4.6), like many other hand tools, come in a variety of sizes. There are, however, four standard sizes, No. 1 and No. 2 are the ones usually needed. The star-shaped hole in a Philips screw and the tip of a Philips driver must fit together properly so that the walls of the screw head or the tip of the driver or both will not be damaged.

In addition, there are some special types of screw drivers which are not always necessary, but are, on occasion, very handy. For example, in tight situations, an 'offset' screwdriver is especially helpful.

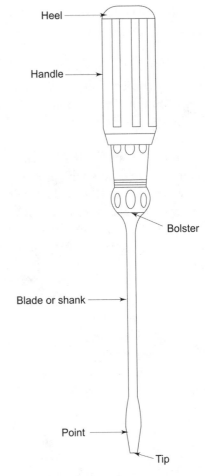

Fig. 4.5   *Various parts of a screw driver*

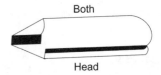

Fig. 4.6   *Philips head screw driver*

The watchmaker's are all metal screw drivers with knurled rotating handle with an easy slip finger tip head. Typical tip sizes are 0.8 mm, 1.4 mm, 2.0 mm, 2.4 mm, 2.9 mm and 3.8 mm.

Ratchet screw drivers have a selector level that allows the screw driver to rotate freely in either the clockwise or anti-clockwise direction and obtain a ratchet driving action in the other direction.

When using screw drivers, the basic rule is to fit the tool to the work. The size of the screw and the type of opening it has determine which driver you use. Screw drivers are the most misused and abused hand tools of all, and care should be taken in their use. A screw driver should not be used at an angle to the screw. Also do not depend on a driver's handle or covered blade to insulate yourself from electricity.

4.1.5 Nut Drivers

Nut drivers (Fig. 4.7) are like screw drivers except that they fit nuts instead of screw heads. They are very useful in mounting a nut on a threaded stud and in holding a nut while its screw is being tightened. Nut drivers are available as individual drivers with separate handles, as individual driver shafts that plug into a common handle or individual sockets that plug into a universal handle/shaft combination. Nut drivers are available with either solid or hollow shafts. The advantage of hollow-shaft drivers is that they allow the user to keep a grip on the nut even though the screw on which the nut is mounted is protruding.

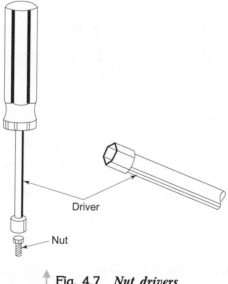

Driver

Nut

Fig. 4.7 *Nut drivers*

There are master driver sets available that include nut driver shafts, blotted screw driver shafts and Philips screw driver shafts.

While selecting drivers, the following factors should be borne in mind:

(a) The handles should be sturdy, made of heavy-duty plastic and preferably have rubber grips for comfort.

(b) Tool shafts should be deeply and firmly embedded into handles in the case of individual drivers.

(c) The shafts of plug-in type drivers should lock firmly into place when inserted in the master handle.

(d) The shafts of the tools should be tempered, plated steel and should have ground tips.

(e) Tools should be well-balanced and comfortable to handle.

(f) One should always buy high quality tools and stick to the established manufacturers of hand tools.

Nut drivers with hexagonal sockets range from 3/16″ Hex to 5/8″ or 3/16″ Hex to 1/2″ Hex in size.

Screw driver sets consisting of Philips bits, nut driver bits and bits suitable for slotted screw heads are common items of an instrument technician's kit. It is always advisable to buy a universal handle that will fit a series of interchangeable blades (Fig. 4.8).

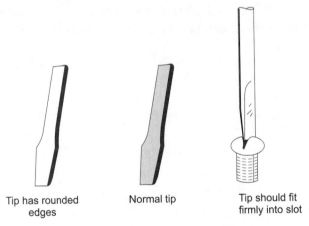

Tip has rounded   Normal tip   Tip should fit
    edges   firmly into slot

Fig. 4.8 *Normal screw driver tips*

4.1.6 Hacksaw

An assortment of metal-working and metal-finishing implements is needed in an electronic repair shop.

A hacksaw is one of such implements which is used to cut and trim the metal parts like shafts of potentiometers and switches, brackets, etc.

Most hacksaws are adjustable so that they can accept blades of several different lengths. More important than the blade length, however, is the coarseness. The number of teeth per unit length determines the coarseness. For general purpose work, a hacksaw blade should have between 14 and 18 teeth per inch.

4.1.7 Drills

Perhaps the most commonly used metal working tool in the electronics shop is the electric drill. For routine work, a drill should have a chuck capacity of 1/4 or 3/8 inch (6.4 or 9.5 mm). Preferably, it should have a variable speed control. Slow speeds are better for drilling through soft materials and for starting holes.

The piece of metal that does the actual cutting of the material being drilled is called a drill-bit. There are two types of drill-bits: those composed of carbon steel and those of tungsten-molybdenum steels. The latter are called high speed drill bits and preferred in electronic shops. A set of drill-bits includes units with diameters ranging from approximately 3/32 inch (2.4 mm) to 1/4 or 3/8 inch (6.4 or 9.5 mm).

The hand drill with a metal case should have a three conductor power cord. In case of a plastic body, a two-conductor cord is adequate.

For drilling and other metal work, several other items like a centre punch, a ball-peen hammer, a reamer, a set of files and a bench vise are required.

A centre-punch is used to make indentations in the material to be drilled at the exact centres of the holes to be made. These indentations will prevent the drill from wandering around the surface when first brought up to speed.

A reamer is used to enlarge holes in sheet metal. It can also be used to remove burrs around the perimeter of a newly drilled hole. A 1/2 inch (1.3 cm) tapered reamer is adequate for most of the work.

4.1.8 Files

Often during repair work, there is a need to use a different sized replacement component, like a potentiometer or fuse carrier or to smoothen scratch marks on surfaces. In such cases, files prove very useful. They are made of hardened high carbon steel. Files are classified according to their length, cut of teeth and cross-section.

Several tapes of files are needed in an electronics shop, out of which the more common are rough, coarse, bastard, second-cut, smooth and dead-smooth depending on the teeth spacing. When selecting a file, it is important to consider the material to be worked on.

Steel or hard metals need less coarse files like second-cut and smooth. Soft materials like aluminium, brass or copper need coarse or bastard files.

The following are some of the commonly used types of files:

Round File: A file with a round cross-section tapering towards a point for finishing or enlarging round holes.

Half Round File: A semi-circular file usually double cut on the flat face and single cut on the curved face. This is one of the most versatile type of files for general purpose work as it can be used for smoothing curved edges or holes of large diameter as well as flat surfaces.

Three Square File: A three-sided double cut file that tapers towards a point. This is mostly used for smoothening internal angular surfaces.

Special files, called 'needle files', are available in smaller sizes for fine work.

When filing soft materials, there is a possibility that the waste metal may clog the file teeth. In that case, the file teeth should be cleaned using file cleaners. As a matter of caution, never do filing near open electronic/electrical equipment as the filings may enter the equipment, causing electrical shorts.

4.1.9   Other Workshop Tools

A heavy duty bench vise is useful in holding metal pieces as they are drilled or sawed. The jaws of the vise should be at least two inches long, and the vise should be a swivel unit which rotates in the horizontal plane. This allows its jaws to form any desired angle with respect to the edge of the workbench.

Kits are available which include a hand-held power tool and a whole complement of drilling, grinding, polishing, sanding and cutting attachments. One such kit is necessary for every electronics workshop.

It may be remembered that worn-out or broken tools are not safe. They are also not efficient. They can be dangerous to the user and those working near him. Worn tools should be repaired or replaced as required. Remember, the cost of a new tool is small when compared to an injury, lost time or sub-standard performance.

A considerable assortment of tools must be stocked in an electronics shop. Some items are absolutely essential, others are for convenience and enhance work efficiency. Many of these implements can be obtained from hardware stores, while others are special items that must be obtained from more exotic sources. Remember that work will be more enjoyable with the right kind of tools.

4.2 SOFT TOOLS (CHEMICALS FOR WORKBENCH)

Most of us are familiar with the traditional or 'hard' tools such as screw drivers, soldering irons, wrenches, pliers, etc. However, some chemicals or 'soft tools' are available which are as important as the screw driver and other hard tools in the servicing field. They are 'soft' tools because they are consumable, i.e. part of the tools is used up when it performs its function, unlike the hard tools which have indefinite life when handled with care. The chemical tools which are generally employed include:

- Solvents;
- Adhesives; and
- Lubricants.

4.2.1 Solvents

Solvents are either aerosol (spray solvents) type or bottled products. They are used mostly as cleaners.

Spray solvents used in servicing usually contain either chlorinated hydrocarbons or fluorocarbons.

Chlorinated hydrocarbons are compounds of hydrogen, carbon and chlorine. The best known example is carbon tetrachloride.

Fluorocarbons are compounds of hydrogen, carbon and fluorine. The most popular compound is freon. Freon TF is one form of this compound, which can be compounded with a variety of hydrocarbons or other solvents to make many specialised cleaning agents. Spray cleaners can be divided into the following groups.

Non-lubricating Cleaners (no-residue cleaners): They are used to remove grit and dust from circuit boards, mechanical assemblies and parts like magnetic tape heads and ganged tuning capacitors (such as used in radio receivers). Lubricating cleaners should not be used for such applications in which the lubricant is likely to cause more dust to accumulate quickly and would do harm.

Lubricating Cleaners: Lubricating cleaners are preferred for such items as potentiometers and switches, that is because those cleaners leave behind a coating of silicone that provides protection from oxidation and the resulting contact degradation.

Foam-type Cleaners: Foam-type cleaners are specially formulated viscous compounds (spray-cleaners) which come out of the can thick. These compounds contain mild abrasives such as jeweller's rouge, in a viscous base. They are generally used to clean and restore most wiping contacts, including those on wafer and slide switches. What those products do is to adhere to the metal contacts and continually clean and burnish the switch parts. However, be sure to remove the abrasive completely, using a no-residue solvent, once the cleaning and burnishing is done, otherwise the switch action will suffer.

Most spray solvents are safe when used inside a unit, but they should be kept off any of the exterior parts such as cabinets, dial scales, clear plastic window, etc. Generally, plastic such as nylon, delrin and similar substances will not be harmed by most solvents.

Bottled Solvents: For cleaning the rubber drive surface in tape recorders, turntables and the like, bottled rather than spray type solvents are generally used. Most of the people use alcohol-type cleaners. They are not recommended because they tend to leach out the stabilizers in the rubber, making it soft and sticky. The ketones such as acetone or methyl-ethyl ketone (MEK) are preferred because they evaporate very quickly and thus have little time in which to damage the good rubber.

However, acetone and MEK are flammable. They should not be used near sparks or open flames. Secondly, they are notorious for dissolving anything made of plastic and therefore, one should be careful when using them. They are also potent solvents and will destroy the rubber if too much is used. The best way to use these chemicals is to apply them sparingly to the rubber's surface with a cotton swab. That will allow them to take off the top layer of rubber and residue, without harming the good rubber beneath it.

To summarize, the following cleaners are commonly used:
- Contact cleaner (in spray can): used for switches and relays;
- Control cleaner (in spray can): used for potentiometers; includes some type of non-drying lubricant;
- WD40: For cleaning or freeing rusted screws; for cleaning, removing labels and label gum; coating tools to prevent rust etc.;
- Flax Remover: Isopropyl alcohol available in spray can;
- Isopropyl alcohol: for general cleaning; (i) used with Q-tip swabs (cotton buds) for cleaning of everything, but for video heads on VCRs and other helical scan tape transport (ii) chamois covered cleaning sticks for video heads;
- Tuner cleaner (spray can); and
- De-greaser (spray can).

4.2.2   Adhesives

Adhesives are used to fasten two pieces of almost anything together reliably and quickly. They can be useful if parts or the broken pieces are available. This often happens with such things as cabinet assemblies, decorative parts and other plastic pieces, especially if the unit being repaired is old. Although it is generally advisable to replace a broken part with a new one, there are situations when that is not possible and use of adhesives to bond broken parts is the only solution. The various types of adhesives that are commonly encountered in practice are:

Epoxy: Expoxies are universal adhesives that have a variety of setting times and viscosity. The viscosity and the setting time depend on the particular formulation used. They are two-component adhesives that must be mixed before application. At room temperature, heat or pressure is not required for setting.

Epoxy will bind most materials, including metals, glass, ceramics, cardboard, wood, rubber, fibre and most plastics (except nylon and similar compounds), with a bond strength which is generally greater than the strength of the bonded materials themselves.

Viscosity tends to be related to setting time. Quick setting epoxy is thinner than the slower setting ones. The choice of an epoxy for a particular application is determined by what is to be joined as well as how many items are involved. If only one item is to be bonded and their surfaces closely match, one of the thin, quick-setting types epoxy is preferred. If there are gaps in the two surfaces to be joined and possibly more working time is needed, the slower-setting epoxies are better in such situations.

Epoxies are generally used as electronic sealers and potting compounds to insulate a high voltage connection or to encapsulate a circuit for reliability. Hysol epoxy is an example of a potting compound that is normally used to totally encapsulate the circuits.

De-potting an electronic assembly is pretty easy. All that is needed is the judicious application of a small welding torch flame to locally heat the epoxy above its glass temperature and a little patience.

Solvent-Release Adhesives: Solvent-release adhesives include a wide variety of resins and polymers that harden when the solvent either evaporates or is absorbed. The tan-coloured compounds will provide a flexible, waterproof bond between virtually any two materials. The adhesives are used by coating each piece to be joined with a thin layer, waiting for the thermoplastic to become tacky and then pressing the pieces together. Drying time ranges from 15 to 30 minutes.

Silicone-Rubber Adhesive: It is best used to seal against air or moisture or when a flexible bond is needed. It will adhere to almost any clean surface and dries when it is exposed to the moisture in the air. This adhesive can be easily removed with a scraper if necessary, which gives it an advantage over other adhesives. Also, it offers non-running consistency, one-component convenience and resilient bonding. It is a handy-compound for potting components and sub-assemblies. However, its tensile strength is lower than that of either epoxy or other adhesives.

Cyanoacrylate Adhesives: They are single component, solvent-free glues that have the property to set very quickly. They will bond any substance provided good surface contact can be maintained. The biggest advantage these adhesives have over epoxy is that they can be used to join such hard-to-glue materials as polyethylene, teflon, vinyl and silicon rubber. They have an excellent tensile strength. Although their setting time is specified as a few seconds, it is better not to stress the joint for 10 to 15 minutes.

4.2.3 Lubricants

Parts that slide against each other do require occasional lubrication. However, when lubricating small mechanisms, it may be remembered that the lesser the quantity of lubricant used, the better.

For gear trains, bearings and the like, a good grade of light-bodied greases made of polymerised oil provides long-lasting lubrication. This grease will not oxidise or become gummy for several years under normal conditions. For parts that slide against each other, use a good molybdenum di-sulfide or lithium-based grease. Silicon lubricant, which comes in the form of spray is ideal for lubricating such items as hinges and sliding plastic parts. Those sprays are especially useful for restoring proper operation to a slide potentiometer. To do so, the pot is first cleaned with a very small amount of tuner cleaner or Freon TF. This should be done until the cracking sound disappears. Next, an ample amount of silicone lubricant is applied and the control is worked until slider moves smoothly.

Seized bearings can often be freed by a good penetrating lubricant under pressure. These are sold under many trade names like LPSI and WD 40.

4.2.4 Freeze Sprays

'Freeze sprays' are used to cool suspected noisy or intermittent components. They are of several types but in composition all are either fluorocarbons or chlorinated hydrocarbons.

The sprays which evaporate quickly are best for components with good thermal conductivity, such as metal-cased transistors, small electroytic capacitors and metal film resistor. On the other hand, a long evaporating time (20 to 30 seconds) is more suitable for plastic or epoxy encased semiconductors, mylar insulated capacitors or other parts with a heavy insulation. Special anti-static freeze sprays are also available for use with MOSFET and other static charge sensitive components.

Freeze sprays do not harm the plastic used on circuit boards. However, their use should be avoided around clear plastic or cabinet plastic or any painted finish.

Another use of freeze sprays is to assemble or dis-assemble tight fitting metal parts, as cooling the inside piece will reduce its size temporarily.

Soft tools are among the most useful items in the workshop. You may take time to learn their proper use, but it will be observed that in the long run, repairs undertaken with soft tools are faster, easier and less expensive.

5

Soldering Techniques

5.1 WHAT IS SOLDERING?

Quite often, for locating a problem in the functioning of the circuit, it is necessary to remove a component from the printed circuit board and carry out the requisite tests on it. The process of repair usually involves:

  (a) Dis-assembly of a particular component;
  (b) Testing of the component;
  (c) Replacement of the component found defective; and
  (d) Testing the circuit for performance check.

In this exercise of removal and replacement of electronic components, a good soldering practice is very essential.

A soldered connection ensures metal continuity. On the other hand, when two metals are joined to behave like a single solid metal by bolting, or physically attaching to each other, the connection could be discontinuous. Sometimes, if there is an insulating film of oxides on the surfaces of the metals, they may not even be in physical contact. Soldering is an alloying process between two metals. In its molten state, solder dissolves some of the metal with which it comes into contact. The metals to be soldered are more often than not covered with a thin film of oxide that the solder cannot dissolve. A flux is used to remove this oxide film from the area to be soldered. The soldering process involves:

  (a) Melting of the flux which, in turn, removes the oxide film on the metal to be soldered;
  (b) Melting of the solder which makes the lighter flux and brings the impurities suspended in it to the surface;
  (c) The solder partially dissolving some of the metal in the connection; and
  (d) Cooling and fusing with the metal.

The soldering process entails an understanding of:

   (a) Soldering tools;
   (b) Soldering material;
   (c) Soldering procedure;
   (d) Replacing components;
   (e) Special consideration when using MOS and micro-electronic circuits;
   (f) Good and bad soldering joints; and
   (g) De-soldering techniques.

Soldering is a skill that needs to be acquired for efficient repair work. The use of a proper soldering technique is critical for reliability and safety.

5.2   SOLDERING TOOLS

Various tools are necessary to facilitate soldering work. The most essential tools used in soldering practice are discussed in detail below.

5.2.1   Soldering Iron

A soldering iron should supply sufficient heat to melt solder by heat transfer when the iron tip is applied to a connection to be soldered. The selection of a soldering iron is made with regard to its tip size, shape, operating voltage and wattage. The soldering iron temperature is selected and controlled according to the work to be performed. The temperature is normally controlled through the use of a variable power supply and occasionally by tip selection.

   There are two general classes of soldering irons, viz. pencils and guns.

Soldering Pencils: Soldering pencils (Fig. 5.1) are lightweight soldering tools which can generate as little as 12 watts or as much as 50 watts of heat. A 25 watt unit is well suited for light duty work such as soldering on printed circuit boards. Modular soldering irons use inter-changeable heating elements and tips which mate to a main pencil body. Such elements screw into a threaded receptacle at the end of the pencil. A variety of tips (Fig. 5.2) are available to handle most soldering tasks. Very fine, almost needle-like tips are used on printed circuit boards with IC component foil pads which are closely spaced. Larger, chisel and pyramid tips can store and transfer greater amounts of heat for larger, widely spaced connectors. Bent chisel type tips can get into difficult-to-reach areas. Regardless of the type of the tips, it is best to use plated, as opposed to raw copper tips, as these have a much longer life.

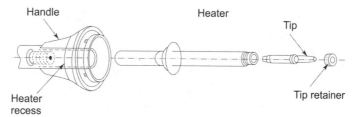

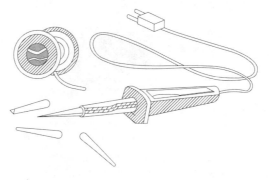

Fig. 5.1 *Soldering iron–pencil type*

Iron clad, chrome plated, pre-tinned tips

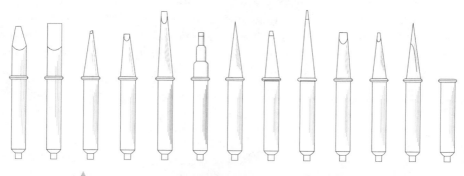

Fig. 5.2 *Different shapes of bits for soldering iron*

A pencil type soldering iron takes a few minutes to attain working temperature and it is better to keep it continuously powered even for interrupted type of soldering work. This would need to keep the iron secured in a safe place at the working temperature. One method is to keep it in a special soldering iron holder which may be a coiled steel form into which the hot soldering iron can be inserted. Most stands of this type also include a sponge which is kept moistened and used periodically to clean the soldering tip.

Soldering Gun: A gun is usually heavier and generates more heat than the average pencil. Soldering of heavy duty conductors or connectors requires the use of a gun because it can generate enough heat to quickly bring a heavy metal joint up to the proper soldering temperature. These soldering tools are called guns simply because the resemble pistols. The gun's trigger (Fig. 5.3) is actually a switch that controls the application of AC power to the heating element. The working temperature is reached instantaneously. Some guns provide for selection of different heat levels through a multi-position trigger switch.

Fig. 5.3   *Soldering gun*

Soldering Stations: Soldering stations (Fig. 5.4) contain an iron and a control console that offers switch selectable temperatures, marked low, medium and high. Obviously, this is more convenient than waiting for a modular pencil's heating element to cool, unscrewing it from the holder and then replacing it with another heater tip combination. The tip temperature is controlled by using a heat sensor and closed-loop feedback control to gate power to the heating element. Obviously, soldering stations are expensive as compared to basic soldering pencils.

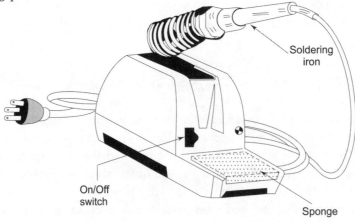

Fig. 5.4   *Soldering station*

Battery Operated Irons: Sometimes, it is inconvenient to depend on the mains power supply for operating a soldering iron. Battery-operated soldering irons are available which depend upon rechargeable batteries as a power source. Recharging is done automatically when the iron is placed in its charger which is built on the stand, and is connected to an AC power source. In these soldering irons, the tips attain working temperature in 5–8 seconds and cool off to an ambient temperature in about one second. Typically, about 125 connections can be made on one charge. For a standard iron, a typical charging interval of approximately 14 hours is required to return the cells to full strength. Of course, there are quick change irons also. Sometimes, the soldering irons have built-in light to illuminate the work area whenever battery power is applied to the heating element.

Soldering irons are best used along with a heat-resistant bench-type holder, so that the hot iron can be safely parked in between use. Soldering stations generally have this feature. Otherwise, a separate soldering iron stand, preferably one with a holder for tip-cleaning sponges, is essential.

Electronics catalogues often include a range of well-known brands of soldering iron. The following factors should be kept in mind when selecting a soldering iron for a particular application.

Voltage: Most soldering irons run from the mains supply at 230V. However, low voltage (12V or 24V) type irons are also available and they generally form part of a soldering station.

Wattage: Typically they may have a power rating ranging between 15–25 watts which is adequate for most work. A higher wattage does not mean that the iron runs hotter. It simply means that there is more power in reserve for coping with larger joints. Higher wattage irons are required for heavy duty work because it would not cool down so quickly.

Temperature Control: The simplest and cheapest type irons do not have any form of temperature control. Unregulated irons form an ideal general purpose iron for most users as they generally cope well with printed circuit board, soldering and wiring.

A temperature controlled iron has a built-in thermostatic control to ensure that the temperature of the bit is maintained at a fixed level, within pre-set limits. This is desirable especially during more frequent use, since it helps to ensure that the temperature does not over-shoot in between times and also that the output remains relatively stable. Some versions have a built-in digital temperature read-out and a control knob to vary the temperature setting. A thermo-couple may be used to measure the temperature of the tip and the heating rate is controlled by means of a thyristor. Thus, the temperature can be boosted for soldering larger joints.

Anti-static Protection: For soldering static-sensitive components such as CMOS and MOSFET transistors, special soldering iron stations having static-dissipative materials in their construction are required. These irons ensure that static-charge does not build up on the iron itself. These irons are ESD-Safe (Electrostatic Discharge Proof).

The general purpose irons may not necessarily be ESD-safe, but can be safely used if the usual anti-static precautions are taken when handling CMOS components. In this case, the tip would need to be well-grounded.

BITS: It is useful to procure a small selection of manufacturer's bits (soldering iron trips) with different diameters and shapes, along with the soldering iron. They can be changed depending upon the type of work in hand.

Spare Parts: It is preferable to ensure that spare parts are available for the iron. So, if the element blows, you don't need to replace the entire iron. This is especially relevant in the case of expensive irons.

5.2.2   Strippers

Strippers are used to remove insulation from the wires. The most usually employed strippers are those of the cutting type (Fig. 5.5). These strippers are so designed that they can accommodate various sizes of wire normally used in electronic equipment. In order to prevent damage to the wire by nicking, it should be ensured that the specific wire size hole is selected in the cutting stripper.

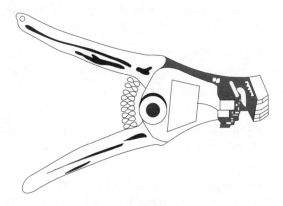

Fig. 5.5   *Wire stripper*

In the thermal strippers, the wire to be stripped is placed between two electrodes. The electrodes get heated when electric current is passed through them. The resulting heat melts the insulation. When using thermal strippers, toxic fumes emanating from compounds such as polyvinyl chloride or polytetrafluoroethylene must be properly exhausted by using some type of fan ventilation system.

5.2.3 Bending Tools

Bending tools are pliers having smooth bending surfaces so that they do not cause any damage to the component.

5.2.4 Heat Sinks

Some components such as semiconductor devices, meter movements and insulating materials are highly heat-sensitive. They must be protected from damage due to heat while soldering. Devices such as a set of alligator clips, nose pliers (Fig. 5.6), commercial clip-on heat sinks, felt-tipped tweezer, anti-wicking tweezers and other similar devices are usually placed or clamped at the site of soldering so that they prevent the heat from reaching the components.

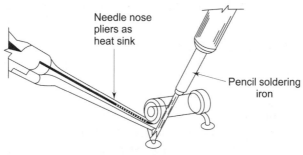

Needle nose pliers as heat sink

Pencil soldering iron

Fig. 5.6 *Use of pliers as heat sink in soldering*

5.2.5 General Cleaning Tools

Before the soldering process is actually performed, the surface on the printed circuit board or the component leads must be properly cleaned. The tools or devices most commonly used for general cleaning are alcohol dispenser, camel hair brush, small wire brush, synthetic bristle brush, cleaning tissue, pencil erasers (Fig. 5.7), typewriter erasers, braided shielded tool, sponge with holder, tweezers and single-cut file.

A very useful soldering aid consists of a plastic or wood wand with a pointed metal tip at one end and a notched metal tip at the other. The blunt end of the aid is used to clear solder from holes in printed circuit boards and from solder lugs. The notched end can be used to make right-angle bends in component leads, to hold leads and wires while the solder joint is made, and to keep leads away from printed circuit boards and lugs during de-soldering operations.

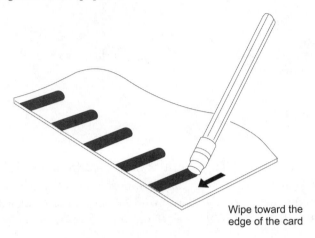

Wipe toward the
edge of the card

Fig. 5.7 *Use of a pencil eraser as a cleaning tool*

5.3 SOLDERING MATERIAL

5.3.1 Solder

The soldering material or solder usually employed for the purpose of joining together two or more metals at temperatures below their melting point is a fusible alloy consisting essentially of lead (37%) and tin (63%). It may sometimes contain varying quantities of antimony, bismuth, silver or cadmium which are added to vary the physical properties of the alloy.

The continuous connection between two metals is secured by soft solder by virtue of a metal solvent or inter-metallic solution action that takes place at a comparatively low temperature. Figure 5.8 shows the tin–lead fusion diagram which explains the alloy or solvent action on molten solder.

Pure lead melts at 621°F while pure tin melts at 450°F. When tin is added to lead, the melting point of lead gets lowered and follows the line PR. Similarly, when lead is added to tin, its melting temperature falls along QR. At point R, where the two lines PR and QR meet, an alloy of the lowest melting point is obtained. The point 'R' represents 63% tin and 37% lead. The alloy at this point is known to have eutectic composition and has a melting point at 361°F. The solder is abbreviated as SN63.

The most common type of solder used in electronics work is an alloy consisting of 60% tin and 40% lead. The alloy is drawn into a hollow wire whose centre is filled with an organic paste-like material called rosin. The resulting product is called 60/40 rosin-core solder. Its melting temperature is 375°F (190°C) and it solidifies as it cools. This alloy is available in wire form in several gauges. Thinner gauges are preferred over thicker ones. Fine solder is easy to position on the joint and requires less heat for the formation of a joint.

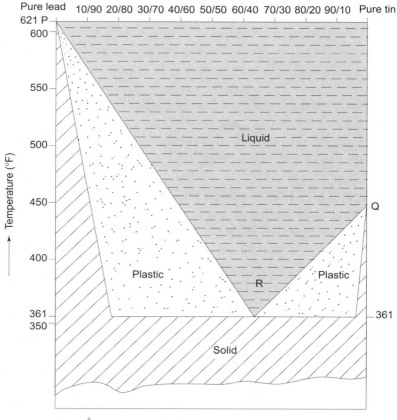

Fig. 5.8 *Lead–tin fusion diagram*

The solder alloy wires are commercially available in different diameters ranging from 0.25 mm to 1.25 mm. Usually, 20–22 SWG is of 0.91–0.71 mm diameter and is fine for most work. 18 SWG may be chosen for larger joints requiring more solder. The volume of the flux in the wire is about 25% corresponding to a mass of about 3%.

It is found that with SN63, the stress resistance of the solder joint is the maximum, i.e. at the lowest melting point, the alloy has the highest pull strength.

Several other alloys exhibit eutectic behaviour. However, they suffer from certain disadvantages; particularly when they have silver (tin 62.5%, lead 36.1%, Silver 1.4%), as they tend to be more expensive. However, for specialized applications, special wires with high or low melting points and some with 1.5 to 5% silver are required. It is a good practice to buy a solder wire from a reputed manufacturer because both the alloy composition and flux used may vary, which often proves detrimental to the product.

The lead present in the solder does not cause any health hazard. However, when handling lead dust, a mask must be used. Smoking during soldering may cause the lead smoke to be inhaled and this must be avoided. Also, one should thoroughly clean one's hands after soldering before eating or smoking.

5.3.2 Flux

Sometimes another substance know as 'flux' is used to aid the soldering process. Flux is needed to scrub away the microscopic film of oxides on the surfaces of metals to be soldered and it forms a protective film that prevents re-oxidation while the connection is heated to the point at which the solder melts. Flux is helpful in the case of a stubborn joint that would not accept solder. Most metals tend to form compounds with atmospheric oxygen which leaves a coating of oxide even at room temperature. The oxides are re-moved by fluxes which remain liquid at soldering temperature, react chemically with the oxides and disperse the reaction products. Fluxes are applied before or during soldering.

The solder used in most electronic work contains this flux as a central core which has a lower melting point than the solder itself. When the molten flux clears the metal, it accom-plishes two things:

(a) It allows the solder to wet the metal; and

(b) It holds the oxides suspended in solution.

The molten solder can then make contact with the cleaned metal and the solvent action of solder on the metal can take place.

5.4 SOLDERING PROCEDURE

The soldering procedure basically involves the preparation of the following three items:

(a) The soldering iron selection;

(b) The component to be soldered; and

(c) The circuit board.

5.4.1 Selection of Soldering Iron

The soldering iron selected should be of the precision type, small but powerful enough to reliably solder components to printed circuit boards. An iron between 25 and 40 watts with a nickel plated tip, or one of the miniature irons capable of a tip temperature of 400°F, is the most suitable. Hotter temperatures run a real risk of spoiling the adhesive bond that holds the copper foil to the board. Do not use a higher temperature to make up for an improperly-tinned tip.

The ideal tip is a single flat or chisel tip of about 2.5 mm. The old style unplated copper tips are not very suitable, as they wear away very quickly.

The soldering iron should be examined carefully every time it is to be used. The soldering iron tip should be properly connected or screwed into the holder and it should be free from oxides. The shape of the tip must meet the requirements of the task to be performed. If any one of these items is not as good as it should be, the following steps are adopted:

(a) The oxides from the tip surface are removed by using an abrasive cloth or sand paper.

(b) The tip is given its proper shape generally by filing. This is normally done on the unplated copper tip.

(c) The iron is heated to the minimum point at which the solder melts. Before using the iron to make a joint, the tip is coated or tinned lightly by applying a few millimetres of solder.

(d) For keeping the tip clean, after it has been prepared, the heated surface of the tip should be wiped with a wet sponge. This is to remove dirt, grease or flux which, if allowed to remain, can become part of the joint and make the joint dry and defective.

If during soldering, excessive heat is generated at the soldering iron tip and the component gets heated beyond its maximum temperature, the component may be permanently damaged, weakened, or drastically affected in value or characteristics. Such effects may not be noticed during the assembly or test but may show up later when the equipment is in use.

The tip temperature to be selected must be based on the temperature limitation of the substrate. The circuit boards which have a substrate of fibre-glass epoxy of 280°C should not be heated for more than 5 minutes. Hotter temperatures reduce the time in inverse relationship; the higher the temperature, the less time the boards will stand it before being damaged.

Further, heat transmitted along the leads may cause unequal expansion between leads and packages, resulting in cracked hermetic seals. In general, for hand soldering, the recommended soldering iron wattage is 20–25 W for fine circuit board work, 25–50 W iron for general soldering of terminals and wires and power circuit boards, 100–200 W soldering gun for chassis and large area circuit planes. With a properly sized iron or gun, the task will be fast and will result in little or no damage to the circuit board plastic switch housings, insulation, etc. For iron temperatures ranging between 300°C and 400°C, the tips of the soldering iron should be in contact with the lead for not more than 5 seconds. Particularly, the ICs and transistors should be soldered quickly and cleanly.

5.4.2 Component Preparation

Cleaning: Before any component lead, wire or terminal is soldered in a circuit, it is essential to clean it with some braided cleaning tool followed by brushing the cleaned surface with a stiff bristle brush dipped in alcohol. The surface is dried with paper or lint-free rag.

It must be noted that the solder will just not take to dirty parts. They should be free from grease, oxidation and other contamination. Old components or copper board can be notoriously difficult to solder because of the layer of oxidation which builds up on the surface of the leads. This repels the molten solder and the solder will form globules, which will go everywhere except where needed. While the leads of old resistors and capacitors should be cleaned with a small hand-held file or a fine emery paper, the copper printed circuit board needs to be cleaned with an abrasive rubber block or eraser. In either case, the fresh metal underneath needs to be revealed.

Component Forming: In order to ensure that the components fit properly into the circuit, in which they are to be installed, they must be properly formed (Fig. 5.9). Forming of the component has two main functions:

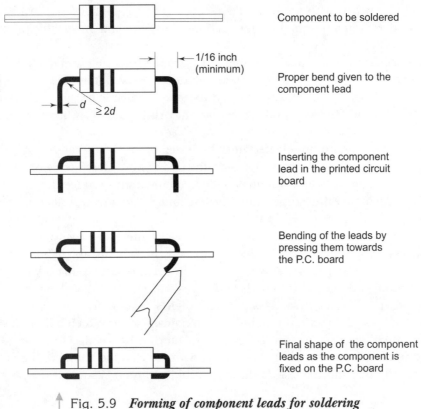

Component to be soldered

1/16 inch (minimum)

Proper bend given to the component lead

Inserting the component lead in the printed circuit board

Bending of the leads by pressing them towards the P.C. board

Final shape of the component leads as the component is fixed on the P.C. board

Fig. 5.9 ***Forming of component leads for soldering***

(a) To secure the lead to the circuit; and

(b) To provide proper stress relief. The relief is necessary to prevent rupture of the component lead from the component or in case of a wire, to prevent a stress pull on the solder joint and rupture of the wire strands.

The following steps need to be taken to properly bend the leads of the components:

(a) The bend should be attempted no closer than 3–5 mm from the component body.

(b) The radius of the bend should be equal to twice the thickness of the lead wire.

(c) Centre the component between its solder connections.

(d) Bend the protruding lead 45° after insertion into the circuit board with the help of a bending tool.

(e) Cut the lead so that no portion, when bent, exceeds the perimeter of the pad. Press the cut lead firmly against the lead.

(f) On a joint, when the lead is not bent, cut the lead to the thickness of SWG 20.

Forming is not necessary in the case of integrated circuits. They are simply soldered to the board without cutting their leads.

Lead bending is the most important factor in forming. A few sharp back and forth bends in a component lead can easily cause it to break or crack. Bending a lead too close to the component encapsulation may result in excessive stress at the lead entrance, and cause cracks in the encapsulation. Such cracks allow moisture to enter inside the component, causing gradual degradation of the component subsequently resulting in its premature failure.

5.4.3 Circuit Board

Although the principle of general cleaning also applies to circuit boards, it is, however, necessary to follow certain precautions. This is because circuit boards may contain some components that may be spoiled if a braided brush is used. In case of circuit boards, a sharply pointed typewriter eraser may be used to remove dirt, contaminants or other foreign substances from the pad to be soldered. It is then cleaned with alcohol and a brush and left to dry.

It may be noted that in the whole process of replacement of components by soldering, the single-most crucial factor is cleanliness, which should be scrupulously followed. With dirty surfaces, there is a tendency to apply more heat in an attempt to force the solder to take. This will often do more harm than good because it may not be possible to burn off any contaminants anyway, and the component may become over-heated. In the case of semiconductors, the temperature is quite critical and they get damaged when excessive heat is applied.

5.5 SOLDERING TECHNIQUE

A good soldering technique must ensure that:

(a) The solder forms a firm joint.

(b) The solder should cover all elements of the joint.

(c) The shape of the elements in the joint is not obscured.

(d) The solder, when it solidifies, appears as a bright solid and flake-free surface.

To best meet these requirements, proceed as follows:

(a) Place the iron at an angle of 45° with the tip touching as many elements of the joint as possible.

(b) Place the solder near the iron and let it flow. Pass it around the joint till you come back near the iron.

(c) Remove the iron and let the solder flow into the area from where the iron has been removed. All the elements of the joint should get covered with the solder.

(d) Never solder when the equipment you are working on is switched on; better unplug it before soldering.

(e) When the solder has successfully flowed into the lead and track, take the solder away and then remove the iron. Many people make the mistake of removing the iron first and this will nearly always result in a dry joint, due to the solder taking heat from the joint prematurely.

Temperature: An important step to successful soldering is to ensure that the temperature of all the parts is raised to roughly the same level before applying the solder. Heating one part but not the other will produce an unsatisfactory solder joint. The melting point of most solders is in the region of 188°C (370°F) and the iron tip temperature is typically 330–350°C (626°–662°F). Figure 5.10 shows the temperature range for ideal soldering

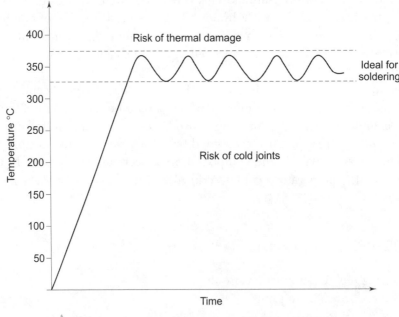

Fig. 5.10   *Ideal temperature range for soldering*

work. Above this temperature range, there is a risk of thermal damage whereas below this range, cold junctions are likely to develop.

Time: Next, the joint should be heated with the bit for just the right amount of time. Excessive time will damage the component and perhaps the circuit board copper foil. The heating period depends on the temperature of your iron and size of the joint. Larger parts need more heat than smaller ones while some parts (semiconductor devices) are sensitive to heat and should not be heated for more than a few seconds. In such cases, thermal shunts or heat sinks are used to protect heat-sensitive components from damage due to heat while soldering. These devices are placed or clamped in place to prevent the heat from reaching the component while its leads are being soldered.

Solder Coverage: In order to achieve a successful solder joint, it is essential to apply only an appropriate amount of solder. Too much solder is an unnecessary waste and may even cause short circuits at the end joints. Too little solder may not fully form a successful joint or may not support the component properly. How much solder to apply really comes only with practice.

5.6 REPLACEMENT OF COMPONENTS

Printed circuit boards used in modern equipments are generally the plated-through type consisting of metallic conductors bonded to both sides of an insulating material. The metallic conductors are extended through the component holes or inter-connecting holes by a plating process. Soldering can be performed on either side of the board with equally good results. Before a component replacement is attempted, the following precautions should be observed:

(a) Avoid unnecessary component substitution. It can result in damage to the circuit board and/or adjacent components.

(b) Do not use a high power soldering iron on etched circuit boards. Excessive heat can dislodge a conductor or damage the board.

(c) Use only a suction device or wooden toothpick to remove solder from component mounting holes. Never use sharp metal object for this purpose as it may damage the plated-through conductor.

(d) After soldering, remove excess flux from the soldered areas and apply a protective coating to prevent contamination and corrosion.

The following steps need to be followed for replacing a component:

(a) Carefully read the replacement procedure from the service manual of the instrument.

(b) Switch-off the power, if applicable

(c) Remove any assemblies, plugs and wire that will facilitate repair work.

(d) Label the component to be removed.

(e) Observe carefully how the component is placed before removing it. Record information regarding polarity, placement angle, positioning, insulating requirements and adjacent components.

(f) Be careful to handle the printed circuit board by the edges only. The finger prints, which though invisible, can cause an accumulation of dirt and dust on the boards, resulting in low impedance bridges in portions of the circuit board, which normally should have a very high impedance. Gloves should preferably be worn to prevent finger print problem if the boards must be handled.

(g) Remove the dry film or the hermetic sealer from the solder joint that is to be worked on. This is done by using a cotton tipped applicator dipped in the recommended chemical. Large quantities of solvents should not be allowed to drip on the board because the impurities will then only be shifted from one place to another on the board. This cleaning is necessary because it can be difficult to burn through a layer of dry film with a soldering iron. In addition, if the dry film is not removed before heating, the appearance of the board will be badly changed.

(h) Heat the solder fillet on the solder side of the printed circuit board. Using the de-soldering tool (suction device), gently and carefully remove the component. Too much soldering iron heat should not be used otherwise the foil is lifted or plated-through holes get removed.

In case of the multi-lead component, the vacuum de-soldering tool must be used to remove almost all the solder from the component leads before the latter can be removed from the board. This procedure must be carefully followed because multi-lead components multiply the probability of printed circuit board damage during repairs.

(i) Some components are difficult to remove from the circuit boards due to a bend placed in each lead during machine insertion of the components. The purpose of the bent leads is to hold the component in position during a flow solder manufacturing process which solders all components at once. To make removal of machine inserted components easier, straighten the leads of the components on the back of the circuit board using a small screw driver or pliers while heating the soldered connections.

(j) After removing the component from the printed circuit board, the area around the removed component must be cleaned up by using the cotton tipped applicator in a solvent. Also, there may be solder in the plated-through holes or other areas of the board that must be removed in order to allow easy insertion of a new component.

(k) Clean leads of new component or element with a cleaning tool, such as a braided tool. Use abrasives if required. In case of a wire lead, the insulation must be removed. The secret to a good solder joint is to make sure everything is perfectly clean and shiny and not depend on the flux alone to accomplish this. In case of multiple strands, form the strands. Tin to about 3 mm from insulated part.

(l) Shape the leads of the replacement component to match the mounting hole spacing. Insert the component leads into the mounting holes and position the component as originally positioned. Do not force leads into mounting holes because sharp lead ends may damage the plated-through conductor.

(m) Start with a strong mechanical joint. Don't depend on the solder to hold the connection together. If possible, loop each wire or component lead through the hole in the terminal. If there is no hole, wrap them once around the terminal. Gently anchor them with a pair of needle nose pliers.

(n) Heat the parts to be soldered, not the solder. Touch the end of the solder to the parts, not the soldering iron or gun. Once the terminal, wires or component leads are hot, the solder will flow via capillary action, fill the voids and make a secure mechanical and electrical bond. Apply the soldering iron to the joint and feed solder into it. The solder should be applied to provide a complete seal covering all elements. Be careful with the amount of solder and the amount of heat. Check the component side of the board for good solder flow. Remember that SN63 is the best type of solder for soldering electronic components. SN60 is acceptable.

(o) Remove the soldering iron and allow the solder to cool and solidify. Do not disturb the board for a while, otherwise you will end up with a bad connection or what is called a 'cold solder joint'.

(p) Clean the area of splattered rosin flux and residue using isopropyl alcohol. Be careful not to leave cotton filaments on the printed circuit board. Allow the circuit board to air dry completely.

(q) Apply protective coating, if possible, on the repaired area and allow this to air dry.

(r) It is always advantageous to check the integrity of the joint that has been soldered or repaired. This check can be performed with an ohmmeter (multimeter) by measuring the resistance between the solder and the component lead. Any reading except a short reveals a defective joint. Recognize that defective solder joints, that are cracked, pitted, cold stressed, have excessive flux or the impure solder.

(s) When working with semiconductor devices and micro-electronic IC circuit components, a heat sink may always be used, while soldering. Also, when working on equipment having components like these, the specifications of allowable soldering iron sizes, voltage ranges and other factors must be studied. This is essential to understand the investment damage that one can do if a unit is repaired improperly.

While replacing components, it may be noted that mechanical shocks can seriously damage the components. For example, semiconductors can get damaged by the high impact shock if dropped on a concrete floor even from a table height. Cutting of leads can also cause shock waves which may damage delicate or brittle components. Therefore, cutting or scratching of surfaces of components by the careless use of tools or sharp test probes should always be avoided.

It is always wise not to remove or replace any component while the power is on. This may well produce voltage or current surges that could damage the component itself and other sensitive components in the circuit.

5.7   SPECIAL CONSIDERATIONS FOR HANDLING MOS DEVICES

MOS (Metal Oxide Semiconductor) devices are highly sensitive devices and get damaged easily by accidental over-voltages, voltage spikes and static-electricity discharges. The human body can build up static charges that range up to 25000 volts. These build-ups can discharge rapidly into an electrically grounded body or device, and particularly destroy certain electronic devices. The resultant high voltage pulse burn out the inputs of integrated circuit devices. This damage might not appear instantly, but it can build up over time and cause the device to fail.

The most common causes of electro-static discharge (ESD) are: moving people, low humidity (hot and dry conditions), improper grounding, unshielded cables, poor connections and moving machines. When people move, the clothes they are wearing rub together and can produce large amounts of electro-static charges in excess of 1000 volts. Motors in electrical devices, such as vacuum cleaners and refrigerators, generate high levels of ESD. Also ESD is most likely to occur during periods of low humidity, say below 50%. Any time the charge reaches around 10,000 volts, it is likely to discharge to grounded metal parts.

An important point to remember is that 10,000 to 25,000 volts of ESD are not harmful to human beings whereas 230 volts, 1 amp current produced by the mains power supply is lethal. The reason for this is the difference in current-delivering capabilities created by the voltage. The ESD voltages, though in the kilovolts range, produce currents only in the microampere range, which are not harmful for human beings. However, the level of static electricity on your body is high enough to destroy the inputs of CMOS (complementary metal-excide semiconductor) device if you touch its pins with your fingers.

Special care is needed in storing, handling and soldering MOS devices. The following precautions must be observed when using such devices:

(a) While storing and transporting MOS devices, use may be made of a conductive material or special IC carrier that either short circuits all leads or insulates them from external contact.

(b) The person handling MOS devices should be connected to the ground with grounding strap as shown in Fig. 5.11. These anti-static devices can be placed around the wrists or ankle to ground the technician to the system being worked on.

These straps release any static charge or the technician's body and pass it harmlessly to ground potential.

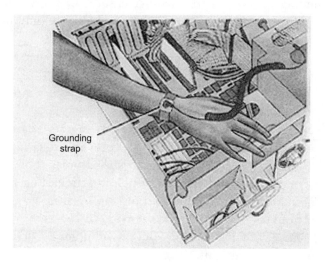

↑ Fig. 5.11   *Use of anti-static strap*

Anti-static straps should never be worn while working on high voltage components, such as monitors and power supply units. Some technicians wrap a copper wire around one of their wrists or ankles and connect it to the ground side of another. This is not a safe practice because the resistive feature of a true wrist strap is missing.

The work areas should preferably include anti-static mats (Fig. 5.12) made of rubber or other anti-static materials they stand on while working on the equipment. This is particularly helpful in carpeted work areas because carpeting can be a major source of ESD build-up. Some anti-static mats have ground connections that should be connected to the safety ground of an AC power outlet.

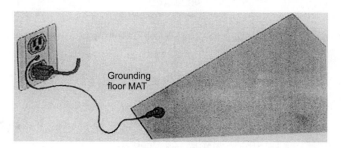

↑ Fig. 5.12   *Use of anti-static mat in the work area*

(c) Before touching any component inside the system particularly containing MOS devices, touch an exposed part of the chassis or the power supply housing with your finger. Grounding yourself in this manner ensures that any static charge on your body is removed. This technique, however, works safely only if the power cord is attached to a grounded power outlet.

(d) Mount MOS integrated circuits on printed circuit boards after all other components have been mounted.

(e) When replacing a defective IC, use a soldering iron with a grounded tip to extract the defective IC and while soldering the new IC in place.

(f) After the MOS circuits have been mounted on the board, proper handling precautions should still be observed. To prevent static charges from being transmitted through the board wiring to the device, it is recommended that conductive clips or conductive tape be put on the circuit board terminals.

(g) To prevent permanent damage due to transient voltages, do not insert or remove MOS devices from test sockets with the power on.

(h) Avoid voltage surges as far as possible. Beware of surges due to relays, and switching electrical equipment on and off.

(i) Signals should not be applied to the inputs while the device power supply is off.

(j) All unused input leads should be connected to either the earth or supply voltage.

(k) Personnel handling MOS devices are advised to wear anti-static clothing, synthetic fibre clothing should particularly be avoided.

(l) Do not insert PC boards into connectors that have voltages applied to them.

(m) Work stations should have non-conductive table tops, non-conductive trays, grounded soldering irons, etc.

The switching action of some controlled-output soldering implements can generate voltage spikes, which can be transmitted to and adversely affect MOS devices. Care should be taken in selecting soldering irons so that they have low voltage spikes.

5.8 SOLDERING LEADLESS CAPACITORS

In some equipment, leadless capacitors are used. Special techniques are required to successfully solder such capacitors to circuit boards. The following steps will minimize the problems that may be encountered when soldering leadless capacitors (Fig. 5.13):

(a) Tin the capacitor by using a small soldering iron with low heat and holding the capacitor down by weighting the edge of it with a silver coin.

(b) Tin the area of the circuit board where the capacitor is to be attached.

(c) Place the capacitor on the board in the desired location.

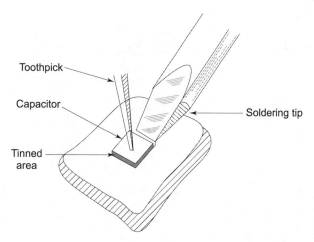

Toothpick

Capacitor

Tinned area

Soldering tip

↑ Fig. 5.13 *Special techniques for soldering leadless capacitors*

(d) Apply heat to the board adjacent to the capacitor without touching the capacitor. Do not attempt to effect a bond by applying heat on top of the capacitor as this will permanently damage the capacitor.

(e) Press down lightly on the capacitor using a toothpick or other small wooden stick, until it settles down on to the board indicating that the solder has melted underneath. Remove the heat and allow it to cool.

5.9 GOOD AND BAD SOLDERING JOINTS

It is with experience that one learns the difference between a good or bad soldered joint. However, the following points should be kept in mind:

(a) The solder should be uniformly distributed over the elements and base metal. All solder joints, particularly in the high voltage circuit paths, should have smooth surfaces. Any protrusions may cause high voltage arcing at high altitudes.

(b) The quantity of the solder should be only so much that it does not obscure the shape of the element.

(c) No residue such as flux or oxide should be left on the surfaces.

(d) No solder should reach the shield of the wire.

A good solder connection will be quite shiny, not dull gray or granular. If your result is less than perfect, re-heat it and add a bit of new solder with flow to help it re-flow.

The examples of bad solder are given in Fig. 5.14.

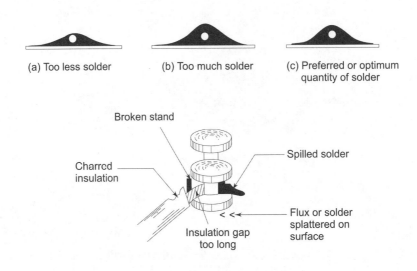

Fig. 5.14 *Bad soldering practices*

5.10 DE-SOLDERING TECHNIQUE

De-soldering means removal of solder from a previously soldered joint. Two techniques are common in soldering:
(a) Wicking; and
(b) Sniffing.

5.10.1 Wicking

In the wicking process, a heated wick, well-saturated with rosin is placed on top of the joint to be de-soldered. The solder will flow rapidly into the rosin area due to capillary action leaving the joint to which it was previously affixed.

A wicking solder remover may consist of a braided shield wire with the core removed or it may be a piece of multi-strand wire. Wicks are available commercially which are suitable for desoldering work. The de-soldering technique using the wicking process is as follows (Fig. 5.15):
(a) Place the wick on top of the solder joint to be de-soldered.
(b) Position the iron tip on top of the wick. The heat of the iron will melt the solder. The solder will readily flow into the wick.
(c) Cut off the wick containing the removed solder. Repeat the process until all the solder is removed from the joint.

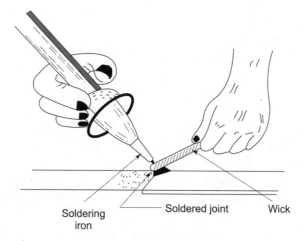

Fig. 5.15 *Wicking process for desoldering*

Take extreme care to ensure that the solder is not allowed to cool with the braid adhering to the work, otherwise you run the risk of damaging PCB copper tracks when you attempt to pull the braid off the joint. This technique is more effective specially on difficult joints wherein a de-soldering pump, described below, may prove unsatisfactory.

5.10.2 Sniffing

In sniffing, a rubber ball (Fig. 5.16) is employed as a solder sucker (sniffer). The sniffer uses the forced air pressure to accomplish the sniffing (removal of solder) action.

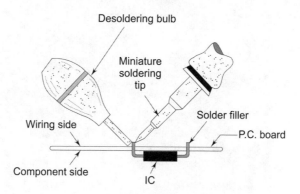

Fig. 5.16 *Sniffing technique in desoldering*

Another vacuum type sniffer uses a spring loaded plunger:

The following steps are adopted in sniffing:

(a) The air is first squeezed out of the rubber ball.

(b) With the ball depressed, the pointed end of the sniffer tube is placed next to the solder to be removed.

(c) The joint is heated with the soldering iron. The tip of the iron should be kept in the solder and not on the sniffer.

(d) The pressure on the sniffer ball is slowly released to allow air to enter the ball through the sniffer tube. As the air enters, it pulls the molten solder into the tube with it.

(e) After the solder has been completely pulled into the tube with it, the sniffer is removed from the joint. By depressing the ball again, the collected solder can be forced out.

A de-solder pump is another device for solder removal. It uses a spring-loaded mechanism. To use the device (Fig. 5.17 (a) and (b)), the spring is cocked and the tip of the vacuum pump is held against the solder joint. When the solder melts, the trigger is operated which releases the spring, creating a powerful vacuum action. Some of these devices can generate a static charge. Be sure to get a type that is specified as 'anti-static.'

(a)

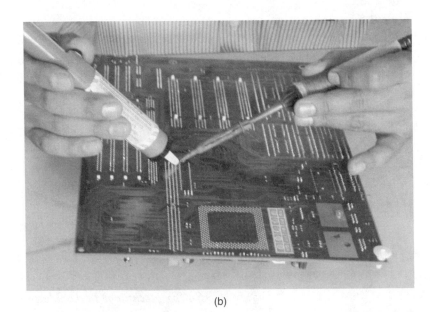

(b)

Fig. 5.17   *(a) Vacuum desoldering pump; and (b) Use of a vacuum pump for use in desoldering*

For stubborn joints or those connecting to the power planes (surface or multi-player boards), you may need to add some fresh solder and/or flux and then try again. Generally, if you only get part of the solder off the first time, repeated attempts will fail unless you add some fresh solder.

A very important consideration which must be kept in mind while de-soldering is that the heat required may damage the base materials and adjoining components. De-soldering should be carried out by using appropriate tools so that a minimum amount of heat is used during the de-soldering process.

During any repair work, it is well worth taking time and care so as not to damage or lift the copper back from the printed circuit board, as the printed circuit board is usually a very expensive item.

Do not use a sharp metal object such as a twist drill for removing solder from component mounting holes. Sharp objects may damage plated-through conductor.

Removing multi-lead components such as integrated circuits presents a special problem. If the component to be removed is still functional, it must be unsoldered quickly lest it can be damaged by heat. Alternatively, if the device is defective, it also needs to be removed fairly quickly to avoid lifting of printed circuit foil conductors by excessive heat.

Specialized devices are needed to solve this problem. One such device is a special DIP-shaped soldering iron tip (Fig. 5.18) and a spring-loaded IC extractor tool. The tool is placed above the IC to be removed and locked into position. When the tip is hot, it is applied to all the dual-in line IC pins or the foil side of the board. The extractor tool lifts the IC off the board as soon as the solder holding it melts. Special de-soldering tools are available for use with other IC and transistor cases.

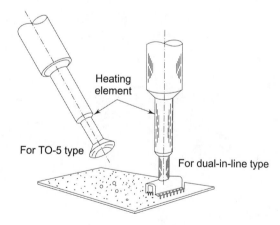

Fig. 5.18 *Special desoldering tip for integrated circuits*

5.11   SAFETY PRECAUTIONS

If you happen to receive burns during soldering/de-soldering operations, the following first aid steps are necessary:

(a) The affected area should be immediately cooled with running water or ice for 10–15 minutes.

(b) Remove any rings before swelling starts.

(c) Apply a sterile dressing to protect against infection.

(d) Do not apply lotions or ointments.

(e) Do not prick any blisters which may form later on.

(f) Seek professional medical advice wherever necessary.

It may be noted that the work of soldering/de-soldering should be carried out carefully to avoid any possibility of electric shock or burns. To avoid electrical shock, disconnect the equipment under repair from the AC power before removing or replacing any component or assembly. It is a good practice to periodically inspect the grounding lead, the state of the handle insulation and cord insulation. The ground terminal must be the last line to be disconnected while pulling the plug.

Mechanical and Electro-mechanical Components

MECHANICAL COMPONENTS

In electronics systems, lights, fuses and switches are used in abundance. These are very vital components because they show up the operational capability and status of the equipment. The presence of these components often speeds up the detection of failure. It is thus important to understand how these are built and installed. The description which follows includes a study of the types of fuses normally found in electronic systems, the types of fuse holders and typical replacement procedures used to restore the circuits to their normal operation; an analysis of the switches used in electronic systems, their replacement procedures and finally various types of connectors.

6.1   FUSES AND FUSE HOLDERS

6.1.1   Fuses

A fuse is used as an indicator for the technician to alert him to a defect or malfunction in the system though the primary function of any fuse is to protect an electronic unit from damage by excessive current or voltage. The fuse is thus used to protect both the wiring from heating and possible fire due to a short circuit or sever overload and to prevent damage to the equipment due to excess current resulting from a failed component or improper use.

The following points are important regarding the use of a fuse in the circuit:

Location: The fuse is normally placed in an input power line. The source of voltage may be AC or DC. The fuse may be located within the cabinet of the equipment or may be remotely placed. However, in the latter case, the fuse is wired to the circuits of the system.

Therefore, it is important to know the location of the fuse both physical as well as in the wiring diagram.

Symbolic Representation: The fuse is represented on a wiring diagram as a symbol like those shown in Fig. 6.1. It is usually designated as F followed by the number that the fuse represents.

The fuse itself carries information regarding its voltage and current rating. Some of them carry information on its characteristics, i.e. whether it is normal blow or slow blow type. A typical nomenclature adopted for fuses is given in Fig. 6.2.

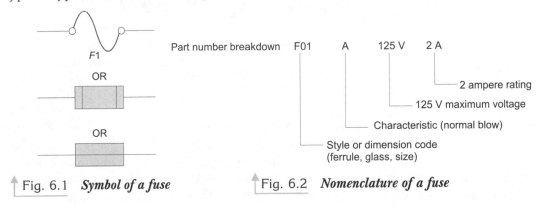

Fig. 6.1 **Symbol of a fuse**   Fig. 6.2 **Nomenclature of a fuse**

Principle of Operation: The fuse basically consists of a current-carrying element which melts when heated to temperatures above 170°F. The melting action opens the circuit thereby removing the source of voltage from the circuit. The fuse will heat and remain intact provided the applied current does not produce heat that exceeds the melting point of the metal. It is, therefore, important that the material used as a fusing element is engineered to accurate thickness, lengths and widths to carry specific voltage and current loads. Obviously, the basic laws of electricity are applied for selection of fuse ratings for a specific circuit taking into consideration the current required to maintain the normal circuit operation, total energy dissipation and voltage requirements. A short circuit in the equipment will melt a circuit fuse instantaneously. When a fuse needs to be replaced, the exact replacement of a fuse with both proper voltage and current ratings must be used. In addition, the exact replacement of a fuse with both proper voltage and current ratings must be used. In addition, exact replacement types, such as fast blow or slow blow, must be installed.

Fuses found in electronic equipment are usually of the cartridge type consisting of a glass or sometimes ceramic body and metal end caps. The most common sizes are 20 mm × 5 mm. Some of these have wire leads to the end caps and are directly soldered to the circuit board but most snap into a fuse holder or fuse clips. Miniature types of fuses include: Pico (tm) fuses that look like green 0.25 W resistors or other miniature cylindrical or square varieties, little clear plastic buttons, etc. Typical circuit board markings are F or PR.

Specifically designed miniature fuses are used as IC protectors to ensure a highly rapid response to prevent damage to sensitive solid state components including integrated circuits and transistors. These are usually encased in TO92 plastic cases but with only two leads or little rectangular cases about 0.1″ W × 0.3″ L × 0.2″ H. These may be designated as ICP, PR or F.

A slow blow or delayed action fuse allows for instantaneous overload, such as normal motor starting, but will interrupt the circuit quickly for significant extended overloads or short circuits. A large thermal mass delays the temperature rise so that momentary overloads are ignored.

The blown-up fuse can indicate the type of service problem in a non-functional electronic equipment. This is possible if the inside of the fuse is visible as is the case with a glass cartridge type. Usually if the glass case of the fuse appears clear and there are broken pieces of the centre conductor, the problem has been caused by a slow, gradual overload on the power supply. If the glass cover of the fuse is discoloured and the centre conductor is almost missing, it shows that the fault was caused due to a short circuit or other problem that produced a lot of current to flow very quickly and destroying the fuse violently with the production of lot of heat (Fig. 6.3).

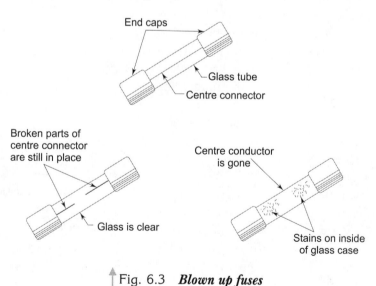

Fig. 6.3 *Blown up fuses*

Sometimes a fuse which has an element that looks intact but tests open may have just become tired with age. Even if the fuse does not blow, continuous cycling at currents approaching its rating or instantaneous overloads results in repeated heating and cooling of the fuse element. It is quite common for the fuse to eventually fail when no actual fault is present.

The fuse is generally checked, when in doubt for its continuity, using an ohmmeter.

6.1.2 Fuse Holders

Figure 6.4 shows different types of fuse holders normally found in electronic systems. All of them carry contact points for making connection with the rest of the circuit. When the fuse holder is to be replaced, the contact at these points needs to be unsoldered and resoldered.

Fuse holders often become damaged, mostly due to heat and mechanical abuse. The holders will get heated if they are corroded, dirty or holding the fuse loosely. The heat in turn can damage the insulating material in the holder and the holder will then have to be replaced. While replacing the fuse holder, the power to the system must be turned off.

High performance space-saving fuses, designed for either direct soldering onto the PCB or for use with PCB mounting fuse holders provide improved reliability due to direct solder connection. They are available in quick blow or anti-surge types with current ratings between 100 mA to 4A. Figure 6.5 shows the two types of packages for this type of fuses.

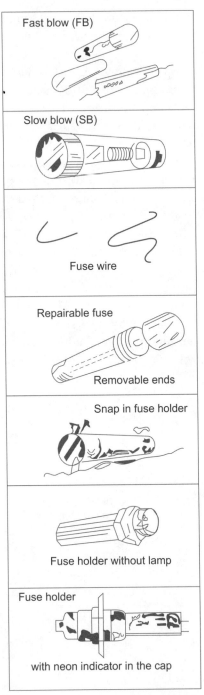

Fast blow (FB)

Slow blow (SB)

Fuse wire

Repairable fuse

Removable ends

Snap in fuse holder

Fuse holder without lamp

Fuse holder

with neon indicator in the cap

Fig. 6.4   *Fuse holders of various shapes*

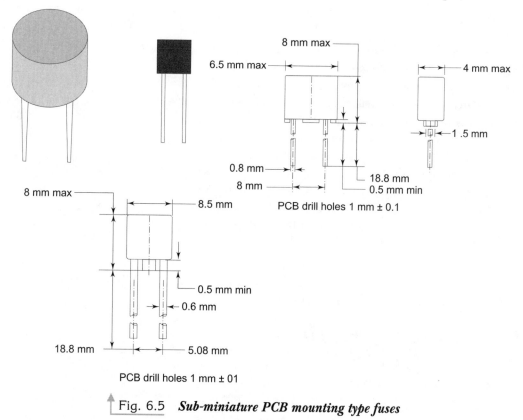

Fig. 6.5 *Sub-miniature PCB mounting type fuses*

6.1.3   Fuse Replacement

Although fuse replacement is not a very difficult task, yet the following parameters must be taken into consideration when replacing a fuse:

(i) *Current Rating:* This should not be normally exceeded. A smaller current rating can safely be used but depending on how close the original rating was to the actual current.

(ii) *Voltage Rating:* It is safe to use a replacement with equal or high voltage rating.

(iii) *Type:* Normal, fast blow, slow blow. It is safe to substitute a fuse with a faster response characteristic but there may be consistent or occasional failure mostly during power on. The opposite should be avoided as it risks damage to the equipment as semiconductors tend to die quite quickly.

(iv) *Mounting:* To meet the requirements of space and provide for proper holding of the fuse.

In case an exact replacement of the fuse is not available, you can use the wire of a suitable gauge which can be carefully soldered to the two metal end caps of the cartridge to create a suitable replacement.

6.1.4   Circuit Breakers

Circuit breakers may be thermal, magnetic or a combination of the two. Small push button type circuit breakers for electronic equipment are most often thermal. In these devices, the metal heats up due to the current flow and breaks the circuit when its temperature exceeds a set value. The mechanism is often the bending action of a bi-metal strip or disk. This action is similar to the operation of a thermostat.

Flip type circuit breakers are normally magnetic. An electro-magnet pulls on a lever held from tripping by a calibrated spring. These are most commonly used at the electric service panels, but rarely in electronic equipment.

6.1.5   Thermal Fuses/Protectors

Like a normal fuse, a thermal fuse/protector provides a critical safety function. Therefore, it is not advisable to just short it out if it fails. However, for testing, it is acceptable to temporarily short out the device to see if the equipment operates normally without over-heating. Thermal fuses are incorporated in the transformers or motors to provide protection from over-heating. In case of failure of the thermal fuse, it should be replaced with the same type or the entire transformer, motor, etc. This is especially critical for unattended devices.

6.2   SWITCHES

Switches are used in electronic systems for power 'ON' and 'OFF'. The common types of switches used in electronic equipment are:
  (a) Toggle-switch;
  (b) Push-button (spring – loaded) switch, with or without lamp; and
  (c) Micro switch.
  Figure 6.6 shows various types of switches used in electronic equipment.

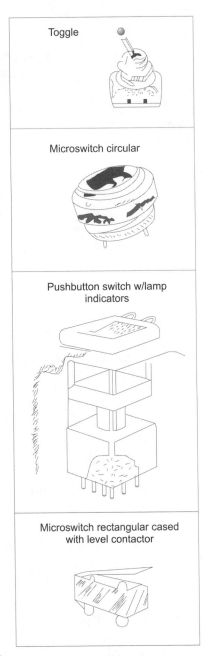

Toggle

Microswitch circular

Pushbutton switch w/lamp indicators

Microswitch rectangular cased with level contactor

Fig. 6.6   *Various types of switches*

6.2.1   Toggle Switch

The most common switch is the toggle switch. The various versions of toggle switch are (Fig. 6.7):

SPST—Single pole single throw;
DPST—Double pole single throw; and
DPDT—Double pole double throw.

The toggle switch can be spring-loaded where it can be turned ON and will spring back to OFF when released. Such type of switches are used where a re-set pulse is required or a pre-set pulse must be employed.

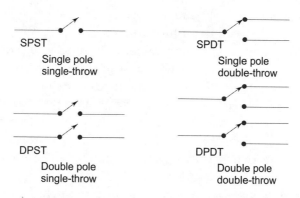

SPST
Single pole
single-throw

SPDT
Single pole
double-throw

DPST
Double pole
single-throw

DPDT
Double pole
double-throw

Fig. 6.7   *Various versions of toggle switches*

6.2.2   Push-button Switch

A push-button switch is similar in action to the toggle switch and has two positions, ON and OFF. Most of the push-button type of switches incorporate a lamp indicator under the pushing surface. This type of switch is available in two types:

(a) The holding type: It employs a holding coil which latches the switch in the ON position when depressed. The second pressing of the switch releases the voltage from the holding coil and returns the switch to its OFF condition.

(b) Non-holding or spring release return: The parameters for selection of the push button switches are: mechanical life (the number of contacts, usually in hundreds of thousands), contact bounce, electrical contact life based on contact wear, size and cost.

6.2.3   Microswitch

The microswitch is a spring loaded device and is often placed in a drive unit so that closing its contacts causes a cycle to repeat or stop. In other words, when the switch is operated, it causes a change in operation.

Causes of Malfunctioning of Switches: The defects occurring in various switches used in electronic equipment most often can be characterised as follows:

(a) Broken contact;
(b) Burned contact;
(c) Shorted contacts;
(d) Defective spring; and
(e) Burned switch body.

These defects manifest themselves, when checks are made, as:

(a) No continuity between the contacts where expected;
(b) Loose toggle;
(c) Sticking spring-loaded switch; and
(d) Non-operating holding coil in push button type switch.

A defective switch is normally not repairable. It must be replaced. The following precautions must be observed when replacing the switch:

(a) The power must be shut-off before attempting replacement.
(b) Some switches may have connections. The exact replacement of the wires is an absolute most. Therefore, carefully label all wires before desoldering.
(c) After the switch is replaced, solder all the wires in their proper positions and check the working of the switch with the power applied. The operation should be as specified.

A VOM can be used to make an in-circuit check on a switch. Touch the meter prods to both sides of the switch as shown in Fig. 6.8. When the switch is set in the OFF position, the meter should show infinite resistance. When it is put in the ON position, the meter should show zero resistance indicating that the two sides of the switch are electrically connected.

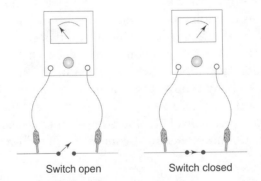

Switch open   Switch closed

Fig. 6.8   *Use of a VOM for in-circuit test of a switch*

6.2.4   Keyboard Switches

A keyboard is one of the commonest devices to be connected into all form of data processing and controlling systems. A keyboard can be as simple as a numeric pad with function keys, as in a calculator, or merely the range selection keys on a measuring system. On the other hand, there are complex alpha-numeric and typewriter keyboards for use with computers and other data entry equipment.

Most available keyboards have single-contact switches. The switching technology could be any of the following types (Fig. 6.9):

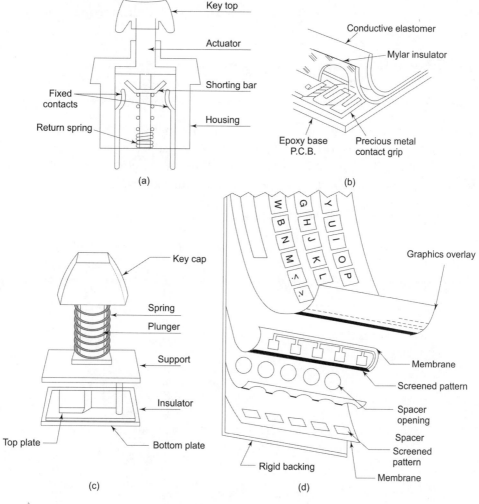

Fig. 6.9 *Keyboard switches (a) Mechanical key switch; (b) Elastomeric type; (c) Capacitive type; and (d) Thin membrane key switch*

Mechanical Switches: These are the most common types. They provide metal-on-contact and therefore ensure good tactile feedback. They usually suffer from wear and corrosion.

Elastomeric Switches: They are either pressure-sensitive or based on resistance change.

Capacitor Switches: They make use of two pads on the underlying circuit board to act as capacitor plates. Depressing the key increases the capacitance between the two pads and creating a signal in the sense circuit.

Membrane Switches: These are made by screen printing conductive silver contacts on two sections of a thin, flexible polyester sheet. This sheet is folded over so that the two sections form the upper and lower switch circuits.

The other types of keyboard switches include Hall-effect, optical sensing and proximity switches.

6.3 WIRES AND CABLES

6.3.1 Wires

Wires carry the electrical current from one point to another. Most wires are made of copper, which are usually coated with tin or silver to reduce copper oxidation and increase solderability. The wires are connected with vinyl and nylon insulations which melt at low temperatures. Care must therefore be taken while soldering. The most commonly used wire types are (Fig. 6.10):

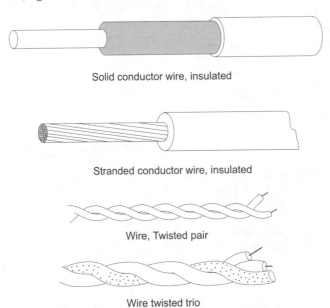

Solid conductor wire, insulated

Stranded conductor wire, insulated

Wire, Twisted pair

Wire twisted trio

Fig. 6.10   *Different types of wires used in electronic equipment*

- *Solid conductor wire (non-insulated):* used for bus bars, grounding of other currents;
- *Solid conductor wire (insulated):* used extensively for wire wrap applications;
- *Standard conductor wire (insulated):* flexible wire, most widely used for wiring in electronics; and
- *Wire twisted pair:* used in signal applications

6.3.2 Cables

A cable is usually a set of insulated wires for carrying electricity or electronic signals. The commonly used cables are (Fig. 6.11):

- *Shielded cable:* used where it is important to minimize induction between wires in a cable or chassis;
- *Coaxial cable:* used to carry high frequency signals without distortion; and
- *Power cable:* used to carry power from source to required areas.

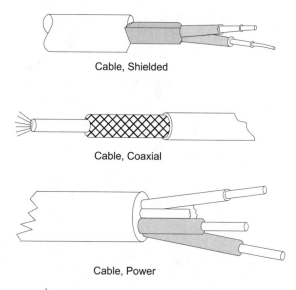

Cable, Shielded

Cable, Coaxial

Cable, Power

Fig. 6.11 *Different types of cables*

Faulty wires and cables are the causes of several problems in electronic equipment and systems. This is because of the mechanical stresses they are subject to particularly those which inter-connect various sub-systems and are placed outside the case or cover of the equipment. A typical example is that of a patient lead in an electrocardiograph machine and multi-wire flat cables used in microcomputers for inter-connecting the various peripherals with the central processing unit.

The cables usually carry very thin conductors inside. These conductors may break when pulled roughly. A cable fails more often at the point when it flexes due to frequent bending stresses at that point. The cables are checked for conductor continuity with a VOM. A high or infinite resistance indicates a problem in the cable conductor.

6.4   CONNECTORS

The function of the connector in an equipment is to connect the circuits through electrical connections to transfer data or voltages. They are manufactured for as few as 1 wire connection to as many as 200–250 wire connections. Each pin in a multi-pin connector is labelled. Labelling can be done with lettering or with numerals. Numbering on the connectors usually follows general reading rules (Fig. 6.12). For example, in circular connectors, the pins may be labelled as *a, b, c* and so on, clockwise.

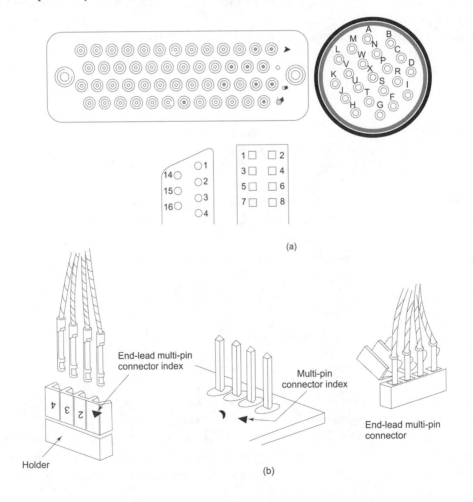

(a)

(b)

Fig. 6.12   *(a) Numbering system in multi-pin connectors*
            (b) Numbering system in end-lead connectors

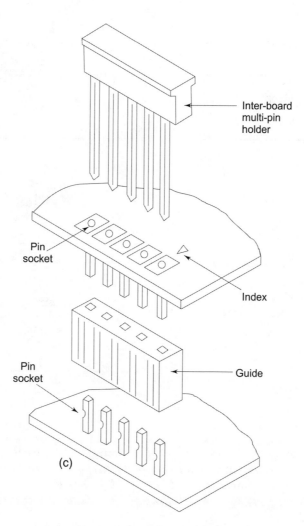

Fig. 6.12 *(c) Numbering system in inter-board multi-pin connector assembly*

A large variety of connectors is available in the market. The common types are rectangular, circular, edge board, coaxial and surface mounted (Fig. 6.13).

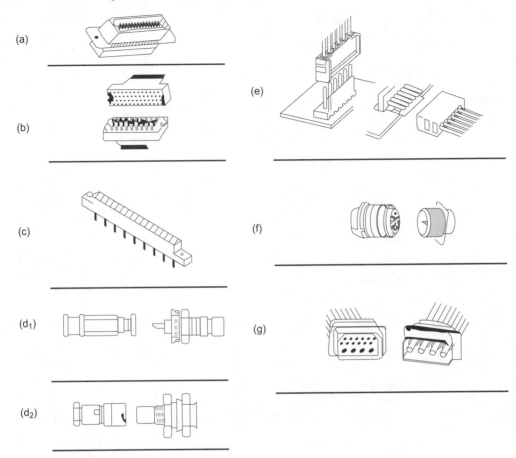

Fig. 6.13   *Common types of connectors used in electronic equipment*
(a) Chassis plug-in unit (b) Chassis connector (c) Printed circuit connector
(d1) Coaxial connector, shield crimp (d2) Coaxial connector, for RG–188/U
(e) Crimp housing and PCB headers (f) Cabling and harness (g) Rack and
panel

The wires may be soldered to the connector pins or the same multiple connectors are designed to hold solderless pins. In the latter case, the wire is secured to the pin by crimping a portion of the pin around the wire with enough pressure to retain the wire. The crimping is carried out by a special 'crimping tool' (Fig. 6.14), which is designed to allow its function to be performed on all sizes of wire and pins. It is desirable to use correct tools for each size of wire and pin used in the system, when removing, replacing or modifying the connectors on equipment. Removal of the pins is facilitated by the use of an extraction tool.

The technician must be aware of the various types of connectors available, termination methods and how to assemble and remove a connector. The connectors are characterized by the following parameters:

- *Contact Resistance:* Should not be more than 0.001 ohms;
- *Breakdown Voltage:* Depends on the contact spacing, geometry, shell spacing, insert and seal materials;
- *Insulation Resistance:* Reflects leakage current that flows not only through the material but also along its surface. The leakage may be due to metallic chip, moisture, etc.
- *Mating Method:* Involves two operations—polarizing and locking. Threaded couplings are useful in extreme environments. Some spring action is required to maintain tension between plug and receptacles.
- *Contact Materials:* Should have good spring qualities, good electrical conduction, corrosion resistant properties. Commonly used materials are beryllium copper, phosphor bronze, spring brass or low leaded brass.
- *Contact Plating:* To improve metal-to-metal contact. Silver plating is used for power contacts with high contact force. Electro-plating is used for low cost applications where few disconnections are anticipated. Lubricated gold plated contacts are used to reduce wear and tear and to improve reliability in cases where frequent removal and insertion are required.
- *Contact Retention Methods:* For insertion and removal of contacts, usually hand pressure is sufficient. But sometimes special tools suitable for the application are required.

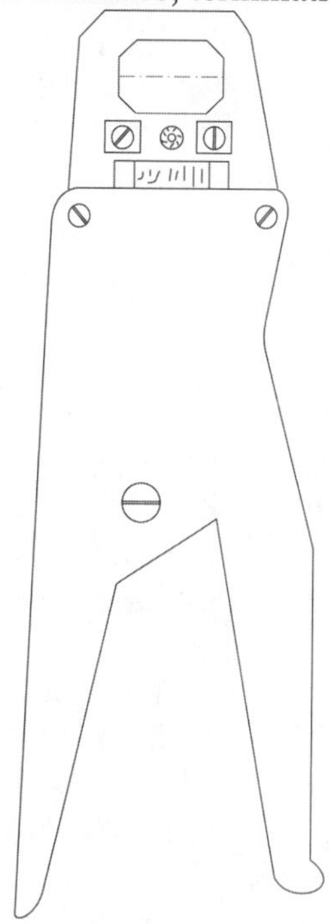

Fig. 6.14 *Crimping tool*

6.4.1 Common Problems in Connectors

Connectors and sockets are one of the major causes for poor reliability of electronic equipment. It is not because the connector quality is bad but because not enough thought has been given to the selection and application of these components. Most of the defects present in a connector will normally become visible on inspection. Some of the common problems in connectors are:

(a) *Loose Fitting Pins:* These will cause intermittent electrical connection or short, which may result in heat build-up. This problem is caused due to improper mating of the pins.

(b) *A Recessed Pin:* This situation results due to the slipping of a male or female pin back into the connector because of a broken clip, a broken plastic recess catch or broken spring tension clip. The recessed pin will not mate properly with the male or female connector resulting in intermittent electrical connection.

Heat built up at the point of contact will cause burning or melting of the plastic body used to hold the pins. Other problems in connectors could be:

(a) Cracked or broken plastic in connectors;
(b) Defective threads on circular connectors; and
(c) Missing guide pins on both circular and rectangular connectors.

6.4.2 Replacement Procedure for Multiple Connectors

Inspecting and replacing a defective connector is a task that requires careful and conscientious effort. It is necessary to properly identify and mark the wires before removing them from the connector to be changed, 100% accuracy needs to be ensured in wire connections when installing a new connector. Most often, in the multiple lead cables, the cable strands are twist-paired and colour-coded. This feature is very helpful in troubleshooting and in marking the wires for identification. The following steps should be followed when replacing a connector:

(a) Turn off the power before starting any repair work.
(b) Carefully mark each wire position and colour code. This may be done in two ways:
   (i) Prepare a chart with a layout of each pin as seen on the connector. Write the wire colour code adjacent to the pinhole number.
   (ii) Use a piece of masking tape on each wire and label each wire with its location letter or number designation taken from the older connector.
(c) Disconnect the connector from the panel, frame or the equipment body.
(d) Replace defective pins using a pin extractor (if applicable) or replace the connector as a whole if it is broken, charred or cracked.
(e) Secure the connection back to the panel or equipment body.
(f) By carefully matching the guide pin, secure the mating unit to the repaired or replaced connector.
(g) Test the equipment by turning on the power.

6.4.3 Flat Cable Termination

An insulation displacement connector (IDC) typically terminates ribbon cables. Cables are trimmed square on an end but not stripped. The cables are inserted into an IDC connector. All connections are then made simultaneously by squeezing the conductors into the connector terminals using special manual or automatic tools. The result is a tight, high pressure solderless connection. The operator requires very little skill to do the job.

It may be emphasized that whenever you are faced with a servicing problem in an equipment, always start by checking the cable and connectors first. The connector should fit snugly into the socket and all the pins of the connectors should be straight and clean. Similarly, all wires in the cable should show continuity when tested with a VOM.

The following precautions are necessary from the point of view of reliability when dealing with connectors:

- The connector selected should be of sound mechanical design and construction to withstand physical and environmental stresses.
- The connectors should be properly mounted and sufficient rigidity maintained.
- Connectors must be installed and maintained by trained personnel.
- All proper tools and accessories must be used.
- Misapplication should be avoided as it could be dangerous.

6.5 CIRCUIT BOARDS

Circuit boards are usually subject to mechanical damage, more often to cracks on the circuit board traces. A cracked board may disconnect the trace entirely or it may cause an intermittent problem that will be sensitive to vibration. This could result in an unreliable operation of the equipment.

Although most of the times, it should be possible to detect disjointed traces by usually inspecting the circuit board, fine cracks can escape notice. In those situations continuity checking can lead to a diagnosis of disconnected traces. Make particular checks on the traces near the edge connectors (Fig. 6.15) as these are subject to more stress while plugging and unplugging the circuit board. Also, the base of each heavy components like transformers and larger capacitors should be checked for cracked traces. The intermittent problems which are sensitive to vibration can be diagnosed by switching on the equipment and tapping or bending the circuit board to see if the fault appears.

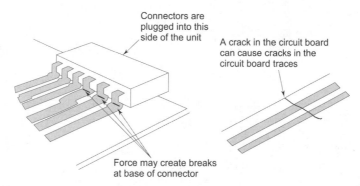

Connectors are plugged into this side of the unit

A crack in the circuit board can cause cracks in the circuit board traces

Force may create breaks at base of connector

Fig. 6.15 *Cracked traces on a circuit board*

Short circuits may be caused in closely located traces by the presence of any one of the following:

(a) Solder ball or whisker, stray piece of wire, loose screw;
(b) Slipping of the metal shield from its position; or
(c) Longer component leads touching adjacent traces.

These types of faults can be spotted by usual inspection. In case of doubt, make checks with a VOM.

A serious problem in circuit boards is that of a dry solder joint. This can occur when the solder is not adequately heated to make a good bond. The connection in that case may appear to be soldered, yet not make a good and reliable contact. By carefully examining the solder joint, the joint may appear rough and grainy rather than smooth and shining. Cold or dry solders are not easy to diagnose by visual inspection and you may have to use a VOM for this purpose.

6.6   TRANSFORMERS

A transformer consists of a laminated iron or ferrite core and two or more insulated windings that are most often not connected to each other directly. If one set of windings is used as the input for AC power (the primary winding), the voltage appearing on each of the other winding/windings will be related by the ratio of the number of turns on each of the windings. The current is related by the inverse of this ratio so that the power does not change.

Transformers are constructed using different cores, depending upon their applications. Some of the commonly used cores are shown in Fig. 6.16. The transformers coils are impregnated in phenolic varnish or other protective materials to protect them from mechanical vibrations, handling, fungus and moisture.

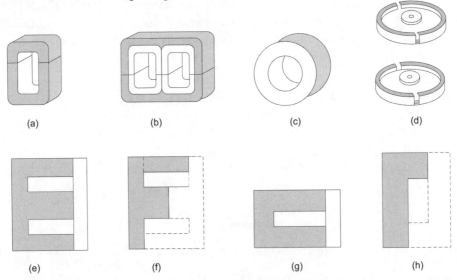

(a)   (b)   (c)   (d)

(e)   (f)   (g)   (h)

Fig. 6.16   *Different types of cores used in transformers: (a) C-core for single phase application; (b) E-core (three phase applications); (c) Tape Wound; (d) Cup-core used in pairs; (e) E-I lamination; (f) F-lamination; (f) U-I lamination; and (h) L-lamination, used in pairs*

6.6.1   Types of Transformers

Transformers are used in nearly every type of electronic equipment both for power and signals. They are most often used to convert the AC line voltage to some other value, lower or higher. The following are the common types of transformers used in electronic equipment:

Low Voltage Transformers   are found in electronic equipment as part of their power supplies or in AC wall adapters to generate one or more DC voltages to run the device or recharge its batteries, etc.

High Voltage Transformers   are found in microwave ovens, neon signs or old TV sets. Their output voltage may be as high as 15 KV or more.

Flyback or Line Output Transformers (LOPTs),   inverter and other specialized transformers are driven by a high frequency oscillator or chopper in various equipment like TVs, monitors, electronic flash units, etc. These transformers do not operate directly from the AC mains, and therefore need to be driven by a proper electronic circuit.

   Other common types of AC line-operated transformers used in repair and servicing laboratories are:

Isolation Transformers   They are wound 1 : 1 so that the output voltage is the same as the input voltage. However, with no direct connection between windings, the equipment can be tested with less risk of shock.

Variable Transformers or *Variac* allow the output voltage to be adjusted between 0 and full (or slightly above) line voltage. *Variac* is useful for testing purposes wherever the behaviour of a piece of equipment is being determined.

6.6.2   Testing of Transformers

Following are some simple tests to be performed for determining the efficacy of a used or new power transformer with no specifications:
   (1) Look for obvious signs of damage or distress. Smell it to determine if there is any indication of previous overheating, burning, etc.
   (2) Plug it in and check for output voltages. They should be reasonably close or probably somewhat higher than what you expect.
   (3) Leave it on for a while. It may get moderately warm and not too hot to touch. It should not melt down, smoke or blow up. If it does any of the latter, the transformer is bad.
   (4) Find a suitable load based on $R = V/I$ from the specifications. Connect it to the relevant secondary winding of the transformer. Make sure it can supply the current

without over-heating. The voltage should not drop excessively between zero and full load conditions.

Transformers are invariably found in mains operated equipment, except in very special cases. They are used to provide the required AC voltage to the circuit by appropriately transforming the mains voltage.

A transformer can be checked (Fig. 6.17) in circuit by using a VOM. The first test should be to check the continuity of the primary winding and that of all the secondary windings. If the primary and secondary windings are not known, proceed as follows:

- Identify all connections that have continuity between them. Any reading less than infinity on the meter is an indication of a connection. The typical values will be between something very close to 0 and 100 ohms.
- Each group of connected terminals represents one winding. The highest reading for each group will be between the ends of the winding; others will be lower. With a few measurements and some logical thinking, it will be possible to label the arrangement, ends and taps of each winding.
- Apply a low voltage AC input (from another transformer driven by a *Variac*) and measure the voltages across each of the connected ends of the winding. This data will enable you to determine voltage ratios and the primary and secondary windings. Often, primary and secondary windings will exit from opposite sides of the transformer.

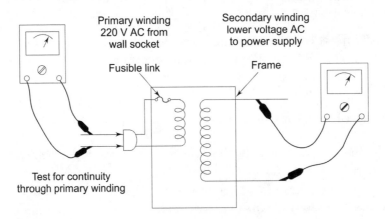

Fig. 6.17   *Testing of transformer using a VOM*

In typical power transformers used internationally, there will be two primary windings (resistance between the two will be infinite). Each of these may also have additional taps to accommodate various slight variations in input voltage for switching between 110 / 115 / 120 volts and 220 / 230 / 240 volts, etc.

For the US (110 VAC) the two primary windings are normally wired in parallel. For 220 volt AC operation, they are wired in series. When switching from one arrangement to the other, make sure that you get the phases of the two windings correct, otherwise you will have a short circuit. This can be tested by applying power—leave one end of one winding disconnected and measure between these points; there should be close to 0 voltage present if the phase is correct. If the voltage is significant, reverse one of the windings and then confirm. Typical colour codes for the primary windings will be black or black with various colour stripes. Almost any colours can be used for secondary windings. Stripes may indicate centre tap connections but not always.

It is best to make voltage measurements on the transformers using a *Variac* so that the voltage is brought up gradually and the mistakes can be detected before anything smokes.

6.6.3 Common Faults in Transformers

The common faults occurring in a power transformer are:

- *Primary Open:* This is usually the result of a power surge but could also be a short on the output leading to over-heating. Since the primary is open, the transformer would be totally lifeless. First confirm that the transformer is indeed beyond redemption. Some transformers have thermal or normal fuses under the outer layer of *insulating tape or paper*.

  In some situations, a transformer winding can short to the frame or core of the transformer. This check should be made again with a VOM.

- *Short in Primary or Secondary:* This may be the result of over-heating or be just due to poor manufacturing. The reason for this could be that the two wires are touching. One or more outputs may be dead and even those that provide some voltage may be low. In this case the transformer may blow the equipment fuse and even if it does not, it probably over-heats very quickly. However, it is first necessary to make sure that it is not a problem in the equipment being powered. Disconnect all outputs of the transformer and confirm that it still has nearly the same symptoms.

6.6.4 Rewinding of Transformers

For determining a suitable transformer as a replacement, the following steps are suggested:

- If the transformer is not totally sealed in epoxy or varnish, dis-assembling it and counting the number of turns of wire for each of the windings would be the best approach. To do so, remove the case and frame (if any) and separate and discard the iron core. The insulating tape or paper can then be peeled off revealing each of the windings. The secondaries will be the outer ones. The primary will be the last—closest to the centre. As you unwind the wire, count the number of full turns around the form or bobbin.

- By counting turns, you will know the precise (open circuit) voltages of each of the outputs. Even if the primary is a melted charred mass, enough of the wire will likely be intact to permit a fairly accurate count. An error of a few turns here and there won't matter.
- To work out the number of turns and size of wire in the windings, it is necessary to know the turns per volt of the transformer. This can be found by counting one of the secondary windings and dividing by its rated voltage. The number of turns then needed can be calculated as the turns per volt multiplied by the voltage you want. The size of wire is determined by the current rating. Use the wire with the same area per amp as the existing winding.
- Measuring the wire size will help to determine the relative amount of current that each of the outputs was able to supply. The overall ratings of the transformer are probably more reliably found from the voltage, current and wattage listed on the equipment name plate. In principle, it would be possible to totally rebuild a faulty transformer. All that is needed is to determine the number of turns, direction, layer distribution, wire size and order for each winding. It should be ensured that proper insulating material and varnish quoting are provided to obtain long-term reliability and safety.
- The ends of the windings must be terminated properly. Use enamelled copper wire. The enamel might need to be scraped of to enable soldering. Usually, there are tags to solder the ends. Also, if the transformer is used in special applications, as in a tape recorder, it most probably needs shielding.
- It should be ensured that the finished transformer is safe. The best way to test the insulation is to test with a high voltage (a few kV) between primary and secondary and then between the core and each winding, and check that there is no leakage current. With the mains applied, check that there is correct voltage at the outputs. Check that the transformer does not get too hot.

6.7   MOTORS

Different motors are used in different equipment, depending upon the application. For example, a synchronous motor may be used in a chart recorder for constant speed operation of paper movement. A stepper motor is used in disk drives in microcomputer systems. Motors are usually checked by testing the continuity of the various windings using a VOM. An infinite resistance indicates an open winding. Also, a short circuit between winding and the motor frame shows a defective motor.

When checking a stepper motor, it must be remembered that it contains four independent sets of windings. Each one of them should be checked independently with a VOM. The value of resistance of each winding is usually indicated on the label of the motor.

6.8 ELECTRO-MAGNETIC RELAYS

A relay is a device by means of which one circuit is indirectly controlled by a change in the same or in another circuit. An electro-magnetically operated relay normally includes an electro-magnet (coil) which controls an armature, which in turn actuates (open or close) electrical contacts.

The relays are characterized by the coil resistance, coil voltage and the number and types of contacts contained in the relay. If a 12-volt relay has $1k\Omega$ resistance, it means that the relay will require 12 milliamperes to turn it on, when placed in a circuit. The relay coil usually has a diode in parallel to bypass the voltage spike generated due to the inductive action of the relay coil.

6.8.1 Solid State Relays

With a sealed construction and no moving parts, the solid state relays are particularly suited to AC switching applications requiring long life and high reliability. The switching is silent, causes no arcing and is unaffected by vibration and corrosive atmospheres. The control input is optically isolated (Fig. 6.18 (a) and (b)) from the zero voltage switching circuit which produces virtually no RF interference. They are operated by a TTL open collector. The output circuit is 'normally open'.

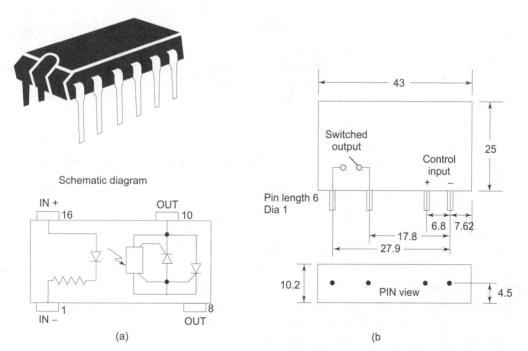

Fig. 6.18   *Solid state relays: (a) with optical isolation of input/output terminals, dual-in-line package; and (b) Single-in-line package*

When an inductive load is to be switched for protection against the effects of transients, the connection of a metal oxide varistor across the relay output is recommended. These relays are available as SIL (single-in-line) or DIL (dual-in-line) PCB mounting type packages.

6.8.2   Reed Relays

Encapsulated reed relays which incorporate a coil (Fig. 6.19 (a) and (b)) to operate the contacts, when energized, are available in DIL (dual-in-line) package. These relays internally incorporate a diode across the coil to protect the driver circuitry from back EMF. These relays normally have open contacts with AC mains (230V) switching capacity. They are normally used with mains Triac Trigger devices.

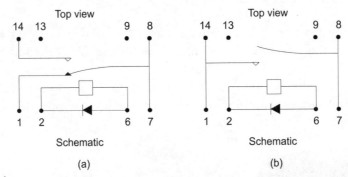

Fig. 6.19   *Reed relays: (a) normal connection between 1-8, changes over to 14–8 when operated; and (b) normally open between 14–8, closes after relay operation*

6.8.3   Faults in Relays

In relays, the following problems may occur resulting in malfunctioning of the circuit in which they are placed:
   (a) The energizing coil goes bad;
   (b) The contacts always remain open;
   (c) The contacts always remain closed; or
   (d) The driver transistor to the relay goes bad.
   The first three problems can be checked by using a multimeter. Although by energizing the coil, the operation of the relay can be visually observed, and clicking sound associated with the change-over operation of the contacts can be heard, it is better to check the integrity of the contacts with an ohmmeter. In the event of any of the above defects found in the relay operation, it should be replaced.

Relays are usually mounted on bases. For testing purposes, they can be removed from the circuit and checked by energizing the coil from a DC voltage power supply. If the relay operates satisfactorily outside the circuit, check the shunt diode and driver transistor or the integrated circuit which drives the relay coil.

6.9   BATTERIES AND BATTERY CHARGERS

The requirement of portable power seems to be increasing rapidly with the proliferation of notebook and palmtop computers, cellular phones, pagers, pocket cameras, camcorders and cassette recorders. Two of the current areas of development related to these products are:

- Development of higher capacity battery technologies (electro-chemical systems) for rechargeable equipment—lithium and nickel metal hydride are among the more recent developments.
- Implementation of smart power management system (optimal charging and high efficiency power conversion) for portable devices. A variety of ICs are now available to implement rapid charging techniques while preserving battery life.

However, most of the devices still use pretty basic battery technologies, and the charging circuits are often very simple and do perform the job which is adequate for many applications.

6.9.1   Battery Basics

Strictly speaking, a battery is made up of a number of individual cells, most often wired in series to provide multiples of the basic cell voltage. The cell voltages that are commonly encountered are 1.2, 1.5, 2.0, or 3.0 volts. Four types of batteries are typically used in most of the portable equipment. These are:

- *Alkaline:* It consists of one or more primary cells with a nominal terminal voltage of 1.5V. Examples are AAA, AA, C, D, N, 9V (transistor) or 6V (lantern batteries). For special applications like clocks, watches, calculators, and some cameras, miniature button cells are used. In general, recharging of alkaline batteries is not practical due to their chemistry and construction.
- *Lithium:* The terminal voltage is around 3V per cell and they have a much higher capacity than alkalines. They are often found in cameras and smoke detectors wherein their high cost is offset by the convenience of long life and compact size.
- *Nickel Cadmium (NiCd):* This is the most common type of rechargeable battery technology used in small electronic devices and these batteries are available in all the popular sizes. Their terminal voltage is only 1.2V per cell as compared to 1.5V

per cell for alkalines. NiCds should not be discharged below about 1V per cell and should not be left in a discharged state for too long. Overcharging is also not good for NiCds and will reduce their ultimate life.

- *Lead Acid:* This technology is similar to the type used in automobiles but such batteries are specially designed in sealed packages which cannot leak acid under most conditions. They come in a wide variety of capacities but not in standard sizes like AA or D. They are used in some camcorders, flash lights, CD players, security systems, emergency lighting, etc. The nominal terminal voltage is 2.0V per cell. These batteries should not be left in a discharged condition as they will quickly become unusable if left that way for any length of time.

Since the nominal voltages for the common battery technologies differ, it is often possible to identify which type is inside a pack by the total output voltage:

- NiCd packs will be a multiple of 1.2 V;
- Lead-acid packs will be a multiple of 2.0 V; and
- Alakline packs will be a multiple of 1.5 V.

It may be noted that these are open circuit voltages and may be slightly higher when fully charged or new.

Although the problem of leakage of chemicals from the batteries is a lot less common with modern battery technologies, yet it can still happen. It is always a good practice to remove batteries from equipment when it is not being used for an extended period of time. Dead batteries also seem to be more prone to leakage than fresh ones. The stuff that leaks from a battery should be thoroughly cleaned. If the contacts are corroded, fine sand paper or a small file may be used to remove the corrosion and brighten the metal. Do not use an emery board or emery paper or steel wool as any of these will leave conductive particles behind which will be difficult to remove.

6.9.2   Battery Chargers

The energy storage capacity (C) of a battery is measured in ampere hours denoted as Ah or mAh (for smaller types). The charging rate is normally expressed as a fraction of C, e.g. 0.5C or C/2.

In most cases, trickle charging at a slow rate—C/100 to C/20—is preferred as a result of which the batteries are likely to give better performance and longer life. Fast charging is harsh to batteries; it generates heat and gases and the chemical reaction may be less uniform. Each type of battery requires a different type of charging technique.

(i) Batteries are charged with a controlled, usually constant current. Fast charge may be performed at 0.5–1.0C rate for the batteries in portable tools and laptop computers. A 0.5C charge rate for a 2 amp hour battery pack would use a current equal to 1.0 amp. Trickle charge is done at C/20–C/10 rate.

Problems with simple NiCd battery chargers are usually pretty easy to find, including bad transformer, rectifiers, capacitors, or possibly regulator. For sophisticated chargers which may be based on microprocessors or custom ICs, the usual troubleshooting techniques are recommended.

(ii) Lead acid batteries are charged with a current limited but voltage cut-off technique. Although the terminal voltage of a lead acid battery is 2.0V per cell nominal it may actually reach more than 2.5V per cell while charging. The charger for a lead acid battery is simply a stepped down rectified AC source with some resistance to provide current limiting. The current will gradually taper off as the battery voltage approaches the peaks of the charging waveform. For small sealed lead acid batteries, an IC regulator may be used to provide current limited constant voltage charging. Trickle chargers used for lead acid batteries are usually of constant voltage type and the current tapers off as the battery reaches the full charge.

Problems with lead acid battery chargers are usually easy to diagnose due to the simplicity of most designs of such batteries.

6.9.3   Problems with Battery-operated Equipment

The usual problem with battery-operated equipment is that the voltage drops when the device is turned on or the batteries are installed. If the batteries are known to be good and putting out the proper voltage, the following causes of problems are possible:

- Corroded contacts or bad connections in the battery holder;
- Bad connections or broken wires inside the device;
- Faulty regulator in the internal power supply circuits; semiconductors and IC regulators need to be checked;
- Faulty DC–DC inverter components; for which semiconductors and other components should be tested;
- Defective on/off switch; and
- Other problems in the internal circuitry.

7

Passive Components and Their Testing

Electronics equipments make use of a variety of components. A thorough knowledge of components and their limitations is thus an essential part of troubleshooting in electronic circuitry. A good understanding of special precautions in terms of handling, soldering and measurement of components would reduce the likelihood of premature failure and help in isolating failures due to design weakness or misuse.

7.1 PASSIVE COMPONENTS

Passive devices include resistors, capacitors and inductors. They always have a gain less than one, thus, they cannot oscillate or amplify a signal. A combination of passive components can multiply a signal by values less than one; they can shift the phase of a signal, they can reject a signal because it is not made up of the correct frequencies, they can control complex circuits, but they cannot multiply by more than one because they basically lack gain.

7.2 ADJUSTABLE CONTROLS

Adjustable controls are used extensively in electronic equipment. Basically, the requirement of adjustable controls falls into the following three categories:

(a) General purpose controls are normally provided on the front panel of the equipment; for example, focus and brilliance controls on a cathode ray oscilloscope.

(b) Precision controls are generally provided on the front panel of the equipment to give a precise control of the output. For example, adjustment of frequency of the pulse output on a pulse generator or calibrated output voltage control on a power supply.

(c) Pre-set or trimmer action is provided by a variable control in the circuit to serve a distinct function on the operation of the circuit. This action may cause increase or decrease in bias voltage or current, a change in frequency or a shift in frequency. This is also employed to provide control of amplification, attenuation, limiting, or distortion control. A typical example is the adjustment of a fixed frequency output from an astable multi-vibrator by means of an adjustable component.

When carrying out repair work on an equipment, it may be noted that before altering any adjustment (pre-set control) in the circuit, its function must be properly understood, otherwise there are chances of completely disturbing the circuit operation and causing it to operate worse than before. Arbitrarily turning the components may result in attenuation or loss of selected frequencies, distortion of waveform or causing a shift in amplifier operation. It is advisable to refer to the 'circuit operation' part in the service manual of the equipment in order to understand the function of the adjustable control, before it is disturbed.

Generally, the adjustable controls in the circuit are provided by the following electronic components:

(a) Variable resistors; and
(b) Variable capacitors.

The testing methods for variable resistors and capacitors are similar to those of fixed resistors and capacitors respectively.

The following description gives details of various types of passive components, adjustable control and their causes of failure, and techniques of function testing.

7.3 RESISTORS

Resistance is the opposition to the flow of current offered by a conductor, device or circuit. It is related to current as follows:

$$\text{Resistance} = \text{voltage/current (Ohm's Law)}.$$

The resistance is expressed in ohms (abbreviated Ω). The most commonly used types of resistors are:

(a) Carbon composition type (made either by mixing finely ground carbon with a resin binder and an insulating filler or by depositing carbon film onto a ceramic rod).

Most carbon film resistors have less stray capacitance and inductance, so they are usable at higher frequencies. However, their accuracy is limited to 1%. In addition, carbon film resistors tend to drift with temperature and vibration.

(b) Metal resistors (made of metal film on ceramic rod), or metal glaze (a mixture of metals and glass) or metal oxide (a mixture of a metal and an insulating oxide).

Metal film resistors are more stable under temperature and vibration conditions having tolerances approaching 0.5%. However, precision metal film resistors with tolerances below 0.1% are commercially available.

(c) Wire wound resistors (made by winding resistance wire onto an insulating former).

(d) Thick film resistor networks comprising precious metals in a glass binding system which have been screened on to a ceramic substrate and fired at high temperatures. These networks provide miniaturization, have rugged construction, are inherently reliable and are not subject to catastrophic failures. Networks comprising 1 to 50 resistors, 5 to 20 being typical, are commercially available. Single inline packages, DIP (dual inline package) and square packages are available.

The value of resistance is either printed in numbers or it is put in the form of colour coded bands around the body. In the colour code, each number from 0 to 9 has been assigned a colour according to Fig. 7.1.

The colour code comes in the form of four-band (Fig. 7.1a). The first band closest to the end of the resistor represents the first digit of the resistance value. The second band gives the second digit and the third band gives the number of zeros to be added to the first two digits to get the total value of the resistor. The fourth band indicates the tolerance. If the fourth band is absent, the tolerance is $\pm 20\%$.

In the five-band (Fig. 7.1b) colour code, the first three bands indicate the value, the fourth band indicates the multiplier factor and the fifth band, the tolerance of the resistor.

In the six-band (Fig. 7.1c) colour code, the sixth band indicates the temperature coefficient of variation of resistance in terms of parts per million per degree centigrade (PPM/ °C).

Four Band Resistors

1st Band		2nd Band		3rd Band		4th Band (tolerance)	
Black	0	Black	0				
Brown	1	Brown	1	Silver	Divide by 100		
Red	2	Red	2	Gold	Divide by 10		
Orange	3	Orange	3	Black	Multiply by 1		
Yellow	4	Yellow	4	Brown	Multiply by 10		
Green	5	Green	5	Red	Multiply by 100		
Blue	6	Blue	6	Orange	Multiply by 1,000		
Violet	7	Violet	7	Yellow	Multiply by 10,000	Red	$\pm 2\%$
Grey	8	Grey	8	Green	Multiply by 100,000	Gold	$\pm 5\%$
White	9	White	9	Blue	Multiply by 1,000,000	Silver	$\pm 10\%$

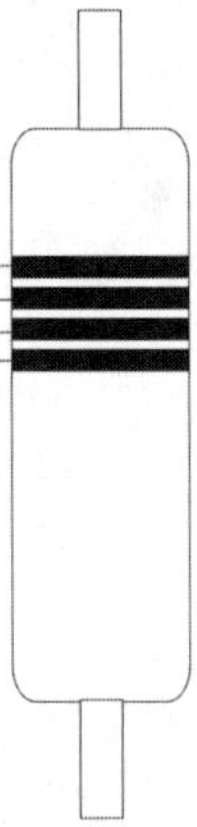

Fig. 7.1   *(a) Colour code of carbon composition and metal film resistor–4 band colour code*

Five Band Resistors

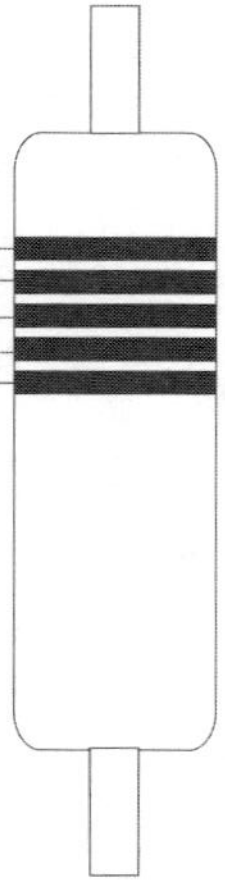

1st Band		2nd Band		3rd Band		4th Band		5th Band (tolerance)	
Black	0	Black	0	Black	0	Silver	Divide by 100	Brown	±1%
Brown	1	Brown	1	Brown	1	Gold	Divide by 10	Red	±2%
Red	2	Red	2	Red	2	Black	Multiply by 1	Gold	±5%
Orange	3	Orange	3	Orange	3	Brown	Multiply by 10	Silver	±10%
Yellow	4	Yellow	4	Yellow	4	Red	Multiply by 100		
Green	5	Green	5	Green	5	Orange	Multiply by 1,000		
Blue	6	Blue	6	Blue	6	Yellow	Multiply by 10,000		
Violet	7	Violet	7	Violet	7	Green	Multiply by 100,000		
Grey	8	Grey	8	Grey	8	Blue	Multiply by 1,000,000		
White	9	White	9	White	9				

Fig. 7.1 *(b) Colour code of carbon and metal film resistors—5 band colour code*

Six Band Resistors

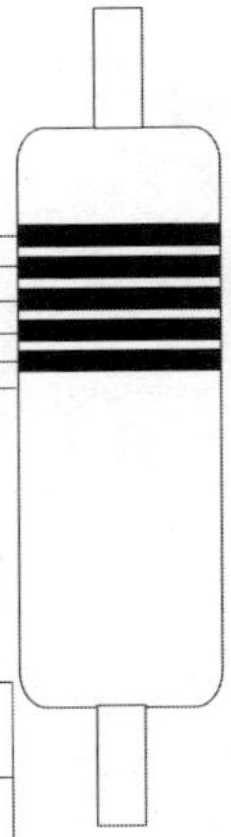

1st Band		2nd Band		3rd Band	
Black	0	Black	0	Black	0
Brown	1	Brown	1	Brown	1
Red	2	Red	2	Red	2
Orange	3	Orange	3	Orange	3
Yellow	4	Yellow	4	Yellow	4
Green	5	Green	5	Green	5
Blue	6	Blue	6	Blue	6
Violet	7	Violet	7	Violet	7
Grey	8	Grey	8	Grey	8
White	9	White	9	White	9

4th Band	
Silver	Divide by 100
Gold	Divide by 10
Black	Multiply by 1
Brown	Multiply by 10
Red	Multiply by 100
Orange	Multiply by 1,000
Yellow	Multiply by 10,000
Green	Multiply by 100,000
Blue	Multiply by 1,000,000

5th Band (tolerance)	
Brown	±1%
Red	±2%
Gold	±5%
Silver	±10%

6th Band (Temp. Coef. PPM/C)	
Brown	100
Red	50
Yellow	25
Orange	15
Blue	10
Blue	6
White	1

(c)

▲ Fig. 7.1 *(c) Colour code of carbon composition and metal film resistors   Six band colour code*

When the value and tolerance of the resistor is printed on the resistor body, the value of the tolerances are coded as follows:

$$F = \pm\ 1\% \qquad G = \pm\ 2\% \qquad j = \pm\ 5\%$$
$$K = \pm\ 10\% \qquad M = \pm\ 20\%^{\pm}$$

To illustrate this code, the following examples are cited:

R 68M is a 0.68 $\Omega \pm$ 20% resistor

5K 6J is a 5.6 k $\Omega \pm$ 5% resistor

82KK is 82 k $\Omega \pm$ 10% resistor

Although it is possible to get resistors of any value, they are generally available in the preferred ranges. The most common series is the E 12 series in which the preferred values are 10, 12, 15, 18, 22, 27, 33, 39, 47, 56, 68 and 82. Much closer values are available in the E 96 series for ±1% tolerance values.

Carbon composition resistors (Fig. 7.2) have poor stability and relatively poor temperature co-efficient, which is of the order of 1200 ppm/°C. Metal film resistors exhibit

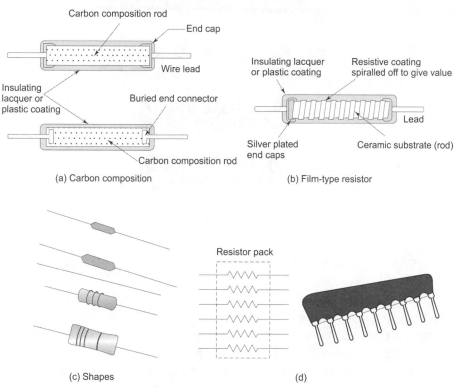

(a) Carbon composition

(b) Film-type resistor

(c) Shapes

(d)

Fig. 7.2 *Fixed Resistors: (a) Carbon composition; (b) Film type; (c) Shapes of fixed resistors; and (d) Resistor pack*

comparatively low temperature coefficient (± 250 ppm/°C) and good stability, both when stored and under operating conditions. Carbon composition and metal resistors are usually available in power ratings of 250mW, 500mW, 1W and 2W. For dissipating more heat, wire wound resistors (Fig. 7.3) are mostly employed, the power ratings being up to 25 watts.

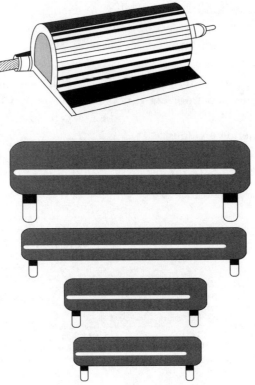

Fig. 7.3   *Wire-wound resistors of high wattage type*

7.3.1   Failures in Fixed Resistors

Resistors are generally very reliable and thus exhibit a very low failure rate. However, failures do take place in resistors. With the passage of time, heat, applied voltage, humidity, mechanical stress and vibrations all act to produce chemical or other changes and lead to a gradual deterioration in resistors.

Any resistor, when placed in an operating circuit, will dissipate power and the maximum power to be dissipated by a resistor depends upon the ambient temperature. Obviously, lower dissipation will give improved stability and lower values of failure rate. In general, the failure rate of a resistor will depend upon its type, the method of manufacture, the operating and environmental conditions and the value of the resistor. The following could be listed as failures of fixed resistors and their possible causes.

Carbon Composition Resistors

(a) *Open circuit*
1. Burning out of the resistor centre due to excessive heat;
2. Fracturing of the resistor due to mechanical stress;
3. Dislodging of end caps; and
4. Wire breaking due to excessive flexing.

(b) *Higher value*
1. Heat, voltage or moisture causing movement of carbon or binder; and
2. Separation of carbon particles due to swelling caused by absorption of moisture.

Metal Film Resistors

(a) *Open Circuit*
1. Scratching or chipping of film during manufacture;
2. High voltage or temperature causing disintegration of film; and
3. Open circuit failure; more likely in the case of higher value resistors due to thin resistance spiral.

(b) *High noise*
1. Bad contact of end connectors due to mechanical stress caused by poor assembly on a circuit.

Wire Wound Resistors

Open Circuit
1. Fracture of wire, especially in the case of fine wire;
2. Corrosion of wire due to electrolytic action caused by absorbed moisture;
3. Slow crystallization of wire (because of impurities) leading to fracture and discontinuity; and
4. Disconnection at the welded end.

The stability of a resistor is defined as the percentage change of the resistance value with time. It depends upon the power dissipation and ambient temperature. The critical temperature in a resistor is the *hot spot temperature* which is the sum of the ambient temperature and the temperature rise caused due to the dissipation of power. Due to the uniform construction of the resistor, the maximum temperature is in the middle of the resistor body and it is this temperature which is known as the *hot spot temperature.*

7.3.2   Testing of Resistors

The normal range of resistors encountered in practice is 1 ohm to 10 meg-ohm. The most convenient method of checking resistances in this range is by using an ohmmeter or analog multimeter (Fig. 7.4) of the moving coil type operated in the resistance measurement mode. The total measurable resistance range in the VOM is divided into three ranges:

$$R \times 1 \quad 0 \text{ to } 2k\Omega$$
$$R \times 10 \quad 0 \text{ to } 200k\Omega$$
$$R \times 100 \quad 0 \text{ to } 20 M \Omega$$

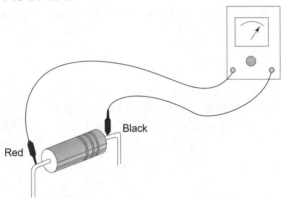

Red   Black

Fig. 7.4   *Testing method for fixed resistors*

The circuit for resistance measurement is so arranged that when the test leads are connected across the unknown resistor, the current through the meter corresponds to the resistance value, which is read on a calibrated scale. Before the unknown resistance is tested, the two meter leads (prods) are shorted together and the zero control corresponding to the range selected is adjusted to read zero ohms on the scale. This, in fact, corresponds to the maximum meter current. With the unknown resistance connected for test, the current falls and therefore, the scale is obviously non-linear.

Digital multimeters are more handy and have the advantage of a linear scale for the measurement of resistance. They make use of a constant current source which is made to develop voltage across the unknown resistance and the measured voltage is proportional to the value of the resistor. The value of the constant current employed depends upon the range in use. For example, it varies from 1mA on the 100 Ω range, to say 1μA, on the 1 MΩ range. Digital multimeters can be used to measure very low resistance values such as the contact resistance of plug and socket connectors making use of the four-terminal (Kelvin) principle. A higher value of current is employed for making these measurements so that an appreciable voltage drop is developed for the convenience of measurement.

7.3.3 Variable Resistors or Potentiometers

Variable resistors basically consist of a track of some type of resistance material to which a movable wiper makes contact. Variable resistors or potentiometers ('pots' as they are popularly called) can be divided into three categories depending upon the resistive material used (Fig. 7.5):

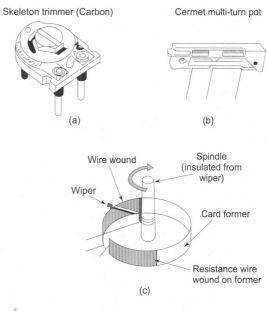

Carbon: Carbon potentiometers are made of either moulded carbon composition giving a solid track or a coating of carbon plus insulating filler onto a substrate.

Cermet: Cermet potentiometers employ a thick film resistance coating on a ceramic substrate.

Wire-wound: Nichrome or other resistance wire is wound on to suitable insulating former for the construction of wire-wound potentiometers.

Potentiometers can be categorized into the following types depending upon the number of resistors and the control arrangement used:

Fig. 7.5 *Three types of variable resistors: (a) Carbon composition; (b) Multi-turn cermet; (c) Wire wound*

Single Potentiometers: Potentiometer control with one resistor;
Tandem Potentiometers: Two identical resistor units controlled by one spindle;
Twin Potentiometers: Two resistor units controlled by two concentric spindles;
Multi-turn Potentiometers: Potentiometers with knob or gear wheel for resistance adjustment, they may have up to 40 rotations of spindle.

Potpack: Rectangular potentiometers, either single or multi-turn.

Potentiometers are typically used for setting bias values of transistors, setting time constants of RC timers, making gain adjustments of amplifiers and carrying current or voltage in control circuits. Therefore, they are packaged in such a way that they are compatible with PCB mounting applications.

Variable resistors can be used either as rheostats or as potentiometers. Figure 7.6 shows the difference in the two applications. When used as a voltage divider, the resistor element is connected to a voltage reference source and the slide arm, which is used as the pick-off point, can be moved to obtain the desired voltage.

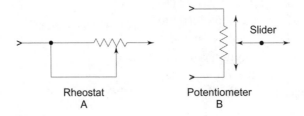

Fig. 7.6   *Difference between a potentiometer and rheostat operation*

For a variable resistor configuration, the resistor element is connected to the circuit at each end and the slide arm is connected to one of the ends. Alternatively, the entire resistance is in series and the slider is connected to an external circuit. Even in this configuration, it may be used like a potentiometer.

Variable resistors can be constructed to follow one of the following laws:

Linear: The resistance of the pot is distributed evenly over its entire length.

Log: The resistance of the pot varies so as to follow the logarithmic law. In these pots, when the wiper is turned, the resistance increases (from zero) very slowly and gradually until about the midway. From then onwards, as the wiper shaft is turned further, the resistance will increase much more rapidly in comparison with the first half of the pot-rotor rotation.

Sine-cosine Potentiometers: As the name implies, the variation of resistance over the track, when the wiper moves follows the sine cosine law. The total operative track length over 360 degrees of rotation is divided into four quadrants of 90 degrees each.

The characteristics of various types of variable resistors are summarised in Table 7.1.

Table 7.1 ▪ *Characteristics of Variable Resistors*

Type	Typical Value	Tolerance	Typical Power Rating	Stability	Construction	Law
Carbon pots	100Ω to 10MΩ	± 20%	0.5 W to 2 W	± 20%	Single-turn Ganged	Linear Log
Wire-wound (i) General type	10 to 100 kΩ	± 5%	3 W	± 5%	Single-turn {Multi-turn {Ganged	Linear
(ii) Precision type		± 3%		± 2%	Helipot(Multi-turn)	Sine-cosine
Cermet	10 to 500 K	± 10%	1 W	± 5%	Single-turn Multi-turn	Linear

7.3.4 Failures in Potentiometers

Potentiometers exhibit a higher failure rate than fixed resistors. This is because they not only jointly have moving parts but also depend for their operation on a good and reliable electrical contact between a wiper and the resistance track. Usually, two types of failures are encountered in practice:

(a) *Complete Failure:* Complete failure usually manifests itself in the open circuit either between the wiper to track or track and the end connections. This can be caused by:
   (1) Corrosion of metal parts by moisture; and
   (2) Swelling and distortion of plastic part (track mouldings) by moisture or high temperature.
(b) *Partial Failure:* Partial failure of a potentiometer results either in rise in wiper contact resistance giving higher electrical noise or an intermittent contact. This is caused by particles of dust, abrasive matter or grease trapped between the wiper and track. A bad potentiometer due to contact problem will give sufficient indications like noise in an audio circuit, irregular behaviour of the controlled parameter, etc.

7.3.5 Testing of Potentiometers

Potentiometers can be checked for correct performance in a similar way as that of fixed resistors using an ohmmeter. The check should be made between the variable contact terminal and each of the two fixed terminals.

Figure 7.7 shows the testing procedure for potentiometers. As the shaft of the pot is turned, the resistance will be observed. If the resistance remains at zero or stays at an infinite reading, the pot may be replaced. When replacing a bad potentiometer, care should be taken to ensure that the potentiometer chosen is of correct resistance value, tolerance, power rating, resolution, shape, size and position of adjustment screw.

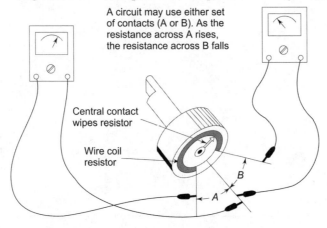

Fig. 7.7 *Testing of a potentiometer with an ammeter*

7.3.6 Servicing Potentiometers

Usually the wiping contacts inside the potentiometer can become dirty or corroded. This can be usually corrected using some spray type non-residue cleaner into the pot. Since most pots are enclosed in a metal case, the cleaner is sprayed either along the shaft or into other openings in the case. After spraying the cleaner inside, the shaft should be turned a few times to complete the action. This would take care of dust that accumulates over the resistance track.

7.3.7 Light Dependant Resistors (LDRs)

LDRs are made of cadmium sulphide. They contain very few free electrons when kept in complete darkness and therefore exhibit very high resistance. When subject to light, electrons are liberated and the material becomes more conducting. When the light is switched off, the electrons are again re-captured and the material becomes less conducting or insulator. The typical dark resistance of LDRs is 1 Mohms to 10 Mohms. Its light resistance is 75 to 300 ohms. The LDRs take some finite time to change this state and this time is called the recovery time. The typical recovery rate is 200 Kohm/sec.

The testing of LDRs is very simple and can be done with a VOM or DMM as it basically involves the measurement of resistance under dark and light conditions.

7.3.8 Thermistors

Thermistors are resistors with a high temperature co-efficient of resistance. Thermistors with negative temperature co-efficient (fall in resistance value with increase in temperature) are most popular. They are oxides of certain metals like manganese, cobalt and nickel. Thermistors are available in a wide variety of shapes and forms suitable for use in different applications. They are available in the form of disks, beads or cylindrical rods. Thermistors have inherently non-linear resistance-temperature characteristics. However, with proper selection of series and parallel resistors, it is possible to get a nearly linear response of resistance range with temperature.

Thermistors with positive thermo-resistive co-efficient are called 'Posistors'. They are made from barium titanate ceramic and are characterized by an extremely large resistance change in a small temperature span.

Thermistors are used for applications such as excess current limiters, temperature sensors, protection devices against over-heating in all kinds of appliances such as electric motors, washing machines and alarm installations, etc. They are also used as thermostats, time delay devices and compensation resistors. Depending upon the application, the thermistor beads need to be properly protected by ceiling the thermistor bead into the tip of a glass tube or inside a stainless steel sheet.

The easiest way to test a thermistor is with a VOM or DMM, wherein the resistance can be measured under ambient conditions and then under heated conditions by bringing a hot soldering iron tip or blow dryer or bringing a heat gun near the thermistor. The resistance should change smoothly—up or down—depending on whether it is PTC or NTC. If the resistance changes erratically, or goes to infinity or zero, the device is bad. However, to determine if the thermistor is operating correctly or not, you will need specification and temperature measuring sensors.

7.4 CAPACITORS

A capacitor (also called 'condenser') consists of two conductors separated by a dielectric or insulator (Fig. 7.8). The dielectric can be paper, mica ceramic, plastic film or foil.

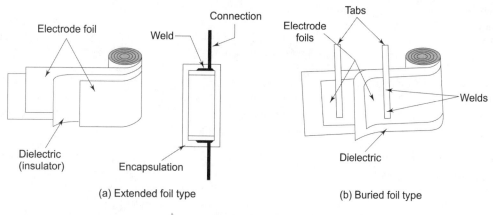

(a) Extended foil type   (b) Buried foil type

Fig. 7.8 *Basic capacitor*

Capacitance is measured in farads. A capacitor has a capacitance of one farad when one coulomb charges it to one volt. The farad is too large a unit. The usual sub-units used are *microfarad* (10^{-16}F) and the *picofarad* (10^{-12}F).

Capacitors find widespread applications in the electrical and electronics fields. Some of the common situations in which capacitors are employed are as follows:

Electrical Field
(a) Power factor correction;
(b) Motor start and run; and
(c) Welding (stored energy in capacitor discharged rapidly).

Electronic Field
(a) Ripple filters in power supplies;
(b) Tuning resonant circuits, oscillator circuits;
(c) Timing elements in multi-vibrators, delay circuits;
(d) Coupling in amplifiers;
(e) Decoupling in power supplies and amplifiers; and
(f) Spark suppression contacts on thermostats and relays.

The value of the capacitor is indicated on the body of the capacitor either in words or in the form of a colour code. Figure 7.9(a) and Table 7.2 shows the capacitor colour code for various types of capacitors.

Table 7.2 ■ *Colour Code for Capacitors*

| Colour | Significant Figures | Capacitors | | | Dipped tantalum voltage rating |
| | | Multiplier (pF) | Tolerance | | |
			Over 10 pF	Under 10 pF	
Black	0	1	± 20%	± 2 pF	4VDC
Brown	1	10	± 1%	± 0.1 pF	6VDC
Red	2	10^2 or 100	± 2%	–	10VDC
Orange	3	10^2 or 1000	± 3%	–	15VDC
Yellow	4	10^4 or 10,000	+ 100% − 0%	–	20VDC
Green	5	10^5 or 100,000	± 5%	± 0.5% pF	25VDC
Blue	6	10^6 or 1,000,000	–	–	35VDC
Violet	7	10^7 or 10,000,000	–	–	50VDC
Grey	8	10^{-2} or 0.01	+ 80% − 20%	± 0.25 pF	–
White	9	10^{-1} or 0.1	± 10%	± 1 pF	3VDC
Gold	–	–	–	–	–
Silver	–	–	–	–	–
None	–	–	± 10%	± 1 pF	–

The value of a capacitor is also sometimes written on the body in the form of numbers. Values beginning with decimal are usually measured in microfarads (μF), all other values are assigned to be in picofarads (pF). Four-digit values are also measured in picofarads but without a multiplier. Some capacitors are coded with a three digit number (Fig. 7.9b)

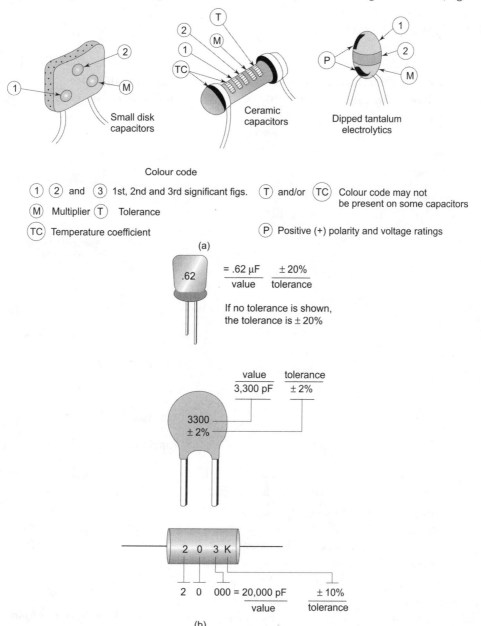

Fig. 7.9 *(a) Capacitor colour code; and (b) Code for numbered capacitors*

which is similar to the colour band system, with a value and multiplier numbers. For example, 203 means that 2 & 0 are attached to 3 zeros and the value of the capacitor would be 20,000 pF or .02 μF. The tolerance letter codes indicate F = ± 1%, G = ± 2%, J = ± 5%, K = ± 10%, M = ± 20% and Z = + 80 to –20%.

7.4.1   Types of Capacitors

Capacitors are categorized into various types (Fig. 7.10) depending upon the dielectric medium used in their construction. The size of the capacitor, their tolerances and the working voltage also depend upon the dielectric used. Some of the common types of capacitors are as follows:

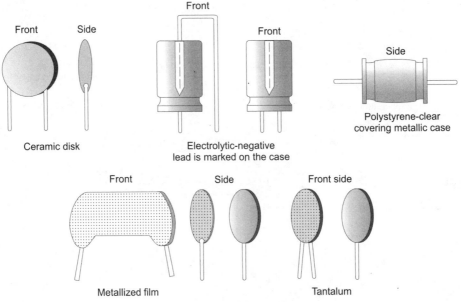

Fig. 7.10   *Various types of capacitors*

Paper Capacitors: Paper capacitors make use of thin sheets of paper wound with thin aluminium foils. To increase the dielectric strength and to prevent moisture absorption, the paper is impregnated with oils or waxes. The capacitor is normally encapsulated in resin. Paper capacitors tend to be large in size due to the thickness of paper and foil. The thickness is reduced in metallized capacitors (Fig. 7.11) by directly depositing the aluminium on the dielectric.

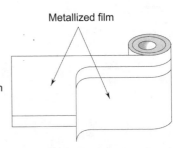

Fig. 7.11   *Construction of metallized capacitor*

Typical range   : 10 nF to 10 μF
Typical dc voltage : 500 V(max)
Tolerance   : ± 10%

Mica Capacitors: A mica capacitor is made by directly metallizing the thin sheets of mica with silver and stacking together several such sheets to make the complete capacitor. The assembly is encapsulated in resin or moulded in plastic.

Typical range   : 5 pF to 10 nF
Typical dc voltage : 50 to 500 V
Tolerance   : ± 0.5%

Ceramic Capacitors: Ceramic capacitors generally employ barium titanate as the dielectric medium. However, low-loss ceramic capacitors use steatite which is a natural mineral. A thin plate of ceramic is metallized on both sides and connecting leads soldered to the metallization. The body is coated with several layers of lacquer. Modern ceramic capacitors in the monolithic type are made of alternate layers of thin ceramic dielectric and electrodes which are fired and compressed to form a monolithic block. These capacitors have a comparatively small size.

Typical range: (a) Low loss (steatite)
                          5 pF to 10 nF
                    (b) Barium titanate
                          5 pF to 1 μF
                    (c) Monolithic
                          1 nF to 47 μF
Typical voltage range: For (a) and (b)
                          60 V to 10 kV
                    For (c): 60V to 400 V
                    Tolerance : ±10% to ± 20%

Plastic Capacitors: The construction of plastic capacitors is very similar to that of paper capacitors and the former are made of both foil and metallized types. Polystyrene film or foil capacitors are very popular in applications requiring high stability, low tolerances and low temperature coefficient. However, they are bulky in size. For less critical applications, metallized polyethylene film capacitors are used. They are commonly referred to as polyester capacitors.

Electrolytic Capacitors: High value capacitors are usually of the electrolytic type. They are made of a metal foil (Figs 7.12 (a) and (b)) with a surface that has an anodic formation of metal oxide film. The anodised foil is in an electrolytic solution. The oxide film is the dielectric between the metal and the solution. The high value of capacity of electrolytic capacitors in a small space is due to the presence of a very thin dielectric layer.

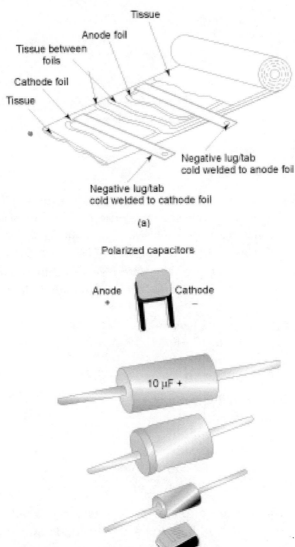

(a) Construction of aluminium electrolytic capacitor
    (b) Shapes of polarized capacitors

Electrolytic capacitors are of the following types:

Aluminium: Plain foil, etched foil and solid,

Tantalum: Solid, wet sintered.

Electrolytic capacitors exhibit a very wide range of tolerance, typically –20% to +50%. They are usually polarized. Care must be taken not to reverse the voltage applied across it. If a reverse voltage is applied, the dielectric will be removed from the anode and a large current will flow as oxide is formed on the cathode. Sometimes, the gases released from the capacitor may build up and cause the capacitor to explode and damage other parts of the circuit.

7.4.2   Performance of Capacitors

The capacitor impedance is a function of frequency: at low frequencies the capacitor blocks signals, at high frequencies the capacitor passes signals. Depending on the circuit connection, the capacitor can pass the signal to the next stage or it can shunt it to ground. The impedance of the capacitor varies with frequency as follows:

$$X_C = \frac{1}{2\pi f C}$$

All capacitors have a self-resonant frequency wherein the parasitic lead and dielectric inductance resonates with the capacitor in a series-resonant circuit. Essentially, the capacitor impedance decreases until it reaches self-resonance where it is minimum impedance.

Aluminium electrolytic capacitors have a very low self-resonant frequency, so they are not effective in high frequency applications above a few hundred KHz. Tantalum capacitors have a mid-range self-resonant frequency, thus they are found in applications up to several MHz. Beyond several MHz, ceramic and mica capacitors are preferred because they have self-resonant frequencies ranging into hundreds of MHz. Very low frequency and timing applications require highly stable capacitors. The dielectric of these types are made from paper, polypropylene, polystyrene and polyester. These capacitors have low leakage current, low dielectric absorption and they come in large values.

ESR (Equivalent Series Resistance) is an important parameter of any capacitor. It represents the effective resistance resulting from the combination of wiring, internal connections, plates and electrolytes. Figure 7.13 shows a capacitor equivalent circuit. ESR is the effective resistance of the capacitance at the operating frequency and therefore affects the performance of tuned circuits. It may result in a totally incorrect or unstable operation like switch mode power supplies and deflection circuits in TVs and monitors. A power supply filter design requires low ESR capacitor because voltage is dropped across ESR and the current flowing through the capacitor causes power dissipation resulting in self-heating. Electrolytic capacitors tend to have a high ESR as compared to other types and it changes, not necessarily for the better, with time.

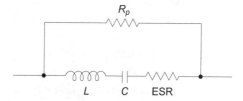

Fig. 7.13   *Equivalent circuit of a capacitor*

An ideal capacitor would only have C and no R. Any R in series with C will reduce the capacitor's ability to pass current in response to a variant applied voltage and it will dissipate heat which is wasteful and could lead to failure of the component. The dissipation factor (DF) is mathematically defined as R/X where R is the resistance in the capacitor and X is the reactance of the capacitor. The higher the R, the higher the DF and poorer the capacitor. From the formula, $DF = R/X$, it is clear that DF is an inverse function of X. As X goes down, DF goes up and vice versa. So DF varies proportionately with frequency, which shows that DF is a function of the test frequency. DF is a measure of capacitor quality and the figure is valid only at the frequency of the test. R_p models the parallel resistance of a capacitor. Its value is usually in hundreds of Mohms except for electrolytic capacitors which have comparatively low value.

The quality factor Q serves as a measure of the purity of a reactance, i.e. how close it is to being a reactance, no resistance. This represents as the ratio of the energy stored in a component to the energy dissipated by the component. Q is a dimensionless unit and is expressed as $Q = X/R$. However, Q is commonly applied to inductors; for capacitors the term more often used to express purity is the dissipation factor (DF). This quantity is simply the reciprocal of Q.

7.4.3 Failures in Capacitors

The failure in capacitors can manifest itself as slow degradation in performance or complete failure of the capacitor:
The degradation in performance results in:
  (a) Gradual fall in insulation resistance or rise in leakage current;
  (b) Increase in dissipation factor, i.e. rise in series resistance or dielectric losses.
In case of a complete failure of the capacitor, the capacitor will have either:
  (a) Open circuit due to end connection failure; or
  (b) Short circuit due to dielectric breakdown.
The failure in capacitors could be due to the following reasons:
  (a) *Environmental factors:*
    1. Mechanical shocks and vibration;
    2. Thermal shock; and
    3. High humidity.
  (b) *Misuse:*
    1. Poor assembly technique;
    2. Prolonged storage; and
    3. Subjecting the capacitor to voltage beyond its stated capability.
  (c) *Manufacturing imperfection:*
    1. Impurities in electrolytics; and .

2. Mechanical damage to the end spray of metallized capacitors resulting in over-heating and open circuit.

Faults in different types of capacitors could be due to different reasons. Some typical faults and their possible causes for various capacitors are detailed below.

Paper Capacitor

(a) Open circuit or intermittent performance: Mechanical/thermal shock or damage during assembly;
(b) Short circuit: Mechanical/thermal shock, leaking of seal.

Mica Capacitors

(a) Open circuit or intermittent performance: Peeling-off silver from the surface of mica;
(b) Short circuit: High humidity causing silver migration.

Ceramic Capacitors

(a) Open circuit: Fracture of connection;
(b) Short circuit: Shock or vibration causing fracture of dielectric; and
(c) Variations in capacitance value; loose lead, not adhering properly to ceramic.

Plastic Film Capacitors

(a) Open circuit: Damage to end spray during manufacture or poor assembly.

Electrolytic Capacitors

(a) Open circuit: Fracture of internal connections;
(b) Short or leaky: Loss of dielectric, high temperature;
(c) Fall in capacitance value: Pressure, thermal or mechanical shock producing leakage in the seal and consequent loss of electrolyte.

7.4.4 Testing of Capacitors

(a) Electrolytic and high value capacitors can be checked for open circuit, short circuit and leaky behaviour by means of an ohmmeter operated on the highest scale. In case the capacitor is good, it will slowly charge to show a high value of resistance. In case of open circuit, the charging action would be absent.

An open capacitor can best be detected with a capacitance meter or by checking if the capacitor passes AC signals. If the capacitor is leaky, the final value of resistance indicated by the ohmmeter will not be very high (below a few kilo-ohms). Do not exceed the voltage rating of the capacitor when testing it by the charging method.

(b) Some DMMs have specific modes for capacitor testing. These work fairly well for approximate μF rating. However, for most applications, they do not test at anywhere near the normal working voltage or test of leakage. However, a VOM or DMM without capacitance ranges can still help to make certain types of tests.

Simple capacitance scales on DMMs just measure the capacitance in μF and do not test for leakage, ESR (Equivalent Series Resistance) or breakdown voltage. If the measurement comes up within a reasonable percentage of a marked value, this is all you need to know. However, leakage and ESR frequently change on electrolytic capacitors as they age and dry out.

For small capacitors (0.01μF or less), what you can really test is for shorts or leakage. It may be noted that on an analog multi-meter on the high ohm scale, just a momentary deflection is seen when you touch the probes to the capacitor or reverse them. A DMM may not provide any indication at all. Any capacitor that measures a few ohms or less is bad. Most should show infinite resistance value even on the highest resistance range.

(c) Capacitance meters are available for measuring the value of capacitance and are more accurate than a DMM for this purpose. They facilitate measurement of the value of capacitance ranging from 1pF to 1000μF. The frequency of the bridge supply is typically 1kHz. The detector is generally an AC amplifier tuned to 1kHz with its output feeding a moving coil meter via a rectifier. Similarly ESR testers, which are good for quick troubleshooting, measure the equivalent series resistance. Some provide only a go/no go indication while others actually display a reading. The reading is usually between 0.01 and 100 ohms. So, they can also be used as low-ohms meters for resistors in non-inductive circuits.

Alternatively, using a DC power supply and series resistor, capacitance can be calculated by measuring the rise time to 63% of the power supply voltage from $T = RC$ or $C = T/R$.

(d) The best way to really test a capacitor, if required, is to substitute a known good one. A VOM or DMM will not test the capacitor under normal operating conditions or at its full-rated voltage. Therefore, substitution is a quick way of finding a faulty capacitor and solving major fault problems.

7.4.5   Precautions While Testing Capacitors

It is essential for your safety and to prevent damage to the device under test as well as your test equipment that large or high voltage capacitors be fully discharged before measurements are made, soldering is attempted or the circuitry is touched in any way. Some of the large filter capacitors commonly found in mains operated equipment store a potentially lethal charge. The capacitors which require discharge are the high value main filter capacitor in TVs, video monitors, switch mode power supplies, microwave ovens and other similar devices.

Generally, for discharging the capacitors, a high voltage resistance of about 5 to 50 ohms / V of the working voltage of the capacitor is recommended. This will prevent the arcing associated with screw driver discharge (normally practised by TV technicians) but will have a short enough time constant so that the capacitor will drop to a low voltage in a few seconds.

7.4.6 Variable Capacitors

Variable capacitors are constructed by using any one of the dielectrics like ceramic, mica, polystyrene or teflon. Basically, a variable capacitor has a stator and a rotor. The area of the stator is fixed and turning the rotor from 0 to 180° varies the amount of the plate surface exposed and thereby the value of capacitance.

In most variable capacitors, the change in capacitance is linear throughout the rotation of the rotor. Figure 7.14 shows a linear increase and decrease in the value of capacitance through 360° of rotor rotation.

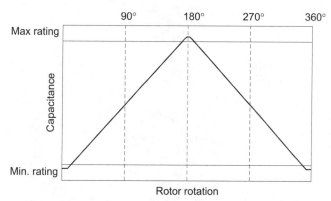

Fig. 7.14   *Linear variation of capacitance with rotation in a trimmer capacitor*

7.4.7 Types of Variable Capacitors

Variable capacitors are generally available in the following two configurations:
  (a) Button type has a variable rotor (Fig. 7.15);
  (b) Tubular type has an adjustable core (Fig. 7.16).
It may be noted that adjustments made with variable capacitors by using a metal screw driver will alter when the screw driver is lifted from the turning screw. This is because placing the metal screw driver on this screw changes the effective area of the metal plated surface of either the stator or, more often, the rotor. In such a case, the use of a non-metallic screwdriver is recommended.

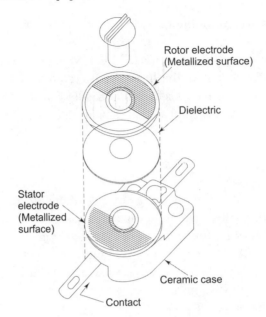

Fig. 7.15   *Variable capacitor*

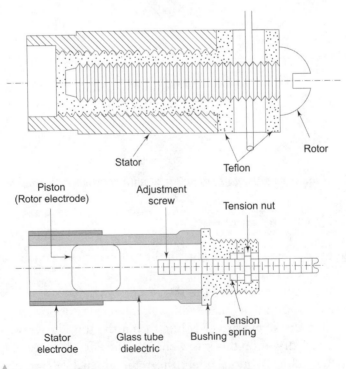

Fig. 7.16   *Tubular type variable capacitors*

7.5 INDUCTORS

Inductance is the characteristic of a device which resists change in the current through the device. Inductors work on the principle that when a current flows in a coil of wire, a magnetic field is produced which collapses when the current is stopped. The collapsing magnetic field produces an electromotive force which tries to maintain the current. When the coil current is switched, the induced emf would be produced in such a direction, so as to oppose the build-up of the current.

$$\text{Induced emf } e = -L \frac{di}{dt} \text{ where } L \text{ is the}$$

$$\text{Inductance and } \frac{di}{dt} \text{ the rate of change of current.}$$

The unit of inductance is henry. An inductance of one henry will induce a counter emf (electromotive force) of one volt when the current through it is changing at the rate of one ampere per second. Inductances of several henries are used in power supplies as smoothing chokes, whereas smaller values (in the milli- or micro-henry ranges) are used in audio- and radio-frequency circuits.

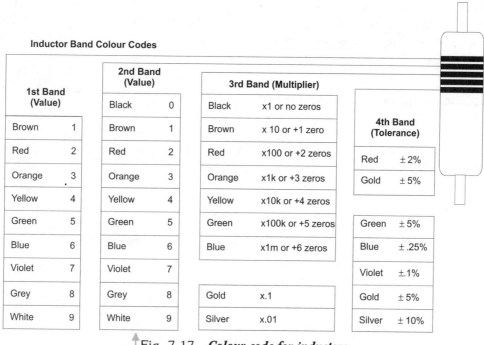

Inductor Band Colour Codes

1st Band (Value)		2nd Band (Value)		3rd Band (Multiplier)		4th Band (Tolerance)	
		Black	0	Black	x1 or no zeros		
Brown	1	Brown	1	Brown	x 10 or +1 zero	Red	± 2%
Red	2	Red	2	Red	x100 or +2 zeros	Gold	± 5%
Orange	3	Orange	3	Orange	x1k or +3 zeros		
Yellow	4	Yellow	4	Yellow	x10k or +4 zeros		
Green	5	Green	5	Green	x100k or +5 zeros	Green	± 5%
Blue	6	Blue	6	Blue	x1m or +6 zeros	Blue	± .25%
Violet	7	Violet	7			Violet	±.1%
Grey	8	Grey	8	Gold	x.1	Gold	± 5%
White	9	White	9	Silver	x.01	Silver	± 10%

Fig. 7.17   *Colour code for inductors*

The value of an inductor may be printed on the component body or it may be printed with colour bands (Fig. 7.17), much in the same way as a resistor. For example, if the first and second bands of an inductor are red (value 2) and the third band is orange (value 3), the value of the inductor is 22000 μH (microhenry). A fourth silver band will indicate its tolerance as ±10%.

The primary use of the inductor is filtering. There are two very different types of filter inductors: the high current inductor wound around a large core is used in power supply filters and low current air core inductors are used in signal filters.

High current inductors require cores to keep the losses within acceptable limits and to achieve high performance. The cores are big and heavy, so they have large weight and size. Switching power supplies require extensive inductors or transformers to control the switching noise and filter the output voltage waveform.

Low current inductors are used for filters in signal processing circuits. An inductive/capacitive filter has sharper slopes than a resistive/capacitive filter, thus it is more effective filter in some applications. In general, inductors are rarely seen outside power circuits.

7.5.1   Testing of Inductors

The inductors can be checked for open circuit by checking continuity with an ohm-meter. Shorted or partially shorted inductors can usually be found by checking the waveform response. When high frequency signals are passed through the circuit, partial shorting often reduces high frequency response (roll-off).

7.5.2   Measurement of Inductance

The measurement of inductance, like the capacitor, is preferably carried out at 1 kHz energizing supply source for measurement. Universal RLC bridges are commonly used for this purpose.

8

Testing of Semiconductor Devices

8.1 TYPES OF SEMICONDUCTOR DEVICES

Semiconductor devices are versatile units employed in a variety of applications in electronic equipment. For example, they are used as rectifiers, amplifiers, oscillators, modulators, voltage and current sources, electronic switches, voltage shifters and variable resistors. They are also employed as energy convertors (for example, light emitting diode) and for generating logarithmic and anti-logarithmic functions.

Semiconductor devices can be broadly classified into the following two areas:

Bipolar: In bipolar devices, the action of the devices depends upon the flow of the type of charge carriers across forward- or reverse-biased *pn* junctions. For example, in an *npn* bipolar transistor, the flow of electronics across the reverse-biased collector-base junction is controlled by the forward-biased base-emitter junction. As electrons move into the base, some re-combine with holes, but the majority diffuse across the base and are swept up by the collector. The commonly used bipolar devices are transistors, diodes, unijunction transistors, thyristors, logic ICs such as TTL and linear ICs.

Unipolar: Unipolar devices use only majority carriers for current flow and this current is controlled by an electro-static field, say between the gate and source, or the gate and substrate. Typical examples of unipolar devices are Junction FETs, MOSFETs, CMOS logic and some linear ICs.

8.2 CAUSES OF FAILURES IN SEMICONDUCTOR DEVICES

Failure in semiconductor devices could take place due to any of the following reasons:

   (a) The device may possess an inherent weakness from the manufacturing process and may fail prematurely in equipment. The failures could be short-circuits and open circuits.

(b) Failures could be caused by misuse or by bad handling during assembly or test.

(c) Exceeding the maximum rated values of voltage, current and power for a particular semiconductor device may result in its failure.

(d) Electrical interference is one of the major causes of premature failure of semiconductor devices. Voltage surges carried along the mains leads caused by switching of heavy machines or relays often cause breakdown of semiconductor junctions.

8.3 TYPES OF FAILURES

Failures are mostly those of an open or short circuit at a junction. For example, in a bipolar transistor, a short or open circuit may develop between the base and emitter or collector and base. A short circuit may also occur between the collector and the emitter.

8.4 TESTING PROCEDURES FOR SEMICONDUCTOR DEVICES

In the repair and servicing situation, it would obviously not be practical to carry out all the possible measurements on devices. Therefore, it is not necessary to be concerned so much with device parameters, but rather with one or two that are of vital importance for correct circuit operation.

8.4.1 Diodes

The cathode and anode ends of metal-encased diodes can be identified by the diode symbol marked on the body. In case of glass-encased diodes, the cathode end is indicated by a stripe, a series of stripes or a dot. For most silicon or germanium diodes with a series of stripes, the colour code identifies the equipment manufacturer's part number. Figure 8.1 shows various shapes of diodes.

A diode can be conveniently checked with an ohmmeter by measuring its forward and reverse resistance. A conventional signal diode or rectifier should normally show a low resistance (typically a few hundred ohms) in the forward direction and a very high value (nearly infinity) of resistance in the reverse direction. It should not read near zero ohms (shorted) or open in both directions. When checking the diode with the ohmmeter, the following factors should be remembered:

(a) The same diode will appear to have different forward resistance values when checked with various types of ohmmeters.

   Similarly, different values of forward resistances will be encountered on different ranges on the ohmmeters.

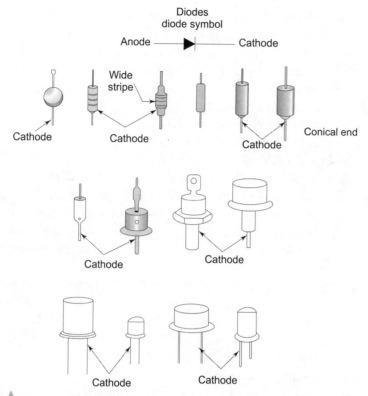

Fig. 8.1 *Various package shapes and basis of diodes*

This variation in resistance is due to the non-linear (Fig. 8.2) voltage/current characteristics of the diode. Different types of ohmmeters and the same ohmmeter in different ranges may not have the same test voltage, resulting in variation in the value of the resistance, which is actually not so.

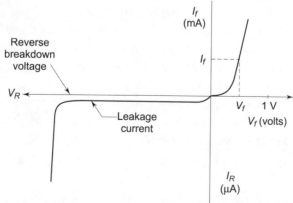

Fig. 8.2 *Voltage-current characteristics of semiconductor diodes*

(b) When checking the diode with an ohmmeter, one end of the diode should always be disconnected from the circuit as in circuit measurements are seriously disturbed by the presence of low values of shunt resistance, if present in the circuit.

(c) The forward resistance of germanium diodes is much less than that of the silicon diodes, whereas a silicon diode normally measures a much higher value of reverse resistance than a germanium diode.

(d) On a digital (DMM) multi-meter, there is usually a diode test mode. Using this, a silicon diode should read between 0.5 to 0.8V in the forward direction and open in reverse. For a germanium diode, it will be lower, 0.2 to 0.4 V or so in the forward direction. Using the normal resistance ranges will usually show open for any semiconductor junction since the meter does not apply enough voltage to reach the value of the forward drop. Note, however, that a defective diode may indeed indicate a resistance lower than infinity especially on the highest ohms range. So, any reading of this sort would be an indication of a bad device but the opposite is not guaranteed.

(e) Sometimes it is necessary to check the values of V_f (forward voltage drop) and R_{BR} (reverse breakdown voltage) to determine whether or not these values are within limits. A typical diode's characteristics are shown in Fig. 8.2. The characteristics can be shown on the oscilloscope or on X-Y plotter by using suitable circuits. The measurement of V_f can be done by passing a constant current, say 5mA, through the diode and reading V_f on a voltmeter.

Diodes are often used in stacks in high voltage rectifier circuits because the peak inverse voltage rating of the stack is equal to the rating of a single diode multiplied by the number of diodes in the stack. When checking the stacked diodes with an ohmmeter, it will be observed that the resistance of two diodes connected in series is greater than twice the resistance of a single diode. Similarly, the resistance of three diodes connected in series will be greater than three times the resistance of the single diode. This discrepancy is again due to the non-linear characteristics of the diode.

Diodes are also often used in parallel to obtain higher current capability. When two diodes are connected in parallel, the combination can provide twice the current capability of a single diode. When making ohmmeter checks on parallel connected diode combinations, the resultant resistance appears to be more than half of the resistance of a single diode, for reasons explained above.

It may be noted, however, that this seeming discrepancy would not be experienced if the ohmmeters work as constant-voltage or constant-current devices. On the other hand, ohmmeters are commonly used in routine repairs and servicing work as varying current-voltage instruments.

The following precautions may be observed when checking diodes:
(a) Do not use an ohmmeter scale that has a high internal current. High currents may damage the diodes under test.
(b) Do not check tunnel diodes with an ohmmeter.

For a VOM, the polarity of the probes is often reversed from what you would expect from the colour coding—the red lead is negative with respect to the black one. DMMs usually have the same polarity as is expected of them. This can be confirmed by using a known diode as a reference.

8.4.2 Special Types of Diodes

Besides the general purpose semiconductor diodes, there are many other types of diodes which have special characteristics. Their function and the maximum current rating of a diode should be known before attempting an ohmmeter test, since the average battery type ohmmeter, when connected to certain detector diodes, specially those used in microwave applications, could easily burn them out. Even if burn-out does not occur, a simple ohmmeter test may not be a true indication of the condition of a diode. For example, the zener diode is designed to be highly conductive in the reverse direction beyond a certain applied voltage. Similarly, the resistance of a tunnel diode varies widely with applied voltage in the forward direction. Following is a description and characteristics of some of the special types of diodes.

Zener Diode: A silicon diode has a very low reverse current, say 1μA at an ambient temperature of 25°C. However, at some specific value of reverse voltage, a very rapid increase occurs in reverse current. This potential is called a breakdown avalanche and may be as low as 1V or as high as several hundred volts, depending upon their construction.

A zener diode has very high resistance at bias potentials below the zener voltage. This resistance could be worth several megohms.

At zener voltage, the zener diode suddenly shows a very low resistance, say between 5 and 100 Ω.

A zener diode behaves as a constant voltage source in the zener region of operation, as its internal resistance is very low. The current through the zener diode (Fig. 8.3) is then limited only by the series resistance R. The value of the series resistance is such that the maximum rated power rating of the zener diode is not exceeded.

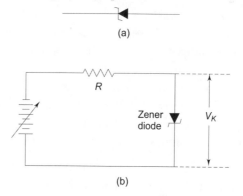

Fig. 8.3 *Zener diode: (a) symbol, (b) use as a constant voltage source*

Zeners most often fail short or more likely, open with short. However, it is also possible, though less common, for the zener voltage to change (almost always to a lower voltage) and for the shape of the I-V curve to change. In that case, it normally becomes less sharp-cornered.

A zener diode can be checked with an ohmmeter below the zener voltage when its operation is similar to that of a silicon diode. At zener voltage, it is tested by measuring the voltage appearing across it, when it is in the circuit.

With a VOM, therefore, a good zener diode should read like a normal diode in the forward direction and open in the reverse direction, unless the VOM applies more than the zener voltage for the device. A DMM on its diode test range may read the actual zener voltage if it is very low but will read open otherwise. The most common failure would be for the device to short-read 0.0 ohms in both directions. Then, it is definitely dead.

Varactor Diode   A varactor diode is a silicon diode that works as a variable capacitor in response to a range of reverse voltage values. Varactors are available with nominal capacitance values ranging from 1 to 500 pF, and with maximum rated operating voltages extending from 10 to 100 volts. They mostly find applications in automatic frequency control circuits. In a typical case, a varactor shows a 10 pF capacitance at reverse voltage of 5 volts and 5 pF at 30 volts. Figure 8.4 shows different shapes of varactor diodes.

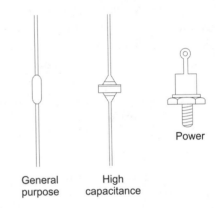

Fig. 8.4   *Varactor diodes—different types*

Varistor   A varistor is a semiconductor device that has a voltage dependent non-linear resistance which drops as the applied voltage is increased. A forward biased germanium diode shows this type of characteristics and is often used in varistor applications, such as in bias-stabilization circuits.

Symmetrical varistor arrangements are used in meter protection circuits (Figs 8.5 (a) and (b)) wherein the diodes bypass the current around the meter regardless of the direction of current flow. If the meter is accidentally overloaded, varistors do not permit destructive voltages to develop across the meter.

Light Emitting Diodes (LED)   A light emitting diode is basically a *pn* junction that emits light when forward biased. LED lamps are available in various types (Fig. 8.6) and mounted with various coloured lenses like red, yellow and green. They are used mostly in displays employing seven segments that are individually energised to form alpha-numeric characters.

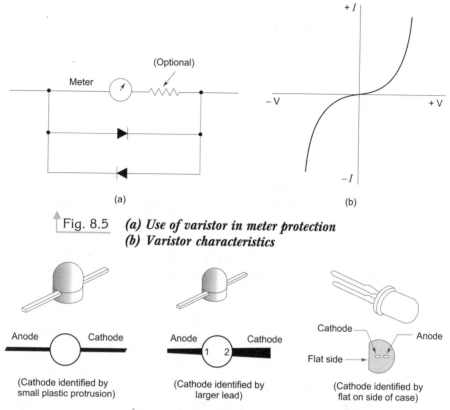

Fig. 8.5   *(a) Use of varistor in meter protection*
             (b) Varistor characteristics

Fig. 8.6   *Light emitting diodes*

LED displays are encountered in test equipments, calculators and digital thermometers whereas LED arrays are used for specific applications such as light sources, punched tape readers, position readers, etc.

Electrically, LEDs behave like ordinary diodes except that their forward voltage drop is higher. For example, the typical values are: IR (infra-red) : 1.2 V, Red : 1.85 V, Yellow: 2 V, Green : 2.15 V. Further, the actual voltages may vary quite a bit depending upon the actual technology used in the LED. Therefore, the LED voltage drop is not a reliable test of colour though multiple samples of similar LEDs should be very close. So, the LED should be tested only for short and open with a multi-meter. On the other hand, an LED may pass the electrical tests, but it can be weak. Therefore, the LEDs should be tested for their proper operation. A defective LED can be easily spotted from its actual operation in the circuit if the other associated circuit components and connections are correct.

Photo-diode: A photo-diode is a solid state device, similar to a conventional diode, except that when light falls on it (*pn junction*), it causes the device to conduct. It is practically an open circuit in darkness, but conducts a substantial amount of current when exposed to light.

A photo-diode can be checked for its performance by measuring the voltage developing across a load (Fig. 8.7) when the photo-diode is reverse-biased and first exposed to light and then to darkness (covering the diode with a dark paper, etc).

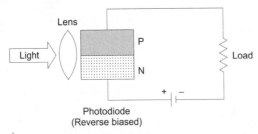

Fig. 8.7 *Basic photodiode arrangement*

Tunnel Diode (TD): A tunnel diode is a *pn junction* which exhibits a negative resistance interval. A tunnel diode voltage current characteristics are shown in Fig. 8.8. Negative resistance values range from 1 to 200 ohms for various types of tunnel diodes.

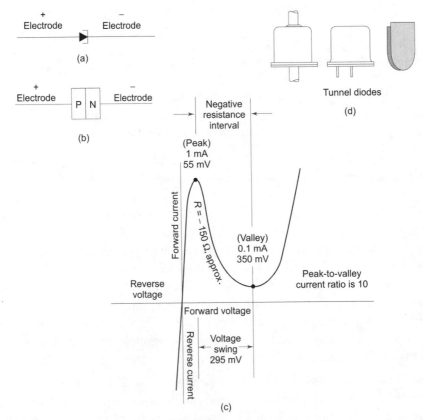

Fig. 8.8 *Tunnel diode: (a) symbol, (b) p and n regions (c) voltage-current characteristics, and (d) different types of housings*

Tunnel diodes can be utilized in switching circuits. A switching circuit has two quiescent points, i.e. it can be driven from its low current quiescent point to its high current quiescent point by means of pulses.

A quick evaluation of the tunnel diode's ability to switch and at the current level at which does switch often helps in the troubleshooting process. The TD may be evaluated using the sawtooth output waveform from an oscilloscope as a current source (Fig. 8.9). A 670 ohm resistor from the sawtooth out connector in series with the TD to ground will give a calibrated current/div horizontally (say 1mA/div.). The sawtooth voltage goes from 0 to 10 volts. Therefore, the horizontal display becomes current/div. Looking at the voltage drop across the diode will give a vertical display of the low/high voltage states of the diodes.

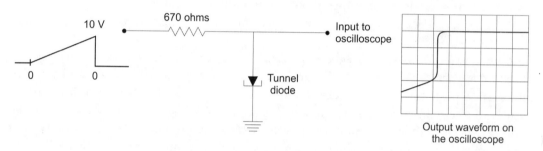

Fig. 8.9   *Testing of tunnel diodes*

The display does not give an indication of switching time but confirms that the device has the ability to switch at the correct current level and will probably perform normally in the circuit.

Bi-polar Transistors

The most commonly used semiconductor device is the transistor. In troubleshooting transistor circuits, the following points should be kept in mind:

(a) *npn* and *pnp* devices are basically 'off' devices while vacuum tubes are basically 'on' devices,

(b) Transistors are made up of two diodes: a base emitter diode and a base-collector diode.

   In normal amplifier operation, the base emitter diode is forward-biased and the base-collector diode is reverse-biased.

(c) Shorting base-to-emitter turns off transistors, while forward biasing base-emitter junctions turns on transistors.

(d) All transistors have leakage current across their reverse-biased base-collector diodes. For silicon transistors, this current is more than several nanoamperes. In germanium transistors, the leakage current may even be several microamperes.

(e) Leakage current may be measured by shorting the base-emitter junction (diode) and measuring the voltage across the load resistance. The leakage current then equals the voltage (V_L) across the load resistor (R_L) divided by R_L. In making this measurement, it should be ensured that the transistor collector is not direct coupled to any other stage.

(f) Leakage current increases with temperature and doubles about every 10°C.

(g) An abnormal increase in the leakage current at room temperature results in a shift in the normal bias (operating) point. Trouble is generally experienced if the driving signal drives the transistor to or near cut-off. The transistor will not properly turn-off and the result may be clipping or distortion due to the residual leakage current flowing through the external resistors. Heating and cooling a transistor aggravates this condition and sometimes shows up marginal operation.

(h) Shorting collector to emitter simulates saturation, as the transistor behaves like a closed knife switch.

Testing Bi-polar Transistors: A good check of transistor operation is its actual performance under operating conditions. A transistor can most effectively be checked by substituting a new piece for it or one which has been checked previously. However, it should be ensured that circuit conditions are not such that a replacement transistor might also be damaged. If equivalent transistors are not available, use a dynamic tester. Static type of testers are not recommended, since they do not check operation under simulated operating conditions.

In transistor circuit testing, the most important consideration is the transistor base-to-emitter junction. The base-emitter junction in a transistor has the same function as a grid-cathode in a vacuum tube. The base emitter junction is essentially a diode and must be forward-biased for the transistor to conduct. Just like the simple diodes, the forward-bias polarity is determined by the materials forming the junctions. Figure 8.10 shows transistor symbols with terminals labelled and they reveal the bias polarity required to forward-bias the base-emitter junction. The figure also compares the biasing required to cause conduction and cut-off in *npn* and *pnp* transistors. If the transistor's base-emitter junction is forward-biased, the transistor conducts. However, if the base-emitter junction is reverse-biased, the transistor is cut-off.

The voltage drop across a forward-biased emitter-base junction varies with the transistors collector current. For example, a germanium transistor has a typical forward-bias, base-emitter voltage of 0.2–0.3V when the collector current is 1–10mA and 0.4–0.5V when the collector current is 10–100mA. In contrast, the forward-bias voltage for a silicon transistor is about twice that for germanium types: about 0.5–0.6V when the collector current is low, and about 0.8–0.9V when the collector current is high. Figure 8.11 shows the relationship between voltage and current for base-emitter junction in germanium and silicon transistors.

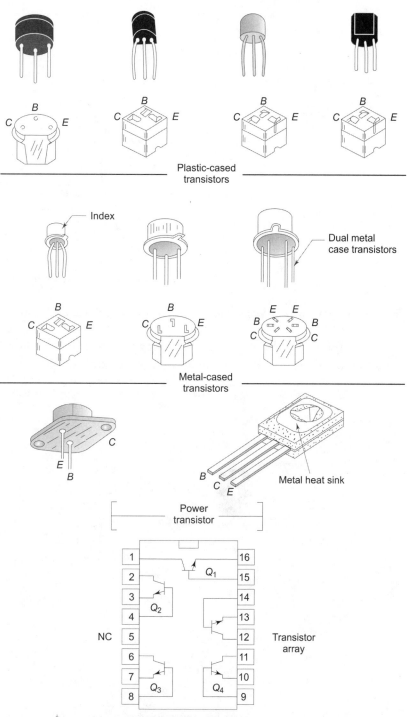

Plastic-cased transistors

Index

Dual metal case transistors

Metal-cased transistors

Power transistor

Metal heat sink

NC

Transistor array

Fig. 8.10 *Transistor symbols and terminals*

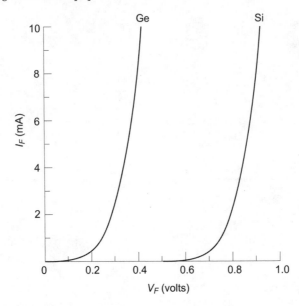

Fig. 8.11   *Forward voltage of base-emitter junction in Ge and Si transistors*

The three basic transistor circuits along with their characteristics are shown in Fig. 8.12. When examining a transistor stage, just determine if the emitter-base junction is biased

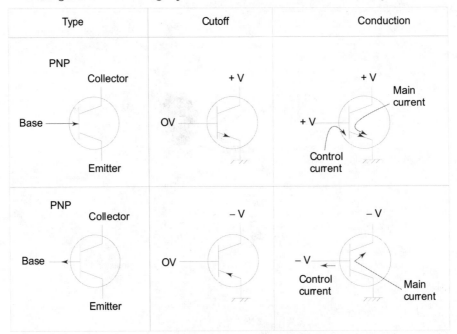

Fig. 8.12   *Three basic biasing arrangements in transistors*

for conduction (forward-biased) by measuring the voltage difference between emitter and base. When using a voltmeter, do not measure the voltage directly between emitter and base because there may be sufficient loop current between the voltmeter leads to damage the transistor. Instead, measure each voltage separately with respect to a common point, say chassis. If the emitter-base junction is forward-biased, check for amplifier action by short-circuiting base-to-emitter, while observing collector voltage (Fig. 8.13). The short-circuit eliminates base-emitter bias and should cause the transistor to stop conducting (cut-off). Collector voltage should then change and approach the supply voltage. Any difference is due to leakage current through the transistor and in general, the smaller this current, the better the transistor. If the collector voltage does not change, the transistor has either an emitter-collector short-circuit or emitter-base open circuit.

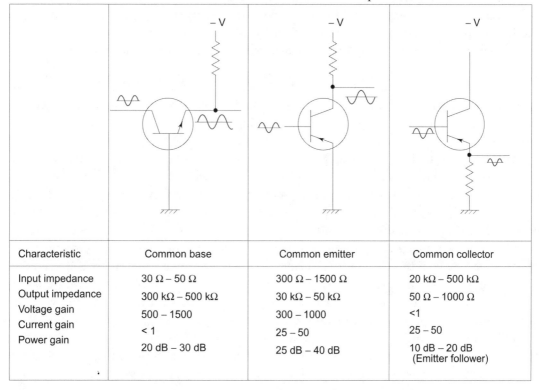

Characteristic	Common base	Common emitter	Common collector
Input impedance	30 Ω – 50 Ω	300 Ω – 1500 Ω	20 kΩ – 500 kΩ
Output impedance	300 kΩ – 500 kΩ	30 kΩ – 50 kΩ	50 Ω – 1000 Ω
Voltage gain	500 – 1500	300 – 1000	<1
Current gain	< 1	25 – 50	25 – 50
Power gain	20 dB – 30 dB	25 dB – 40 dB	10 dB – 20 dB (Emitter follower)

Fig. 8.13 *Transistor amplifier characteristics*

A common problem in transistors is the leakage which can shunt signals or change bias voltages thereby upsetting circuit operation. This problem is particularly serious in direct-coupled or high frequency stages. Leakage current is the reverse current that flows in a junction of a transistor when specified voltage is applied across it, the third electrode (terminal) being left open. For example, I_{CEO} is the DC collector current that flows when a specified voltage is applied from collector to emitter, the base being left open (unconnected). The polarity of the applied voltage is such that the collector-base junction is reverse-based. Obviously, in a transistor, six leakage paths are present (with the third electrode open), as shown in Fig. 8.14.

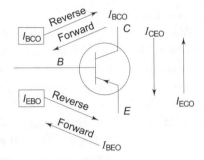

Fig. 8.14   *Leakage paths in a transistor*

Out of these six leakage possibilities, the two most important types which greatly upset the circuit operation are I_{CBO} (collector to base, emitter open) and I_{EBO} (emitter to base, collector open). If I_{CBO} or I_{EBO} occurs, the forward resistance will measure a sub-normal value, which could sometimes be practically zero. In such a case, the transistor is short-circuited.

If the terminal markings in a bi-polar transistor are not known and also, if the identification of the device has been erased, it is possible to identify the base, emitter and collector terminals and the type of the transistor using an ohmmeter. To do so, proceed as follows:

(a) Make resistance measurements between each pair of leads in both the forward-resistance and reverse-resistance directions. A resistance below 250 Ω shows that the ohmmeter is forward-biasing a junction. The higher forward reading is obtained between the emitter and collector leads. The third lead, which is not connected to the ohmmeter, is the base lead.

(b) Next, a resistance measurement is made between the identified base and one of the other leads. If the ohmmeter indicates forward resistance when the negative lead of the ohmmeter is connected to the base, the transistor is a *pnp* type. The transistor will be *npn* type if the forward resistance is indicated with the positive lead of the ohmmeter connected to the base.

(c) To identify which of the two unknown leads is the collector and which the emitter, the two resistance measurements are made between these leads, reversing the ohmmeter polarity for the second measurement. Carefully observe the polarity that gives the lower resistance indication:

  (1) If a *pnp* transistor is under test, the negative lead of the ohmmeter will be connected to the collector lead.

  (2) In the case of a *npn* type transistor, the positive lead of the ohmmeter will be connected to the collector lead.

If none of the six combinations, when measuring resistance, yields a pair of low readings or if more than one combination results in a pair of low readings, the transistor is likely to be bad or it is not a bi-polar transistor.

Figure 8.15 shows typical bi-polar transistor junction resistance readings. The polarity of the ohmmeter to be applied on the various transistor leads are also indicated on the figure.

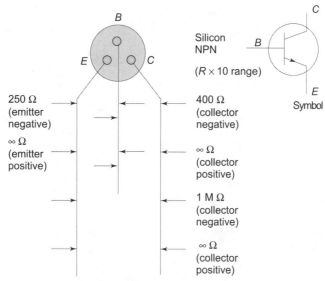

Fig. 8.15   *Bipolar transistor junction resistance values*

Do not hold the transistor under test in your hand. For every degree, the transistor increases in temperature, the base–emitter diode drop (V_{be}) decreases by 2 mV. This is a significant amount when determining the V_{be} and V_{bc}.

Power Transistors: Power transistors are also tested in a similar manner as the small signal transistors. However, it may be noted that the junctions of the power transistors have comparatively large areas. The following points will be observed:

(a) Forward-resistance values are generally lower than those for small signal silicon transistors.

(b) Similarly, lower reverse-resistance values are observed. The test results with an ohmmeter on a good silicon power transistor are shown in Fig. 8.16.

Power transistors are usually mounted on the heat sinks or heat radiators (Fig. 8.16a). They are sometimes mounted on the chassis using a silicone grease to increase heat transfer. When replacing power transistors, the silicone grease should also be replaced. Silicone grease should be handled with care. It should not get into the eyes. The hands must be washed thoroughly after use. After replacing a power transistor in a circuit, check that the collector is not shorted to ground before applying power.

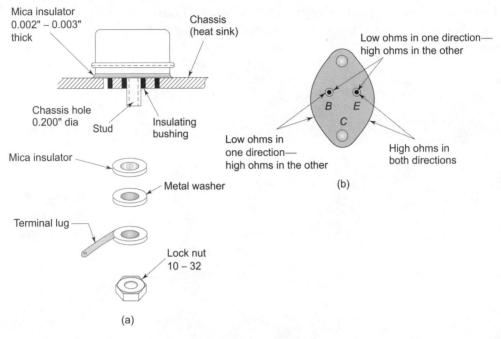

↑Fig. 8.16   ***Power transistor (a) mounting arrangement, and (b) junction resistance values***

Testing Darlington Transistors

A 'Darlington' is a special type of configuration usually consisting of two transistors, fabricated on the same chip or at least mounted in the same package. Darlington pairs are often used as amplifiers in input circuits to provide a high input impedance. Darlingtons are used where the drive is limited and a high gain, typically over 1000, is needed. In this configuration, the emitter base junctions are connected in series and the collector terminals are connected in parallel. A Darlington configuration behaves like a single transistor wherein the current gains (H_{fe}) of the individual transistors it is composed of are multiplied together and the base–emitter voltage drop of the individual transistor it is composed of are added together.

Testing a Darlington pair with a VOM or DMM is basically similar to that of normal bipolar transistors except that in the forward direction, base-emitter resistance will measure higher than a normal transistor on a VOM, but not open. On the diode test range of a DMM, it will show 1.2 to 1.4V due to the pair of junctions in series. It may be noted that 1.2 V may be too high for some DMMs and thus a good Darlington may test open. It should confirm that the open circuit reading on the DMM is higher than 1.4 V or it should be checked with a known good Darlington.

When Darlington pairs are checked with an ohmmeter from the base input to the emitter output terminals, the ohmmeter will show a much higher resistance value than the value of resistance from base to emitter for a single transistor (Fig. 8.17). This apparent discrepancy is due to the non-linear characteristics of a transistor.

Some plastic-cased transistors have lead configurations which may not agree with their replacement transistors made by a different manufacturer than the original. Always check the terminal diagram for correct placement.

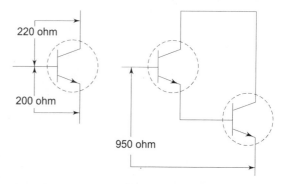

Fig. 8.17 *Darlington pair forward resistance values*

Transistor Testers: The measurement of current gain and leakage are probably the two most important characteristics of transistors which are made by a service technician in typical troubleshooting and maintenance work. The performance of a transistor greatly depends upon the operating conditions and therefore, to determine whether or not a transistor complies with the manufacturer's specifications, the tests should duplicate the voltage and current specified. The operating temperature is usually given as 25°C.

There are several types of transistor testers, ranging from fairly simple 'go–no–go' or 'good–bad' indicting testers to those which incorporate programmed testing. The more commonly used testers, which are often battery-operated provide for the measurement of opens, shorts, DC beta (DC current gain) and facilities for measurement of AC current gain and of leakage currents at various junctions. Some testers utilize a meter as the indicator of various test result values, others simply have a neon-lamp indicator.

It is often easy to construct your own transistor tester particularly for the measurement of DC current gain (DC beta) and AC current gain (AC beta)

$$\beta = \frac{I_c}{I_b} = \frac{\text{collector current}}{\text{base current}}$$

To measure DC beta, a known current is circulated in the input or base circuit. This causes DC current to flow in the output or collector circuit. Beta will be the collector current divided by the base current. Referring to Fig. 8.18 for making beta test, R_2 is adjusted until 5 mA current flows in the input circuit as indicated by millimeter M_1. The current in the meter M_2 will be the collector current. If the collector current is 500 mA, then beta would be 100. This procedure is valid only if the leakage current is small otherwise the leakage current must be first measured with zero base current and subtracted from the collector current when calculating beta.

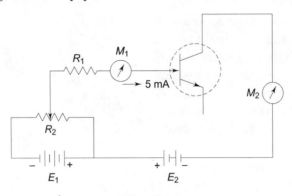

Fig. 8.18   *DC beta test circuit*

AC beta measurement is done with a circuit shown in Fig. 8.19.

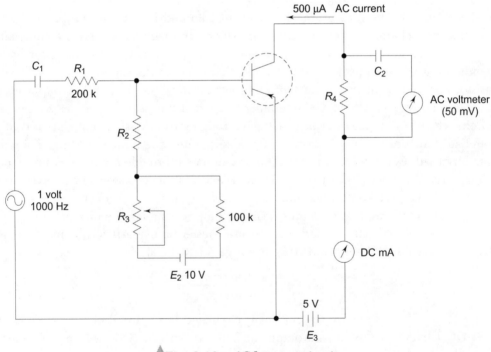

Fig. 8.19   *AC beta test circuit*

A small AC signal is obtained from an oscillator, say 1V amplitude and 1000 Hz frequency. Assuming input resistance of the transistor under test to be small compared to 200 K resistor (R_1), the AC base current will be

$$\frac{1}{200 \times 10^{-3}} = 5\ \mu A$$

DC biasing for the transistor is obtained with battery E_2 and resistors R_2 and R_3. This causes DC quiescent current flow in the collector emitter circuit, indicated by the DC millimeter which may be normally kept as 1 mA.

If the β of the transistor is 100, then the AC collector current will be $100 \times 5 = 500 \ \mu A$. The AC voltage drop across R_4 (100Ω) will be $100 \times 500 \ \mu A = 50 \ mV$.

If any AC voltmeter whose full scale sensivity is 50 mV is connected across R_4, a full scale reading will be indicated by a beta of 100. The voltmeter is then calibrated in terms of beta 0–100.

Normally, the difference between an AC and DC beta test should not exist. However, since most transistors are non-linear, particularly at low and high levels of collector current, discrepancies are found to exist between the two values. Therefore, a small signal beta test will be more representative of the actual gain of the transistor, particularly if the tests are made under typical operating conditions.

Curve Tracer for Testing Transistors: Curve tracers are oscilloscopes used as *X–Y* plotters, wherein conventionally *X* represents the voltage across, and *Y* the current in the device under test. The device may be passive—say a resistor, diode or thermistor or it may be active. However, it is most useful to make quick checks of the operation of small signal semiconductor devices, particularly transistors and FETs. For convenience, the source of voltage or current is built into the instrument. For active devices, an input drive signal is provided which moves the device in steps from a no-input to a high-input condition, so that a family of characteristic curves may be plotted. Figure 8.20 shows a block diagram for testing a transistor with the curve tracer. A curve tracer usually has two sections:

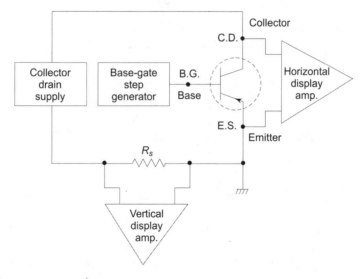

Fig. 8.20 *Curve tracer schematic*

(a) A stimulus section which provides voltages and currents for application to the device under test. This section consists of the collector (drain) supply and the base (gate) step generator.

(b) A measurement section which measures the effects of stimulus. This section consists of the vertical and horizontal display amplifiers.

The collector (drain) supply produces a voltage which is connected to the collector of a bi-polar transistor, the drain of an FET or lead of a diode. The voltage can be either a sweeping voltage or a DC voltage. The sweeping voltage has a triangular wave shape.

The base (gate) step generator produces current steps for application to the base of a bi-polar transistor or voltage steps for application to the gate of a FET. The steps occur at a rate of one step per cycle of the collector (drain) supply.

The vertical display amplifier measures current for display on the vertical axis of the CRT. A resistor in the return path to the collector (drain) supply is used to sense current.

The horizontal display amplifier measures voltage for display on the horizontal axis of the CRT. The voltage measured is V_{CE} for bi-polar transistor, V_{DS} for FET or an anode-cathode voltage for a diode.

The curve tracers are usually available as plug-in units with standard oscilloscope main frame configurations. The resulting display is shown in Fig. 8.21.

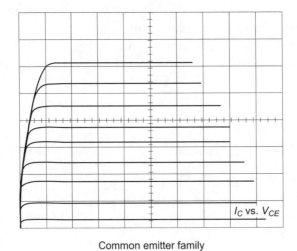

I_C vs. V_{CE}

Common emitter family

Fig. 8.21   *Voltage-current characteristics of a bi-polar transistor*

Curve tracers are usually provided with special sockets to accommodate various lead configurations. They also have an additional DC supply to power such devices as ICs, polarity change switches, calibrated DC offsets and a wide range of accurate vertical and horizontal deflection sensitivities. These measurements are quite helpful for maintenance engineers.

Substitution of Transistors: When an exact replacement for a transistor is not available and you are considering a transistor type for substitution for a particular application in a circuit, the following are important considerations:

(a) Identify whether it is *npn* (using positive supply rail) or *pnp* (using negative supply). Any replacement must be of the same polarity, i.e. *npn* or *pnp*.

(b) It is prudent to replace germanium with germanium and silicon with silicon transistors. This is because leakage current in germanium devices are orders of magnitude higher than in silicon. However, it is usually possible to select a silicon device to substitute for an unobtainable germanium one. This dos not usually work the other way.

(c) Many differently shaped cases are found in transistors. You must know details of the case outline of the transistor intended to be used. Transistor manuals or selectors usually provide such details as terminal identification diagrams.

(d) The transistor is basically a low voltage device. Also, it is catastrophically sensitive to current overloads. Therefore, it is important to check the permissible maximum voltage and collector current to ensure that it is not exceeded under any circumstances.

(e) One of the most important characteristics of a transistor is the current gain (beta, β) H_{FE}. It is necessary to know the current gain for a particular dc bias network for the device. Transistor data sheets usually provide information about the bias current at which the gain is measured.

(f) Power dissipation and high frequency characteristics are the other two important parameters which should be considered before attempting a replacement.

Transistor Type Numbers: Over the years, manufacturers have issued somewhere 50,000 to 100,000 separate transistor type numbers.

A vast majority of these, are, however, no longer in use. Some of the commonly used type numbers are:

1. Joint Electron Device Engineering Council (JEDEC)

The transistor markings in this case take the following form:

Digit, letter, serial number, [suffix]

Where the letter is always 'N', the serial number runs from 100 to 9999 and tells nothing about the transistor except its approximate time of introduction. The (optional) suffix indicates the gain (h_{fe}) group of the device. For example:

A = low gain	B = medium gain
C = high gain	No suffix = ungrouped (any gain)

Examples: 2N904, 2N3819, 2N2221A

The data sheets give information on the actual gain spread and groupings. The reason for gain grouping is that the low gain devices are usually cheaper than the high gain devices, resulting in savings for high volume users.

2. Japanese Industrial Standard (JIS)

These take the following form:

Digit, two letters, serial number, [suffix]

The letters indicate the application area according to the following code:

SA = PNP HF transistor	SB = PNP AF transistor
SC = NPN HF transistor	SD = NPN AF transistor
SJ = P-channel FET/MOSFET	SK = N-channel FET/MOSFET

The serial number runs from 10 to 9999.

The (optional) suffix indicates that the type is approved for use by various Japanese organizations. Since the code for transistor always begins with 2S, it is sometimes omitted. For example: a 2SC733 would be marked C733. The typical examples of JIS based transistor markings are:

2SA1187, 2SB646, 2SC733

3. Pro-Electron System

This European system adopts the following form:

Two letters, (letter), serial number, (suffix).

The first letter indicates the material as follows:

A = Germanium(Ge)	B = Silicon (Si)
C = Gallium Arsenide (GaAs)	R = Compound materials

The biggest majority of transistors are of silicon, and therefore, begin with a B. The second letter indicates the device application:

C = Transistor, AF, small signal	D = Transistor, AF, power
F = Transistor, HF, small signal	L = Transistor, HF, power
U = Transistor, power, switching	

The third letter indicates that the device is intended for industrial or professional rather than commercial applications. It is usually a W, X, Y or Z. The serial number runs from 100–9999. The suffix indicates the gain grouping, as for JEDEC.

Examples: BC108A, BAW68, BF239, BFY51.

4. Old Standards

Some of the old numbers use OC or OD followed by two or three numerals (e.g. OC28) or CV numbers (UK) like CV7. They are no longer used with modern transistors.

5. Manufacturer's Codes

Apart from the above, manufacturers often introduce their own types, for commercial reasons, or to emphasize that the range belongs to a special application. Some common brand specific prefixes are:

TIS = Texas Instruments, small signal transistor (plastic case)

TIP = Texas Instruments, power transistor (plastic case)

MPS = Motorola, low power transistor (plastic case)

MRF = Motorola, HF, VHF and microwave transistor

RCA = RCA

Consulting Data Books on Transistors Transistor data books usually provide information on transistor characteristics on the following points (Table 8.1).

Table 8.1 ▪ *Transistor Characteristics*

Type No.	2N 869	
Polarity and Material PS		$N = npn$ $P = pnp$ $G =$ Germanium, $S =$ Silicon
Case outline	TO 18	Different case outlines for different transistors
Lead identification	LO 1	Lead details provided with the base diagram
V_{CBO} Max	25 V	Maximum permissible collector base voltage with emitter open circuit
V_{CEO} Max	18 V	Maximum permissible collector emitter voltage with base open circuit
V_{EBO} Max	5 V	Maximum permissible emitter base voltage with collector open circuit.
I_c Max	100 mA	Maximum permissible collector current
T_j Max	175°C	Maximum permissible junction temperature
P_{TOT} Max	360 mWF	Maximum permissible device dissipation F= in free air at 25°C. C=with the case surface held at 25°C, H = in free air at 25°C with metal heat sink attached to device.
F_T Min	100 M	Minimum frequency cut-off indicated in K=kilo Hertz, M= MHz, G=GHz F_T=frequency at which common emitter current gain drops to unity

(Contd.)

Type No.	2N 869	
$C_{OB \text{ Max}}$	9 pF	Maximum collector capacitance normally with emitter open circuit
H_{fe}	20 MN	Current gain (normally dc) MN = minimum
$I_c (h_{fe} \text{ bias})$	10 mA	Bias current gain (normally dc) MN = minimum
Use	ALG	Application usage A = audio, L = low current G = general purpose
Supplier	SGI	Possible supplier (abbreviated)

8.4.4   Field-effect Transistors

Field-effect transistors, like bi-polar transistors, have three terminals. They are designated as: source, drain and gate (Fig. 8.22) which corresponds in function to emitter, collector and base of junction transistors. Source and drain leads are attached to the same block (channel of n or p semiconductor material). A band of oppositely doped material around the channel (between the source and drain leads) is connected to the gate lead.

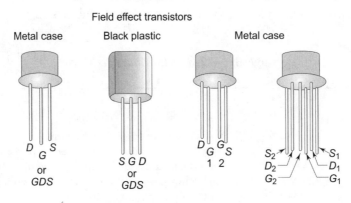

Fig. 8.22   *FET packages and terminals*

In normal junction FET operation, the gate source voltage reverse-biases the pn junction, causing an electric field that creates a depletion region in the source-drain channel. In the depletion region, the number of available current carriers is reduced as the reverse-biasing voltage increases, making source drain current a function of gate-source voltage. With the input (gate-source) circuit reverse-biased, the FET presents a high impedance to its signal source. This is in contrast to low impedance of the forward-biased junction bi-polar transistor base-emitter circuit. Because there is no input current, FETs have less noise than junction transistors. Figure 8.23 shows the schematic symbol and biasing for n-channel and p-channel depletion mode field-effect transistors. Figure 8.24 shows FET amplifier characteristics.

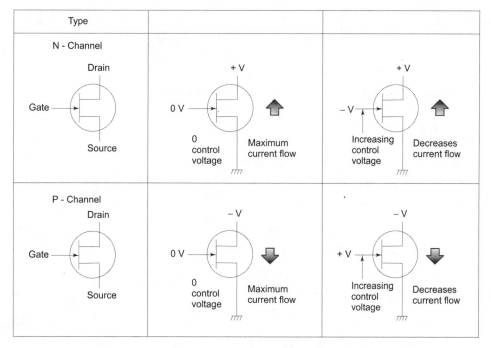

Fig. 8.23   *Biasing arrangement in field-effect transistors*

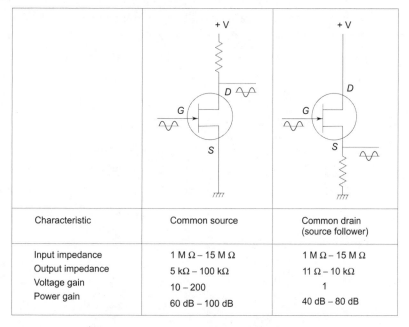

Characteristic	Common source	Common drain (source follower)
Input impedance	1 M Ω – 15 M Ω	1 M Ω – 15 M Ω
Output impedance	5 kΩ – 100 kΩ	11 Ω – 10 kΩ
Voltage gain	10 – 200	1
Power gain	60 dB – 100 dB	40 dB – 80 dB

Fig. 8.24   *FET amplifier characteristics*

Conversely, most MOSFET transistors, including those in the CMOS integrated circuits are 'Enhancement Mode' type devices. With zero gate-to-source bias, these devices are off, and are increasingly turned on by the application of increasing gate-to-source bias (positive for *n*-channel, negative for *p*-channel).

There are three different types of field-effect transistors:

(a) Junction gate;

(b) Insulated gate (non-enhanced type); and

(c) Insulated gate (enhanced type).

Each type comes with either an *n*-channel or a *p*-channel.

Testing field-effect transistors involves slightly greater care and more complex procedure than for bi-polar transistors. Before the FET is tested, the following need to be known:

(a) Whether the device is junction field-effect transistor (JFET) or metal-oxide semiconductor field-effect transistor (MOSFET);

(b) Is the FET an *n*-channel or *p*-channel type?

(c) In case of MOSFET, is it an enhancement type or a depletion type?

Junction gate and non-enhanced type insulated gate FETs are basically 'ON' devices like vacuum tubes. These two devices must be biased-off. On the contrary, the enhanced type insulated gate FET is basically an 'OFF' device and must be biased on.

A junction FET can be checked with an ohmmeter. Figure 8.25 shows junction resistance readings of a JFET. The forward and reverse readings occur between the gate and the source or between the gate and drain only. The resistance between the source and the drain is the same irrespective of the ohmmeter polarity. It may be remembered that the

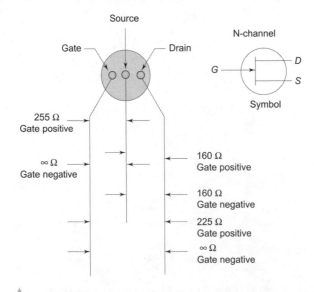

Fig. 8.25 *Junction resistance readings of a JFET*

gate source and gate drain junctions are non-linear and the resistance values will change depending on the range used.

An ammeter can also be used to conveniently check that a junction field effect transistor is able to function correctly. The circuit of Fig. 8.26(a) may be used for an *n*-channel device and that of Fig. 8.26(b) for a *p*-channel device.

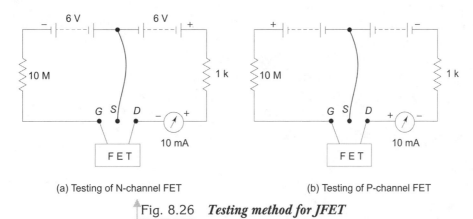

(a) Testing of N-channel FET   (b) Testing of P-channel FET

Fig. 8.26   *Testing method for JFET*

If the gate is initially connected directly to the source (and not as shown) it will be found that the meter provides a reading of a few mA. This current is limited by the 1 K resistor in the drain circuit to a safe value.

If the gate electrode is now connected to the 10 M resistor as shown, the gate to the channel junction is reverse-biased. Thus the channel width decreases and with most devices, the drain current will fall to zero in the circuits shown. As the gate circuit has a very high resistance, voltage can be applied to it through a high value resistor (10 M).

If one wishes to test a device and does not know the connections, one can first find two connections in which a small current will pass in either direction. These are the source and drain connections. A current should pass from the third electrode (the gate) only in one direction to either of the two electrodes. If conduction takes place when the gate is positive, one has an *n*-channel device, whereas if conduction takes place when the gate is negative, the device is of the *p*-channel polarity. One cannot, however, easily determine which electrode is the drain and which is the source, but these electrodes are, to some extent, electrically inter-changeable.

It may be noted that external leakage paths-moisture build-up in soldering fluxes should be watched closely. Such paths make a FET look defective when it is not. Thus, the result may be an external (to FET) leakage path. Therefore, before changing FETs for leakage thoroughly clean the PC board around the FET. The typewriter erasers do an excellent job.

A junction FET or an insulated-gate protected MOSFET can be removed from or inserted into a circuit without any special precautions other than guarding against overheating during soldering and desoldering operations. However, in the case of insulated gate FET, it may be remembered that insulation is between gate and channel and which is in fact a delicate capacitor.

The insulation is so thin and the gate is so small that it can be easily ruined (Fig. 8.27). It is necessary to watch out for the static charge from fluorescent lights if the gate lead is left open. This can happen before the FET is installed, if after removing from its insulated case, is left on a table with its shorting wire removed. A shorting wire is usually a small piece of wire wrapped around all the leads. The following precautions may be observed while handling MOSFETs:

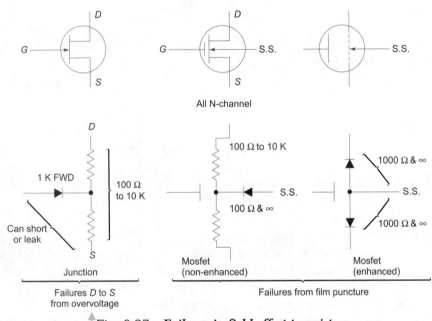

Fig. 8.27   *Failures in field-effect transistors*

(a) Before the MOSFET is inserted into the circuit, the device leads should be kept shorted.
(b) The user should ensure that his body and hand are at the ground potential.
(c) Soldering gun tips should be grounded during soldering procedures.
(d) MOSFETs should never be inserted into or removed from their circuits with the power on.

The ohmmeter can be used to check the MOSFET. The gate-to-source insulation is checked with the highest range of the ohmmeter. The practically infinite reading is normally obtained for both polarities of test voltage. The drain-to-source resistance normally has a comparatively low value and the ohmmeter readings are the same regardless of test voltage polarity. It may be noted that the usual failure mode in a MOSFET is short between the gate and source and the drain and source. In other words, everything is connected together.

Insulated Gate Bipolar Transistor (IGBT)

Prior to the development of IGBTs, power MOSFETs were used in medium or low voltage applications which require fast switching, whereas bi-polar power transistors, and thyristors were used in medium to high voltage applications which require high current conduction. A power MOSFET allows for simple gate control circuit design and has excellent fast switching capability. On the other hand, at 200 V or higher, it has the disadvantage of rapidly increasing on-resistance as the break-down voltage increases. The bi-polar power transistor has excellent on-state characteristics due to the low forward voltage drop, but its base control circuit is complex and fast switching operation is difficult as compared with the MOSFET. The IGBT has the combined advantages of the above two devices.

The IGBT structure is a combination of the power MOSFET and a bi-polar power transistor as shown in Fig. 8.28. The input has a MOS gate structure, and the output is a wide base PNP transistor. The base drive current for the PNP transistor is felt through the input channel. Besides the PNP transistor, there is an NPN transistor, which is designed to be inactivated by shorting the base and the emitter to the MOSFET source metal. The four layers of PNPN, which comprises the PNP transistor and the NPN transistor, form a thyristor structure, which causes the possibility of a latch-up. Unlike the power MOSFET, it does not have an integral reverse diode that exists

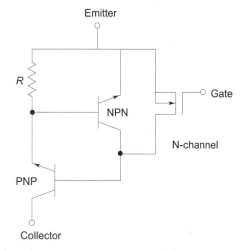

Fig. 8.28   ***Structure of IGBT (Insulated Gate Bi-polar Transistor)***

parasitically, and because of this it needs to be connected with the appropriate fast recovery diode when needed.

8.4.5 Thyristors

Thyristor is the generic name for the solid state devices that have electrical characteristics similar to that of the thyratron. The three types of thyristors which are widely used are:
  (a) Silicon controlled rectifier (reverse blocking triode thyristor);
  (b) Triac (bi-directional triode thyristor); and
  (c) Four-terminal thyristor (bilateral switch).

Thyristors are used extensively in power control circuits. They are particularly suited for AC power control applications such as lamp dimmers, motor speed control, temperature control and invertors. They are also employed for over-voltage protection in DC power supplies.

The thyristor is basically a four layer *pnpn* device (Fig. 8.29) and can be represented as a two-transistor combination structure. The two transistors are cross-connected—one is

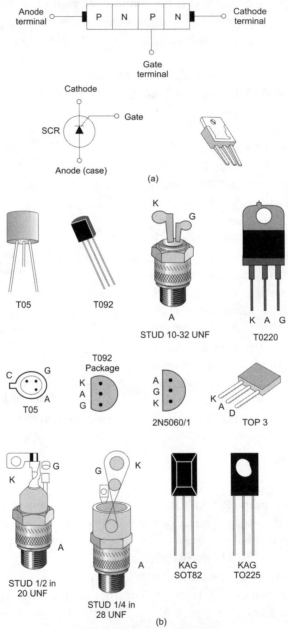

Fig. 8.29   *(a) Silicon-controlled rectifier, device structure and symbol; and (b) Different forms of SCR packages*

NPN and the other is PNP. The base of the NPN is connected to the collector of the PNP and base of the PNP is connected to the collector of the NPN. The device is normally off, but a trigger pulse at the gate switches the thyristor from a non-conducting state into a low resistance forward conducting state. Once triggered in conduction, the thyristor remains on unless the current flowing through it is reduced below the holding current value or it is reverse-biased. This means that the thyristor has extremely non-linear voltage-current characteristics (Fig. 8.30).

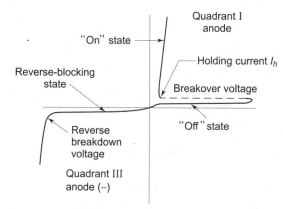

Fig. 8.30   *Voltage-current characteristics of SCR*

The triac is similar in operation to two thyristors connected in reverse-parallel, but with a common gate connection. This means that the device can pass or block current in both directions. Obviously, it can be triggered into conduction in either direction by applying either positive or negative gate signals.

For a motor speed control or light dimmer, the exact time when the thyristor is triggered relative to the zero crossings of the AC power is used to determine the power level. Trigger the thyristor early in the cycle and the load is delivered high power. Similarly, if the thyristor is triggered late in the cycle, only a small amount of power is delivered to the load. The advantage of thyristors over simple variable resistors is that they dissipate very little power as they are either fully 'on' or fully 'off'.

Excessive temperature or high rate of temperature cycling can result in the failure of thyristors and triacs. The failure due to these reasons is mostly manifested in the complete failure of the device, which can be typically open circuit or short circuit between terminals. An ohmmeter can be used to check for these failures. For example, the gate-cathode of a diode. With the gate positive with respect to the cathode, a low resistance (typically below 100 Ω) should be indicated. On the other hand, with the gate negative with respect to the cathode, a high resistance (greater than 100 kΩ) will be indicated. A high resistance is indicated in either direction for the anode to cathode connections. It is advisable to unsolder the device before making measurements with an ohmmeter.

- For SCRs, the gate to the cathode should test like a diode (which it is) on a multimeter. The anode to cathode and gate to anode junctions should read open.
- For triacs (Figs 8.31(a) and (b)), the gate to main terminal (T_2) should test like a diode junction in both directions. T_1 to T_2 and gate to T_1 junctions should read open.
- For diacs (Fig. 8.31(c)), there is no gate terminal. Resistance should be infinite in both directions. With a VOM, they can be tested only for shorts.

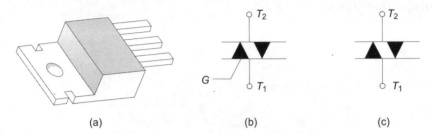

Fig. 8.31 *Triac: (a) Package; (b) Symbolic representation; and (c) Diac symbolic representation*

Partial failures in thyristors result in erratic triggering due to impaired gate sensitively. To check this, the thyristor should be switched on and the voltage measured between anode and cathode, which should be approximately 1V and the voltage between gate and cathode should be 0.7V. In case of faulty conditions, the following will be observed:

(a) *Anode to cathode short circuit:* Thyristor conducts in both directions with no gate signal applied. Voltage between anode and cathode, when measured, will be zero volts.

(b) *Anode to cathode open circuit:* Thyristor does not conduct, no current flows from anode to cathode, voltage between anode and cathode is always high.

9

Linear Integrated Circuits

9.1 LINEAR INTEGRATED CIRCUITS

Linear integrated circuits are characterized by an output that is proportional to its inputs. Such circuits are designed as DC amplifiers, audio amplifiers, RF amplifiers, IF amplifiers, power amplifiers, differential amplifiers, analog multiplexers and comparators, etc.

An important class of linear integrated circuits is operational amplifiers (op-amps). These amplifiers were originally utilized in analog computers to perform various mathematical operations such as addition, subtraction, integration and differentiation. Op-amps are now used to perform precise circuit functions some of which will be discussed in this chapter. There are over 2500 types of commercially available op-amps. Most are low power devices with a power dissipation of IW. However, they differ according to their voltage gain, temperature range, noise level and other characteristics.

9.2 OPERATIONAL AMPLIFIERS (OP-AMP)

An operational amplifier is a complete amplifier circuit constructed as an integrated circuit on a single silicon clip. Inside, it contains a number of transistors and other components packaged into a single functional unit. The op-amp is available as an off-the-shelf item. For troubleshooting, op-amp is treated as a single device rather than attempting to dissect it into its constituent components that go on to make the chip.

An operational amplifier has a balanced arrangement in the input. It is characterized by extremely high (DC-static) and low frequency gain, a very high input impedance, a low closed loop output impedance and a fairly uniform roll-off in gain with frequency over many decades. The linear roll-off characteristic of an operational amplifier gives it the universality and the ability to accept feedback from a wide variety of feedback networks with excellent dynamic stability. The particular application of an op-amp is obviously determined by the device and its external circuit connections.

9.2.1 Symbolic Representation

The op-amp is symbolically represented as a triangle (Fig. 9.1) on its side. In digital circuit symbols, the inverter is represented as a triangle, but the op-amp symbol is much larger. The triangle indicates the direction of signal flow. It is associated with three

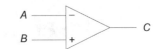

Fig. 9.1 *Symbolic representation of an op-amp*

horizontal lines, two of which (A and B) indicate the signal input and the third (C), the output signal connections.

The signal input terminals are described by minus (–) and plus (+) signs inside the triangle. The (–) input is called the inverting input, because the output voltage is 180 degrees out of phase with the voltage to this input. On the other hand, the (+) input is called the non-inverting input because the output voltage is in phase with this voltage applied to this terminal. The names inverting and non-inverting terminals have been given to indicate the phase of output signal in relation to the voltage applied at the inputs. Figure 9.2 shows the operation of the op-amp as inverting and non-inverting amplifier.

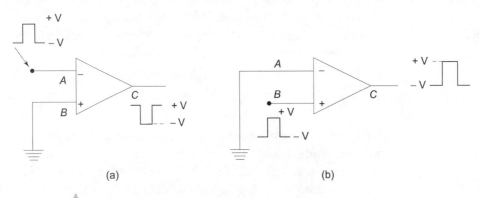

(a)   (b)

Fig. 9.2 *(a) Inverting operation of an op-amp; and (b) Non-inverting operation of an op-amp*

Operational amplifiers are available both in metal as well as dual-in-line epoxy packages. Sockets for both the types of packages are available and are mostly used in the printed circuit boards, though the IC can be directly soldered on the board. The numbering convention for the pin numbers on the IC and the socket are shown in Fig. 9.3a.

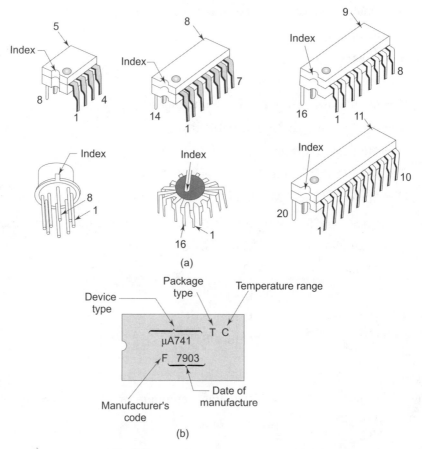

Fig. 9.3 *(a) Commonly available linear IC packages; and (b) Op-amp identification*

9.2.2 Op-amp Identification

In general, the op-amps carry the following three types (Fig. 9.3b) of information on the pack:

uA 741	T
Device type	Package type

C
Temperature

Device Type: This group of alpha-numeric characters defines the data sheet which specifies the functional and electrical characteristics of the device.

Package Type: One letter represents the basic package style:

D = Dual in-line package (Hermetic, ceramic)

F = Flat pack

H = Metal can package

J = Metal power package (TO-66 outline)

K = Metal power package (TO-3 outline)

P = Dual-in line package (Moulded)

R = Mini DIP (Hermetic, ceramic)

T = Mini DIP (Moulded)

U = Power package (Moulded, TO-220 outline)

Temperature Range: The following three basic temperature grades are in common use:

C = Commercial	0°C to + 70°C	
M = Military	−55°C to + 125°C	
	55°C to + 85°C	
V = Industrial	−20°C to + 85°C	
	−40°C to + 85°C	

Examples

(a) uA 710 FM

This code indicates a uA 710 voltage comparator in a flat pack with military temperature rating.

(b) uA 725 HC

This number code indicates a uA 725 instrumentation operational amplifier, in a metal can with a commercial temperature rating capability.

In addition, the year of manufacture batch number and manufacturer's identification are also given on the device.

9.2.3   Power Supply Requirements for Op-amps

Op-amps need to be powered with DC power supply, like any other transistor amplifier. The power supply should be of proper voltage regulation and filtering for correct operation of the op-amp.

The power supply leads on the op-amp are marked + V and −V to which positive and negative supply voltages should be connected respectively, with reference to the ground. The positive and negative supply voltages are usually symmetrical, i.e. the two voltages are equal but opposite in sign. The most commonly used voltage to power op-amps are +15 V and –15 V. However, this is not always the case. Therefore, it is advisable to consult manufacturer's data manuals on the op-amp of interest to determine the power supply requirements.

It may be noted that mostly on the circuit schematics, the power supply leads are not shown on the op-amps. It is assumed that the reader is aware that DC voltage is necessary for operation of the op-amp.

9.2.4 Output Voltage Swing

Just as standard transistor amplifiers are limited in their output voltage swing, so also the op-amp has a limited voltage swing. The limitations on the output voltage swing are generally dependent upon the magnitude of the positive and negative DC power supply voltages of the op-amp. Usually, the output of an op-amp can go to a voltage value no more positive than the positive and no more negative than the negative power supply can provide. When the output of an op-amp goes to +V and –V, it is said to be in positive or negative voltage saturation.

9.2.5 Output Current

Op-amps are designed to provide a limited current in the output which is usually a few milliamperes, less than ten milliamperes in most standard op-amps. If more current is drawn, the output signal begins to change because of the current limiting provisions built into the op-amp output circuit which limit their own output current to a safe operating region.

Some op-amps are designed to deliver larger currents, of the order of amperes, at the output pin, but they can only be characterized as special devices and not as standard op-amps.

9.3 CHARACTERISTICS OF OP-AMPS

An ideal op-amp would have the following characteristics:
  (a) Infinite open-loop voltage gain (A);
  (b) Infinite bandwidth;
  (c) Infinite input impedance;
  (d) Zero output impedance;
  (e) Zero offset (voltage at output when input is zero); and
  (f) Maximum output voltage equals +V and minimum –V.

By carefully examining these characteristics, the following implications are obvious:

(a) The voltage across the input terminals of an amplifier with infinite gain must be zero or negligibly small (input voltage = output voltage/gain of amplifier).

(b) No (zero) current can flow between the input terminals of the amplifier because of the infinite input impedance (Input current = input voltage/input impedance). If one input terminal is at the ground reference potential, then the other must be at the ground potential. The terminal is thus called a 'virtual ground' and the input is called a 'virtual short circuit'. (In virtual short circuit, no potential can exist across it, and no current can flow through it.)

(c) With the infinite voltage gain, we can expect to get a very large voltage output from a very small voltage input. In fact, with a small voltage across the input terminals, the amplifier output is driven into positive or negative saturation very easily. An op-amp is said to be operating properly when the output is in the linear region of operation, unless the op-amp circuit is designed to operate to perform non-linear function.

9.4   TYPICAL OP-AMP CIRCUITS

Before the procedure for troubleshooting op-amps is outlined, it will be worthwhile to consider various circuit configurations in which the op-amp is applied to perform different functions such as precision amplifier, summing device, integration, differentiation, generator, active filter, comparator, etc.

An operational amplifier is normally used with a negative feedback circuit, and the performance of the circuits is then primarily determined by the magnitude of the external components connected to the amplifier. Examples of the basic operational amplifier configurations using different feed-back elements are as follows:

Inverting Amplifier (Fig. 9.4)

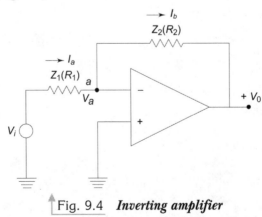

Fig. 9.4   *Inverting amplifier*

$$\text{Gain A} = \frac{V_o}{V_t} = -\frac{R_2}{R_1}$$

where A is the closed loop gain of the amplifier.

This is an important fact that the gain of the inverting op-amp depends only on the feedback-path resistors R_2 and R_1. The minus sign indicates that the output signal is 180° out of phase with the input signal.

When the feedback circuit comprises elements other than resistors, the closed-loop gain of the ideal amplifier is

$$A = -\frac{Z_2}{Z_1}$$

where Z_2 and Z_1 are the impedances of the two feedback elements.

If $\qquad\qquad R_2 = 20\text{k}, R_1 = 2\text{k},$

then

$$A = 10$$

Non-inverting Amplifier (Fig. 9.5): The closed loop gain of a non-inverting amplifier is given by

$$A = 1 + \frac{R_2}{R_1}$$

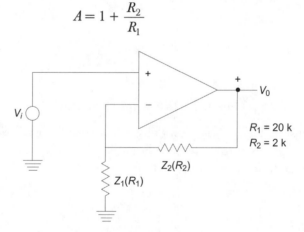

Fig. 9.5 *Non-inverting amplifier*

If R_2 is 20 k and $R_1 = 2$ k, then

$$A = 1 + \frac{20}{2} = 11$$

Voltage Follower: The voltage follower circuit is shown in Fig. 9.6.

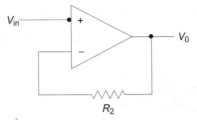

Fig. 9.6   *Voltage amplifier*

Differential Amplifier: A differential amplifier (Fig. 9.7) is one that is sensitive only to the difference between the two input voltages applied to its input terminals. The voltage output in this circuit is

$$V_o = (V_1 - V_2)\,\frac{R_2}{R_1}$$

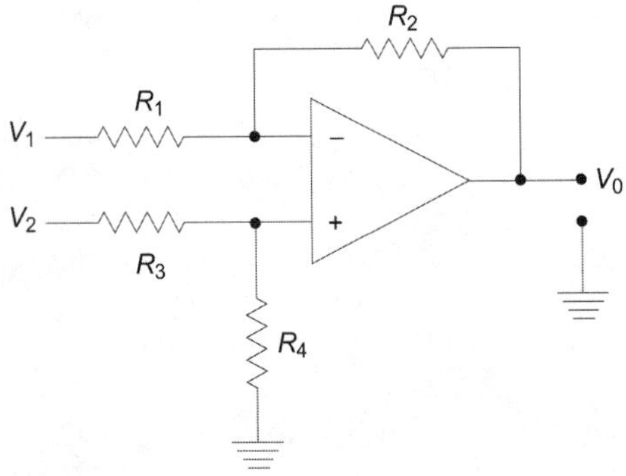

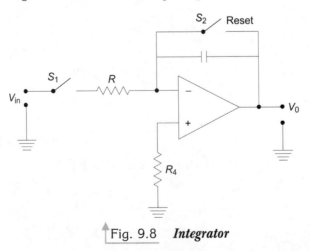

Fig. 9.7   *Differential amplifier*

Integrator: An integrator (Fig. 9.8) is essentially a low-pass filter whose output is proportional to the product of the amplitude and duration of the input, i.e. it encompasses the area under the voltage time curve of the output signal.

Fig. 9.8   *Integrator*

In the integrator circuit shown, voltage output

$$V_o = -\frac{1}{RC} \int V_t \, dt$$

Differentiator: A differentiator is a rate-of-change circuit—it produces an output signal proportional to the rate of change of its input signal.

The operation of the basic differentiator is obvious (Fig. 9.9). *C* and *R* form a divider whose output depends on the relative values of *X* (the reactance of the capacitor) and *R*. The voltage output,

$$V_o = -RC \frac{dVi}{dt}$$

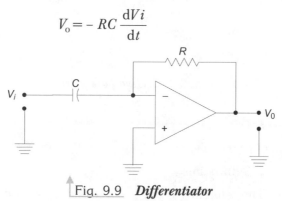

Fig. 9.9 **Differentiator**

Active Filters: The op-amp is widely used as an active filter because reasonably valued capacitors and resistors may be used to produce active filters operating at frequencies as low as 0.01 Hz. With an op-amp, standard filters can be conveniently implemented.

Figure 9.10 shows a second-order low pass filter which will start to roll-off at 12dB/octave (40 dB/inter decade). The high pass filter is implemented simply by interchanging the capacitors and resistors in the (+) channel (Fig. 9.11).

The circuit shown in Fig. 9.12 is a multiple-feedback bandpass filter. The filter bandwidth (sharpness) is a function of its *Q*.

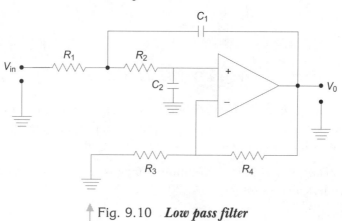

Fig. 9.10 **Low pass filter**

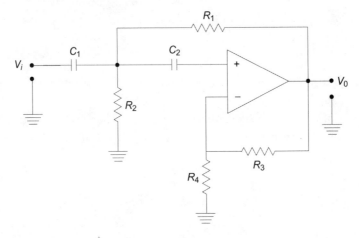

Fig. 9.11   **High pass filter**

Bandwidth $BW = f_o/Q$

where f_o = resonant frequency

BW = bandwidth at 3 dB points

Reference Voltage Source: The reference voltage source is illustrated in Fig. 9.13.

Negative Reference Voltage Source: The negative reference voltage is shown in Fig. 9.14.

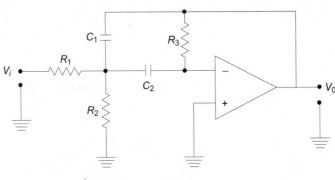

Fig. 9.12   **Band pass filter**

Voltage Regulator (Fig. 9.15): The regulated output voltage may be varied by varying R_2 or R_1. If the load current is more than the output capabilities of the op-amp, a transistor used as an emitter follower can be connected at the output.

Logarithmic Amplifier (Fig. 9.16): The output voltage is constant times the logarithmic of the input current less a constant, i.e.

$$V_o = K_3 \left[\log I_1 - K_4\right]$$

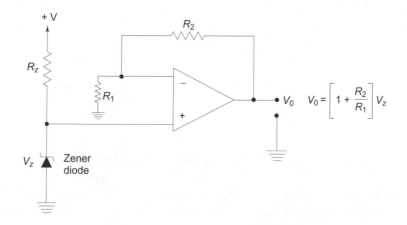

Fig. 9.13 *Reference voltage source*

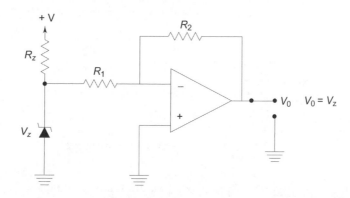

Fig. 9.14 *Negative reference voltage source*

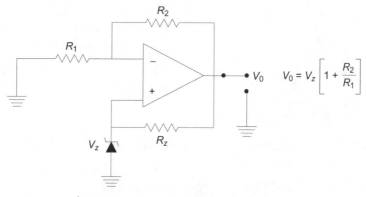

Fig. 9.15 *Voltage regulator circuit*

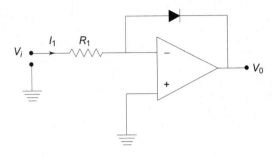

Fig. 9.16   *Logarithmic amplifier*

Precision Diode Circuit (Fig. 9.17): The precision diode circuit overcomes the problem of threshold voltage needed for forward conduction of the diode, thereby permitting the rectification of signals in the millivolt range.

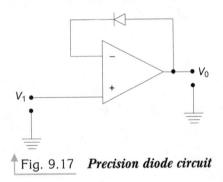

Fig. 9.17   *Precision diode circuit*

Peak Detector Circuit: Figure 9.18 shows a peak detector circuit.

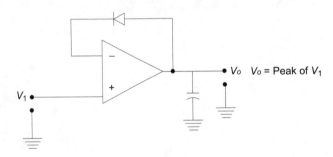

Fig. 9.18   *Peak detector circuit*

Precision Clamp: A precision clamp is shown in Fig. 9.19.

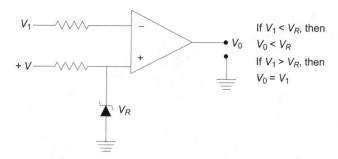

Fig. 9.19 ***Precision clamp circuit***

Comparator: The comparator circuit is used to switch at any given reference level. The output can be made to switch from plus saturation to minus saturation and vice versa with less than one millivolt change across its input (Fig. 9.20)

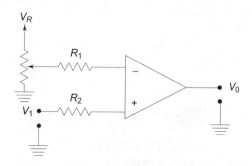

Fig. 9.20 ***Comparator circuit***

Divider Circuit: A divider circuit is shown in Fig. 9.21.

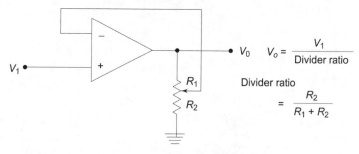

Fig. 9.21 ***Divider circuit***

Instrumentation Amplifier (Fig. 9.22): The gain is adjusted by varying R_1
If $R_2 = R_3$ and $R_4 = R_5 = R_6 = R_7$

Then
$$V_o = \left(1 + \frac{2R_2}{R_1}\right)(V_2 - V_1)$$

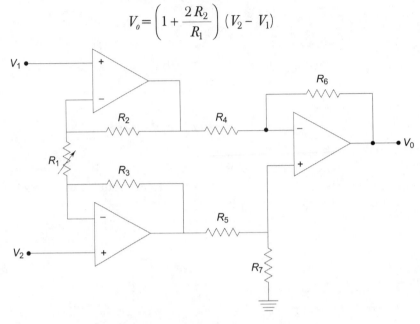

Fig. 9.22   *Instrumentation amplifier*

Precision Wide–Range Logarithmic Amplifier (Fig. 9.23): A logarithmic amplifier is an amplifier whose output is a logarithmic function of its input signal.

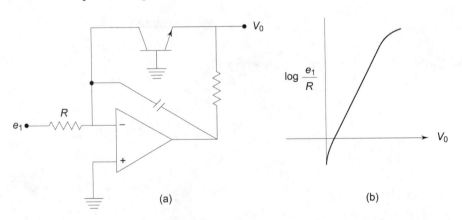

(a)   (b)

Fig. 9.23   *(a) Precision wide-range logarithmic amplifier; and*
(b) Characteristics of precision wide-range logarithmic amplifier

Charge Amplifier: A charge amplifier is shown in Fig. 9.24.

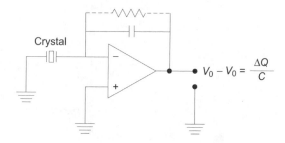

$$V_0 - V_0 = \frac{\Delta Q}{C}$$

Fig. 9.24   *Charge amplifier*

Current-to-Voltage Amplifier (Trans-resistance) (Fig. 9.25): Such an amplifier produces a certain output voltage for a certain input current.

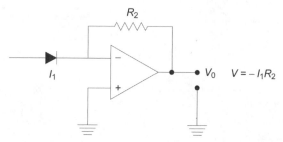

$$V = -I_1 R_2$$

Fig. 9.25   *Current-to-voltage amplifier (trans-resistance)*

Photo-cell Amplifier: Figure 9.26 illustrates a photo-cell amplifier.

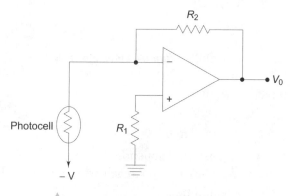

Fig. 9.26   *Photo-cell amplifier*

Voltage-to-Current Amplifier (Trans-conductance) (Fig. 9.27): Such an amplifier produces a certain output current for a certain input voltage.

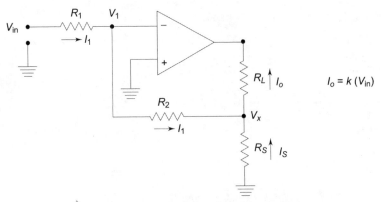

$$I_o = k\,(V_{in})$$

Fig. 9.27   *Voltage-to-current amplifier*

Square Wave Generator (Fig. 9.28): This circuit is a signal generator that generates a square wave output voltage.

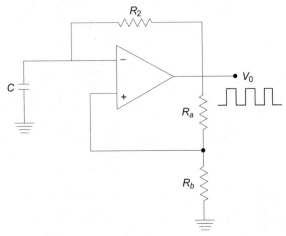

Fig. 9.28   *Square wave generator*

If β $(R_b/(R_a + R_b))$ is set at 0.473, then

$$f = \frac{1}{2\,R_2\,C}$$

The output of the square wave generator depends on the saturation voltages. If two zener diodes are added parallel to R_2, back to back, each with a zener voltage of V_z, then the amplitude of the square wave would be $\pm\,V_z$.

Sine Wave Generator: Figure 9.29 shows a simple Wien bridge oscillator circuit in which potentiometer R_a controls the amplitude of sinusoidal output. With $R_1 = R_2$ and $C_1 = C_2$, the frequency of oscillations is given by

$$f = \frac{1}{2\pi\,R_1C_1}$$

Diodes are used to reduce distortion in the output signal.

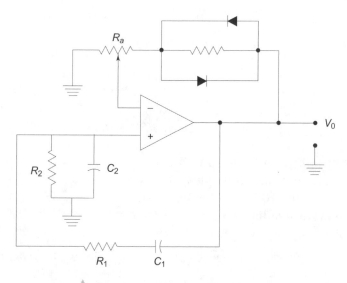

Fig. 9.29 **Sine wave generator**

The above illustrations have shown the op-amp in different circuit configurations depending upon the external circuit components. For troubleshooting such circuits, it is essential to first determine the type of output expected from the circuit. These illustrations should help to do so.

9.5 HOW TO CONSULT OP-AMP SPECIFICATION DATA BOOKS

There are presently in the market hundreds of different types of operational amplifiers with diverse characteristics and prices. In order to understand their use in a particular application or circuit, it is essential to be able to understand the information given in the specification sheets or data manuals. The important characteristics of operational amplifiers are as follows:

9.5.1 Major DC Parameters

Input Offset Voltage (Vos): The input voltage required to make the output voltage zero is called the input offset voltage (Vos) and is typically a millivolt or two (Fig. 9.30).

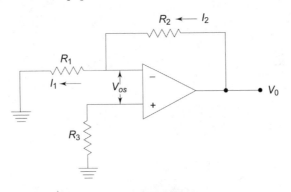

Fig. 9.30   *Input offset voltage*

Input Bias Current (I_B): In order for an op-amp to operate properly, it is necessary to supply a DC bias current (pA to uA) to the two input differential amplifier stages. The average of the two bias currents is called the input bias current. Referring to Fig. 9.31:

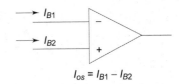

Fig. 9.31   *Input bias currents*

$$I_B = \frac{I_{B1} + I_{B2}}{2}$$

Input Offset Current (I_{os}) In the op-amp, the two input transistors are usually not perfectly matched, and a different bias current will be required by each stage. The difference between the two input bias currents is called the input offset current (Fig. 9.31).

$$I_{OS} = I_{B_1} - I_{B_2}$$

Voltage Gain: The gain of an op-amp as with any other amplifier, is the ratio of a change in the output voltage to change in the input voltage. The gain is usually specified in dB.

The symbol AVOL is used to indicate open loop voltage gain, the gain of the amplifier without feedback.

Common Mode Rejection Ratio (CMRR): Common mode rejection ratio is defined as the ratio of common mode voltage to differential mode voltage which gives the same output. Typical op-amps have 80–100 dB CMRR value.

Input Impedance: The input impedance of an op-amp consists of that impedance between the two input terminals in parallel with the impedances of each input and ground, as shown in Fig. 9.32a. The impedance between the inverting and non-inverting input is called the *differential input impedance* (R_{in}) The impedance between both inputs and ground is referred to as *common-mode impedance* (R_c). For Figure 9.32b:

$$R_{in} = 2R$$
$$R_c = 0.5\,R + R_2$$

Usually $R_2 \geq R,$

Therefore $R_c \geq R_{in}$

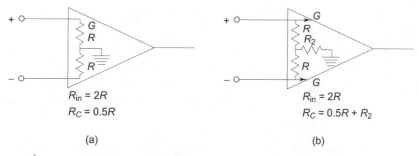

Fig. 9.32 **(a) Differential mode impedance**
(b) Common mode impedance of an op-amp

9.5.2 AC Parameters

Bandwidth:

(a) One bandwidth is the open loop 3 dB bandwidth which is defined as the frequency where the open-loop voltage gain (A) drops 3 dB from its DC (or low frequency) voltage gain. A drop of 3 dB means a power decrease of 50% which is equivalent to the voltage gain decreasing to 0.707 (or $1/\sqrt{2}$) of its original value. Figure shows a plot of open loop gain versus frequency for a typical op-amp.

(b) The open loop frequency response shown in Fig. 9.33 drops off at 20 dB/decade (6dB/octave) because the ordinate is volts. A 3 dB drop on this curve represents a

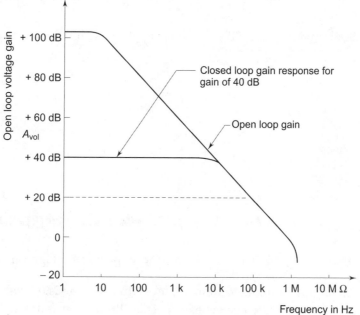

Fig. 9.33 *Typical op-amp frequency response*

drop to 0.707 of maximum value, a 6 dB voltage drop represents a drop to 0.5 value, a 20 dB voltage drop represents a drop to 0.1 value.

(c) Another bandwidth specification is the unity-gain bandwidth factor defined as the frequency at which the open-loop voltage gain reduces to unity.

Note that the 3-dB bandwidth is based on a ratio, whereas the unity-gain bandwidth refers to an absolute value of gain (unity). Thus, there is no direct relation between the two ways of expressing bandwidth.

Slew Rate: Slew rate is the maximum rate of change of output voltage when the amplifier is supplying full rated output. It is expressed in volts/microsecond (V/μs).

9.6   FAULTS DIAGNOSIS IN OP-AMP CIRCUITS

Modern integrated circuit operational amplifiers are highly reliable devices. Even then, it is possible for many different faults to develop inside the IC itself such as internal shorts or open circuits. Therefore, it is essential during fault diagnosis to rapidly isolate the fault to either the IC or one of the other components. External checks such as open circuit or shorted IC pins or tracks sometimes may suggest that the IC itself is defective, but before unsoldering the IC, it is advisable to check for these possibilities.

The techniques of fault diagnosis in op-amp circuits are based on the characteristics of the op-amp presented earlier and some experience and practice. Although representative application circuits have been illustrated, but even if one is not sure about how the circuit functions, it will still be possible to apply the following troubleshooting procedures with good success.

Let us consider an inverting amplifier circuit (Fig. 9.34) with 0.2 volts as input signal.

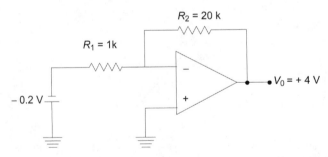

Fig. 9.34   *Typical inverting amplifier with a gain of 20 (R_2 / R_1)*

Measure the output voltage. Since $\dfrac{R_2}{R_1}$ is 20, the expected voltage output should be $0.2 \times 20 = 4$ volts. Supposing on measurement, the output voltage is 0.0 volts. This means that the circuit is not functionally correct as expected. Proceed as follows:

(a) Measure and record the voltage at both (–) inverting and (+) non-inverting pins. The level should not differ by more than ≈ 10 mV. If the voltage level is not within this value, check the external circuitry and components. It, however, does not necessarily mean that op-amp is not defective. The op-amp may be defective, but it is too early in the fault diagnosis process to conclude for certain.

(b) Measure the voltage on the (+) and (–) power supply terminals of the operational amplifier. The measurement should not be just measured on the PC (Printed Circuit) board, but also directly on the op-amp package. In case the supply is present on the PC board, but not on the IC pins, look for the following possible troubles:

   (1) The track on the PC board may be discontinuous or defective.

   (2) Feed-through hole on the PC board is not metallized the entire distance through the board.

   (3) If the device is in the socket, the socket may be defective.

If the voltage is not present on the PC board, start tracing the power supply voltage on the board to determine the cause of its not being present.

(c) If the power supply to the op-amp is correct, it is possible that the output circuit is being excessively loaded. It may be recalled that it was stated that the op-amps can deliver only 1–10 milliamperes to a load connected at its output. To rule out this possibility, isolate the output pin from other connections on the PC board and observe for the following possibilities:

   (1) Isolate just the feedback circuit as shown in Fig. 9.35 and measure the voltage at the output pin. The voltage should be +V or –V because the output will simply swing as far in voltage as the power supply permits. If the output voltage is still 0.0 volts when it is isolated from the feedback path, the op-amp is defective.

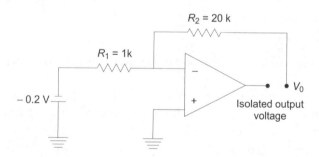

Fig. 9.35 *Op-amp circuit with isolated output of the op-amp*

   (2) If the output voltage is +V or –V, the device may still be defective. In that case, it is essential to determine if the op-amp output is capable of going to +V or –V voltage level as the output may be stuck at one voltage level in case the device is defective. To do so, take the following steps:

(i) Short the resistor R_1 by using a shorting lead. This will connect a DC voltage of –0.2 V on the inverting terminal of the op-amp. In this condition, the output should go to positive saturation.

(ii) In order to make the output go to the –V level, a positive DC signal is injected on the negative terminal (disconnect the shorting lead across the resistor R_1). The signal level should be such that it is safe to apply for the device under test. This information is available in the manufacturer's device data sheets. However, a few hundred millivolts applied at the input terminals will not damage the op-amp. The output will settle at some negative level when a positive signal is injected.

If the output does not reach a positive voltage or negative voltage saturation level as expected in the above two situations, then the device is defective. If the output does switching to + and – voltage levels, the op-amp is possibly good.

(d) To further establish the correctness of the device, isolate the output circuit, but the feedback resistor R_2 should be in the circuit (Fig. 9.36). This will isolate the output circuit completely and the op-amp is no longer overloaded. If the output voltage is now +4.0 volts (as was expected), the op-amp is good. If the op-amp is still not +4.0 volts, then the op-amp is defective.

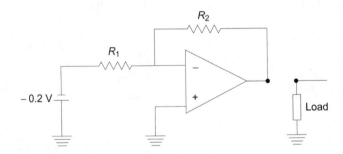

Fig. 9.36   *Isolated load circuit from the op-amp circuit*

(e) If the op-amp circuit works correctly with the output circuit isolated, then the problem is in the load circuit. Check for the output being grounded somewhere on the PC board or some other circuit connected to the op-amp output circuit.

With the above-mentioned simple techniques, it should be possible to troubleshoot and analyse a majority of op-amp circuits encountered in practice. Of course, many circuits with op-amps will not be as simple as an inverting amplifier, these steps will be found useful even in those cases to determine whether a particular op-amp is defective in the circuit or not. This information is highly useful in troubleshooting, particularly when there are a large number of op-amps in a circuit.

10

Troubleshooting Digital Circuits

10.1 WHY DIGITAL CIRCUITS?

Digital integrated circuits are used extensively in all branches of electronics from computing to industrial control, electronic instruments, communication systems and medical equipment. In fact, there does not seem to be any area in electronics where digital circuits in some form are not or will not be used. The basic reason for this is that digital circuits operate from defined voltage levels, which reduces any uncertainty about the resulting output and behaviour of a circuit. Many circuits operate with voltages that can only be either 'on' or 'off', e.g. a light can be switched on or off, a motor can be running or stopped or a valve can be open or shut. All these are digital operations and would need digital circuit elements for their operation and control.

Digital circuits cover a wide range from high current industrial motors to microprocessors. However, the basic elements of all digital circuits are logic gates that perform logical operations on their inputs. Therefore, it is essential to understand the basics of logic gates before discussing troubleshooting techniques.

10.2 BINARY NUMBER SYSTEMS

The binary number system is used practically exclusively in digital computers. This is because of its most remarkable feature, viz. simplicity as it involves just two digits, namely 0 and 1. The binary system has a base of 2 and any number can be expressed in the binary number with powers of 2. For example, the number 10 in the binary system is 1010.

$$10 = 1.2^3 + 0.2^2 + 1.2^1 + 0.2^0$$

$$= 8 + 0 + 2 + 0$$

$$10_{10} = 1010_2$$

The subscripts 2 and 10 are used to indicate the base in which the particular number is expressed. The binary system is also a positional value system in which each digit has its value expressed as powers of 2.

The zeros and ones in the binary notation are commonly called 'BITS'. This is an abbreviated form of 'Binary Digits'. Most frequently, we deal with 8-bit combinations. An 8-bit unit is known as BYTE. Thus, a byte represents numbers in the range 0 to 255. 4-bit units are often referred to as NIBBLES. Thus, one byte consists of two nibbles. 16-bit units are generally known as WORDS. A word thus consists of two bytes and 4 nibbles.

To provide a shorthand notation for the system of logic based on a single valued function with two discrete possible states, Boolean algebra is used. This type of algebra, based on logical statements that are either true or false, is a very useful tool in the design and troubleshooting of digital logic circuits. The validity of a Boolean statement can be verified by drawing a truth table.

10.3   TRUTH TABLES

Truth tables provide a tabular means of presenting the output side of logic devices for any set of inputs. Truth tables contain one column for each of the inputs and a column for the output. In basic truth tables, the column notations are usually H or L (for high and low) or for binary notation, '1' or '0'. For example, in a logic circuit, the truth table can be represented as:

Input States		Output States
A	B	
0	0	0
0	1	0
1	0	0
1	1	1

10.4   LOGIC CIRCUITS

Logic circuits are decision-making elements in electronic instruments. They are the basic building blocks of the circuits that control the data flow and processing of standard signals. In most systems which use logic, the output function represents a voltage level, either high or low.

There are several ways to represent two state 'yes' and 'no' decisions. Some of these are given below.

Table 10.1 *Logic State*

Yes	No
Open	Closed
1	0
Positive	Negative
True	False
High	Low
ON	OFF

10.4.1 Logic Convention

In digital circuits, 0 and 1 are represented by two different voltage levels, often called HIGH and LOW. The logic convention usually employed to relate these two entities is as follows (Fig. 10.1).

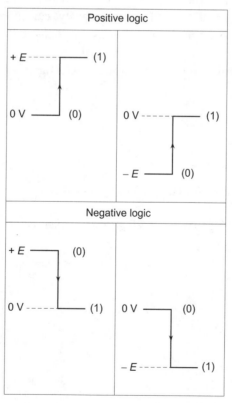

↑Fig. 10.1 *Logic convention usually employed to represent two levels, high and low, in digital circuits*

In the positive logic convention, logic 1 is assigned to the most positive (HIGH) level of the voltage and logic '0' to the least positive (LOW) level.

In the negative logic convention, logic 1 is assigned to the most negative (LOW) level and logic 0 to the least negative (HIGH) level.

This convention is important to understand the interpretation of digital data. For example, suppose 1001 (binary) data is presented on a set of binary coded decimal output lines. In positive logic, this would mean 1001 (binary) = 9 (decimal) while in negative logic, the same would mean 0110 (binary) = 6 (decimal).

10.4.2 The AND Gate

When the presence of two or more factors is necessary to produce a desired result, the AND gate is employed. This implies that the output of the AND gate will stand at its defined '1' state if and only if all the inputs stand at their defined '1' states.

Figure 10.2 shows the functioning of an AND gate which is like that of a set of switches in series. Only when they are all closed simultaneously, can there be an output, i.e. when the power is applied, both switches (A and B) must be closed before the lamp 'X' will light. The logical notation is expressed in Boolean terms as AB=X (A and B equal X). The graphical representation and the truth table of the AND gate are shown in Fig. 10.3.

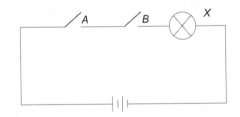

Fig. 10.2   AND gate equivalent circuit

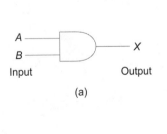

(a)

A	B	X
1	0	0
0	1	0
1	1	1
0	0	0

(b)

Fig. 10.3   AND gate: (a) graphical representation; and (b) truth table

The AND gate is used primarily as a control element with one input regulating the traffic through others. If a word is to be allowed to pass through the gate, a '1' at the control input will open the gate. The '0's' in the word are maintained in the right position at the output because the gate is closed whenever there is at least one '0' input.

10.4.3 The OR Gate

The OR gate provides the means of achieving a desired result with a choice of two or more inputs. This means that the output of the OR gate will stand at its defined '1' state if and only if one or more of its inputs stand at their defined '1' state.

As shown in Fig. 10.4, the functioning of the OR gate is similar to a set of switches connected in parallel. If either or both of the switches (A and B) is closed, power will be applied to the lamp 'X' causing it to glow. The logical notation is expressed in Boolean terms as $A+B=X$ (A or B equal X). The truth table and graphical representation of the OR gate are shown in Fig. 10.5.

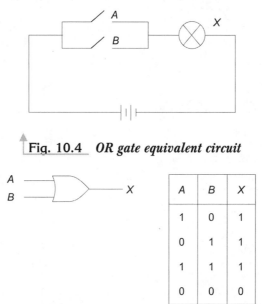

Fig. 10.4 *OR gate equivalent circuit*

A	B	X
1	0	1
0	1	1
1	1	1
0	0	0

Fig. 10.5 *OR gate (a) graphical representation; and (b) truth table*

The OR gate differs from the AND gate in that a '1' at one input OR the other input will give a '1' output; hence the 'OR' gate. However, two '0' inputs give a '0' output and two '1' inputs give a '1' output.

The OR gate is designed to prevent interaction or feedback between inputs.

10.4.4 The INVERTOR (NOT) Gate

An 'INVERTOR' is used if it is necessary to change the state of information before it is used. Therefore, the output of an invertor is always the complement of the input, i.e. zero becoming one and one becoming zero. Figure 10.6 shows the symbol and truth table of an INVERTOR circuit.

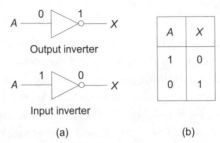

Fig. 10.6 *Invertor: (a) graphical representation; and (b) truth table*

The INVERTOR is never used by itself, but always in conjunction with another logic element. It is then represented by a small circle directly connected to the other logic element.

10.4.5 The NAND (NOT-AND) Gate

When an AND gate has an invertor at the output, the combined circuit is called a NAND gate. This is in effect the opposite of the AND gate. When all inputs are '1' the output is '0'. Typical NAND gate circuitry is shown in Fig. 10.7

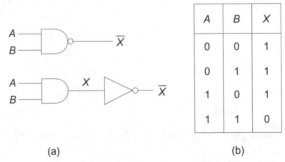

(a) (b)

Fig. 10.7 *(a) Typical NAND gate configuration; and*
(b) Truth table of NAND gate

The functioning of this gate is equivalent to a number of switches in series, in parallel with a lamp (Fig. 10.8). At least one switch must be open in order to make the lamp 'ON'.

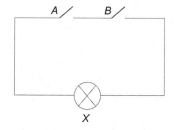

↑**Fig. 10.8** *Equivalent circuit of NAND gate*

10.4.6 The NOR Gate

The NOR gate is an OR gate with an invertor in its output circuit which produces a 'NOT OR'. Thus, when neither one input NOR the other is a '1', the output is '1'; hence the name 'NOR' gate. The symbol and truth table of NOR gate are shown in Fig. 10.9.

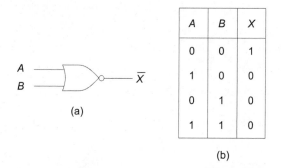

A	B	X
0	0	1
1	0	0
0	1	0
1	1	0

(a)

(b)

↑**Fig. 10.9** *NOR gate: (a) symbol; and (b) truth table*

The functioning of 'NOR' gate is similar to that of a number of switches in parallel with a lamp. All switches must be open if the lamp is to glow (Fig. 10.10).

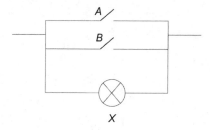

↑**Fig. 10.10** *Equivalent circuit of NOR gate*

10.4.7 The EXCLUSIVE-OR (EX-OR) Gate

The EX-OR gate may be regarded as a combination of AND and OR gates. It produces a 1 output only when the two inputs are at different logic levels.

A (two-input) EX-OR gate as shown in Fig. 10.11 can be regarded as the combination of two AND gates and an OR gate.

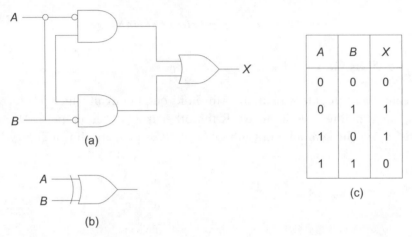

A	B	X
0	0	0
0	1	1
1	0	1
1	1	0

(c)

(a)

(b)

Fig. 10.11 *EX–OR gate: (a) configuration; and (b) symbol; and (c) truth table*

An EX-OR gate has always two inputs, and its output expression is

$$X = A + B$$

The equivalent circuit of an EX-OR gate is given in Fig. 10.12.

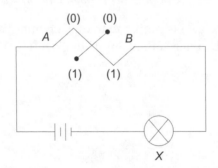

Fig. 10.12 *Equivalent circuit of EXCLUSIVE–OR gate*

10.4.8 THE INHIBIT GATE

The INHIBIT gate is an OR gate with an inhibiting input. In this gate, the output will stand at its '1' state if, and only if, the inhibit input stands at its defined '0' state, AND one or more of the normal OR inputs stand at their defined '1' state.

The symbol and truth table of the INHIBIT gate are shown in Fig. 10.13.

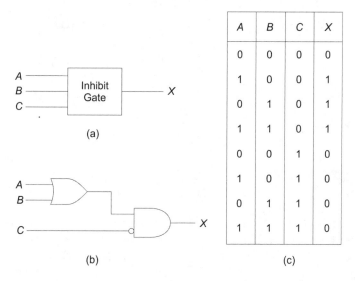

A	B	C	X
0	0	0	0
1	0	0	1
0	1	0	1
1	1	0	1
0	0	1	0
1	0	1	0
0	1	1	0
1	1	1	0

Fig. 10.13 *INHIBIT gate: (a) symbol; (b) configuration; and (c) truth table*

The equivalent circuit of an INHIBIT gate is shown in Fig. 10.14.

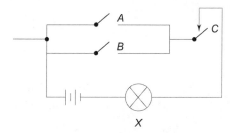

Fig. 10.14 *Equivalent circuit of INHIBIT gate*

This gate is very useful for controlling inputs (A and B) by means of the inhibiting signal C. When the inhibiting signal is present (C = 1), the output is always OFF (X = 0), but when the inhibiting signal is absent (C = 0), the signals A and B can pass to the output X.

10.4.9   The COMPARATOR (EX-NOR) Gate

The COMPARATOR (EXCLUSIVE-NOR) gate can be regarded as a combination of AND and OR gates. For example, a (two input) COMPARATOR gate is a combination of one AND and two OR gates (Fig. 10.15a) and its truth table is shown in (Fig. 10.15b).

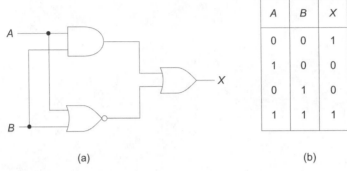

A	B	X
0	0	1
1	0	0
0	1	0
1	1	1

(a)   (b)

Fig. 10.15   *(a) The COMPARATOR configuration; and*
(b) COMPARATOR truth table

The output of the COMPARATOR will stand at its defined '1' state only if *all* the inputs stand at their defined 1 states or if none of the inputs stands at its defined '1' state.

The equivalent circuit of COMPARATOR gate is shown in Fig. 10.16.

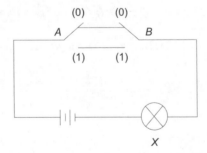

Fig. 10.16   *Equivalent circuit of COMPARATOR*

10.4.10   Graphical Symbols for Logic Elements

The graphical symbols used for logic elements recommended by different organizations are shown in Fig. 10.17.

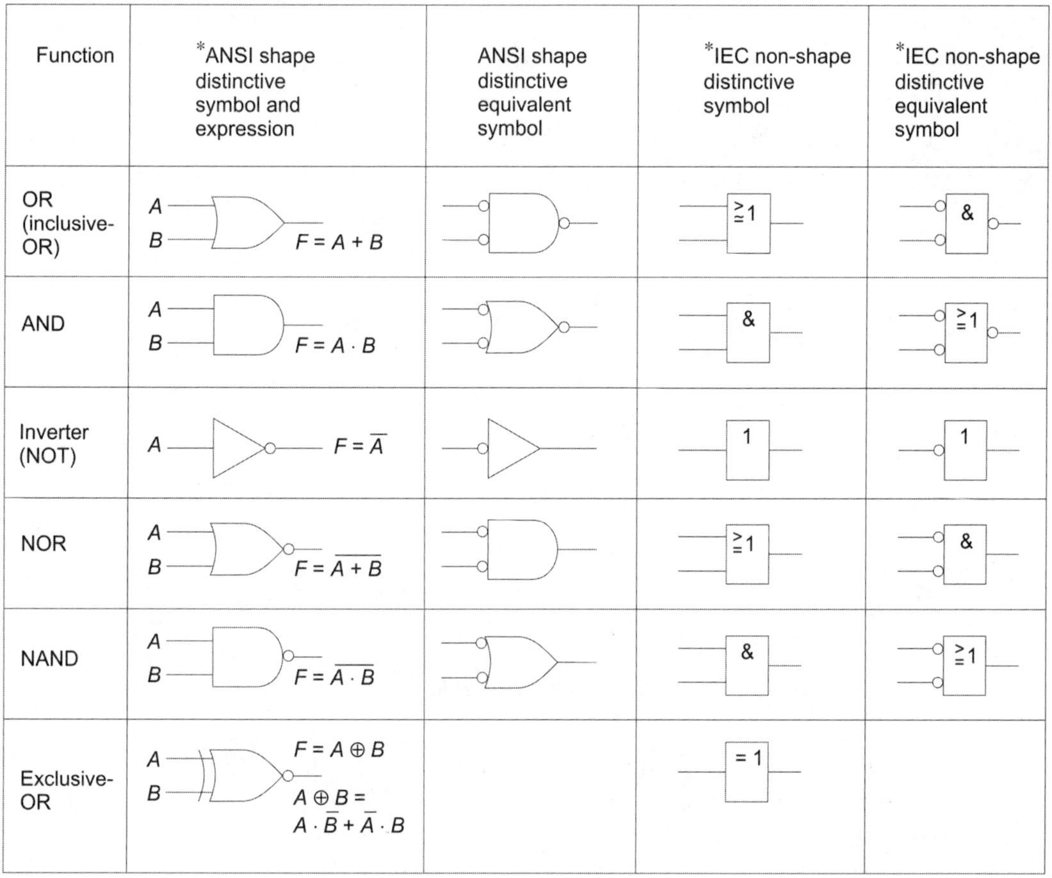

Function	*ANSI shape distinctive symbol and expression	ANSI shape distinctive equivalent symbol	*IEC non-shape distinctive symbol	*IEC non-shape distinctive equivalent symbol
OR (inclusive-OR)	$F = A + B$		≥1	&
AND	$F = A \cdot B$		&	≥1
Inverter (NOT)	$F = \overline{A}$		1	1
NOR	$F = \overline{A + B}$		≥1	&
NAND	$F = \overline{A \cdot B}$		&	≥1
Exclusive-OR	$F = A \oplus B$ $A \oplus B = \overline{A} \cdot \overline{B} + \overline{A} \cdot B$		= 1	

*ANSI : American National Standards Institute

*IEC : International Electro-technical Commission

Fig. 10.17 *Graphic symbols for logic elements recommended by different organizations*

10.5 CHARACTERISTICS OF INTEGRATED CIRCUIT LOGIC GATES

The various families of logic gates are associated with different characteristics. This means that one of them may prove to be the best suited for a particular application. The important characteristics are as follows:

Speed of Operation: The speed of operation of a logic gate is the time required by it to pass from one state to another. This is generally expressed in terms of the propagation delay. The propagation delay of a gate takes place on account of the switching time of a transistor and the rise time of the switching input voltage. The rate at which a flip-flop can switch from one state to the other is called its clock rate.

Threshold Value: The threshold voltage of a gate circuit is defined as the input voltage at which the gate just switches from one state to the other. For TTL logic family, the threshold voltage is 1.4V (Fig. 10.18). However, the maximum input voltage that will definitely

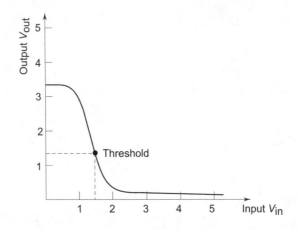

Fig. 10.18   *Positive logic threshold level in a TTL gate*

give logic 0 is 0.8V, whereas the minimum input voltage giving a definite logic 1 is 2.0V. For the correct operation of a gate, specific voltage levels must be applied. For example, for a TTL gate, logic 1, has a typical voltage of 3.3V and a minimum value of 2.4V (Fig. 10.19). On the other hand, logic 0 is typically 0.2V to 0.4V.

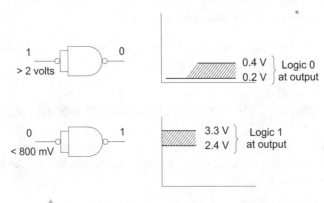

Fig. 10.19   *Logic levels at gate outputs*

Noise Margin: In order to avoid errors in a logic system due to parasitic voltages like spikes, logic devices should have a wide voltage swing between the two states, i.e. logic devices should have a wide noise margin. Figure 10.20 shows noise margins in high and low states.

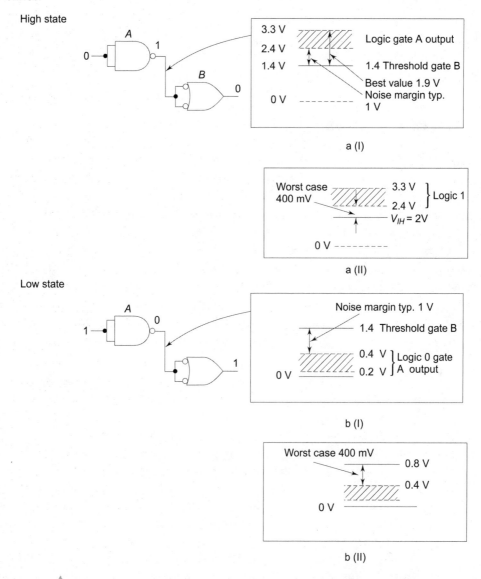

Fig. 10.20 *Noise margin in TTL gate: (a) High State (i) typical 1V (ii) worst case 400 mV; and (b) Low state (i) typical 1V (ii) worst case 400 mV*

Power Dissipation: Power dissipation generally implies the power required for operation of the logic device. As the circuit complexity increases, the power dissipation per gate must decrease so as to limit the amount of heat which may be dissipated in the semiconductor junctions.

Fan-in and Fan-out: The *fan-in* of a gate is the number of inputs which can be connected to the gate without seriously affecting its performance. Similarly, the *fan-out* is defined as the maximum number of circuits that can be connected to its output terminals without the output falling outside the specified logic levels.

10.6   THE LOGIC FAMILIES

The nature of the basic logic elements depends on the properties of the electrical components used to realize them. In the early days of digital techniques, when diodes were largely used in the circuitry, it was natural to take the AND or OR gates as the basic elements. Later, when transistors came to the fore, it became natural to basic logic circuits on the NAND and NOR gates. This is because the output signal of the transistor is opposite to the sign of its input. The most popular and most widely used circuits in modern digital equipment are the transistor–transistor logic and complementary metal-oxide semiconductor logic families. Logic circuits have become increasingly complex. The developments in integrated circuit technology have solved the problem of bulk along with the possibilities of obtaining several functions on one chip. It may be noted that the logic function of any IC gate is the same irrespective of the technology employed in fabricating the gate.

10.6.1   Transistor–Transistor Logic (TTL)

The most popular and the most widely employed logic family is the transistor–transistor family. The various logic gates are manufactured in the integrated circuit form by most manufacturers of semiconductors. The basic element in TTL circuits is the bi-polar transistor. TTL technology makes use of multiple-emitter transistors for the input devices. TTL gates use a Totem-pole output circuit. Another type of output circuit is the open collector output in which an external pull-up resistor is required to get the proper HIGH and LOW level logic outputs.

The popularity of the TTL family rests on its good fan-in and fan-out capability, high speed (particularly Schottky TTL version), easy inter-connection to other digital circuits and relatively low cost. The main characteristics of TTL logic are: propagation delay 10 ns, flip-flop rate 20 MHz, fan-out 10, noise margin 0.4V, and dissipation per gate 10 mW. The standard TTL gates are marketed as 74 series which can operate up to 70°C. However, 54 series are operatable up to a temperature of 125°C. Most IC packages contain more than one gate. For example, IC 7400 is a quad 2-input NAND gates whereas 7420 is a dual 4-input NAND gates. There are various types of TTL families mostly differing only in speed and power dissipation.

10.6.2 Schottky TTL

The gates in the family are faster than standard TTL and consumes much less power. Schottky TTL logic gates are available in the integrated form as 74S/54S series. A low power Schottky TTL series is also commercially available as 54LS/74LS.

10.6.3 Emitter-coupled Logic (ECL Family)

The ECL family provides another means of achieving higher speed of the gate. This differs completely from the other types of logic families in that the transistors, when conducting, are not saturated with the result that logic swings are reduced. For example, if the ECL gate is operated from 5V, the logic is represented by 0.9V and logic 1 by 1.75V.

10.6.4 CMOS Logic Families

The complementary metal-oxide semiconductor (CMOS) logic families offer significant advantages over bi-polar transistor-based logic circuits, particularly since they feature very low power dissipation and good noise immunity.

The great advantage of CMOS technology is the possibility of high density packing of a large number of devices. The technique is most suitable for the construction of large-scale integrated circuits rather than simple gates and flip-flops. Commercially available CMOS gates are available as 4000 series. For example, quad 2-input AND gate in CMOS comes as 4081 (7408 TTL) while quad 2-input NOR gate as 4001 (7402 TTL).

10.7 CMOS DIGITAL INTEGRATED CIRCUITS

Modern electronic equipment making use of logic circuits generally employs CMOS circuits. CMOS stands for complementary symmetry metal oxide semiconductor and alternatively termed COS/MOS. To make effective troubleshooting possible, a fundamental understanding of CMOS operation and characteristics is necessary. The following description gives the service technician a general overview of CMOS logic circuitry.

CMOS, from a black box point of view, operates fundamentally the same way as the conventional, bi-polar TTL family of logic-inverters, flip-flops, NAND, AND, OR, NOR circuits, etc. Figure 10.21 shows the configurations of TTL and CMOS inverters. A typical TTL is made of transistors and registers whereas CMOS is totally semiconductor material resulting in greater simplicity and much less power consumption. The CMOS inverter is made from an *n*-channel and a *p*-channel MOSFET connected as shown and their working is as follows:

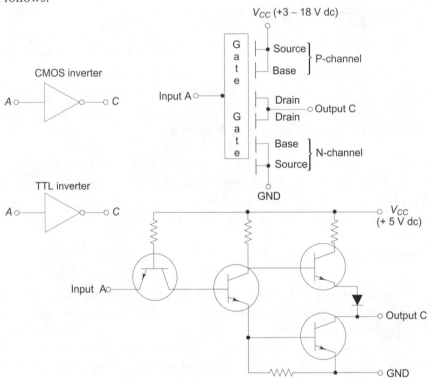

Fig. 10.21   *Configuration of TTL and CMOS invertors*

n-channel MOSFETs are turned on by a positive gate voltage;
p-channel MOSFETs are turned on by a negative gate voltage.
'ON' source-to-drain resistance equals 1000 ohms, typically.
'OFF' source-to-drain resistance equals 10,000 megohms.

Placing +12V (logic 1) on the input turns the *n*-channel 'ON' and *p*-channel MOSFET 'OFF'. The output is then essentially at ground or a logic '0' level, which completes the inversion. On the other hand, placing a ground level (logic 0) on the input turns the *p*-channel MOSFET 'ON' and the *n*-channel MOSFET 'OFF'. The output is then essentially at V_{DD} or 12V or logic 1 level, which completes the inversion.

An important CMOS device is the transmission gate or bilateral switch. The transmission gate operates as a single-pole, single throw switch operated by one-bit binary control input. When the control input is high, the switch is closed and when the control input is low, the switch is open (Fig. 10.22). This switch makes use of the analog properties of CMOS. The signal being switched can be a waveform of any shape, within the frequency limitations of CMOS, and any voltage within the V_{DD} to V_{SS} supplies of the particular device.

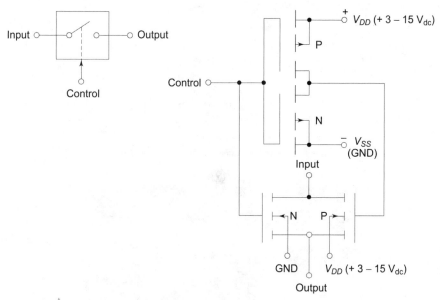

Fig. 10.22 *CMOS transmission gate or bilateral switch*

10.8 CATEGORIES OF INTEGRATED CIRCUITS BASED ON PACKING DENSITY

- *SSI (Small-scale integration)* means integration levels typically having up to 12 equivalent gates on chip. Available in 14 or 16 pin DIP or Flat packs.
- *MSI (Medium-scale integration)* means integration typically between 12 and 100 equivalent gates per IC package. Available in 24 pin DIP or Flat pack or 28 pin ceramic chip carrier package.
- *LSI (Large-scale integration)* implies integration typically up to 1,000 equivalent gates per IC package. Includes memories and some microprocessor circuits.
- *VLSI (Very large-scale integration)* means integration levels with extra high number of gates, say up to 100,000 gates per chip. For example, a RAM may have more than 4,000 gates in a single chip, and thus comes under the category of VLSI device.

10.9 LOGIC IC SERIES

The commonly used logic IC families are:
  (a) Standard TTL (Type 74/54);
  (b) CMOS (Type 4000 B);
  (c) Low power TTL (74L/54L);
  (d) Schottky TTL (Type 74S/54S);
  (e) Low power Schottky TTL (Type 74LS/54LS); and
  (f) ECL (Type 10,000).

10.10 PACKAGES IN DIGITAL ICS

Digital ICs come in four major packaged forms. These forms are shown in Fig. 10.23.

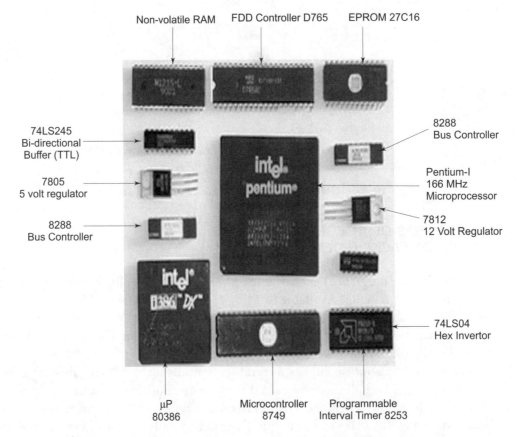

Fig. 10.23 *Typical packaging systems in digital integrated circuits*

Dual-in-Line Package (DIP): Most TTL and MOS devices in SSI, MSI and LSI are packaged in 14, 16, 24 or 40 pin DIPs.

Mini Dual-in-Line Package (Mini DIP): Mini DIPs are usually 8-pin packages.

Flat Pack: Flat packages are commonly used in applications where lightweight is essential requirement. Many military and space applications use flat packs. The number of pins on a flat pack varies from device to device.

TO-5, TO-8 Metal Can: The number of pins on a TO-5 or TO-8 can vary from 2 to 12.

All the above styles of packaging have different systems of numbering pins. For knowing about how the pins of a particular package are numbered, the manufacturer's data sheet on package type and pin numbers must be consulted.

10.11   IDENTIFICATION OF INTEGRATED CIRCUITS

Usually the digital integrated circuits come in a dual-in-line (DIP) package. Sometimes, the device in a DIP package may be an analog component—an operational amplifier or tapped resistors and therefore, it is essential to understand as to how to identify a particular IC.

In a schematic diagram, the ICs are represented in one of the following two methods:
(a) IC is represented by a rectangle (Fig. 10.24) with pin numbers shown along with each pin. The identification number of the IC is listed on the schematic.

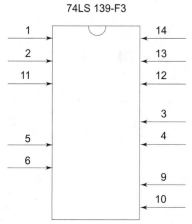

Fig. 10.24   *Representation scheme for digital ICs*

(b) Representation of the IC in terms of its simple logic elements. For example, IC 74 LS 08 is quad 2-input AND gate and when it is represented in a schematic, it is listed as ¼ 74 LS 08 (Fig. 10.25).

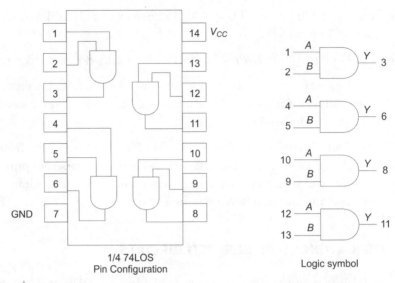

1/4 74LOS
Pin Configuration

Logic symbol

Fig. 10.25 *Representation of IC in terms of its logic elements*

An IC can be identified from the information given on the IC itself. The numbering system, though has been standardized, has some variations from manufacturer to manufacturer. Usually an IC has the following markings on its surface (Fig. 10.26).

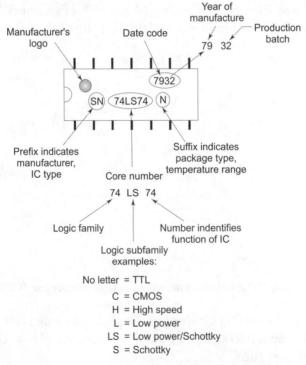

Year of manufacture

Manufacturer's logo　　　Date code　　　　　Production batch

79　32

7932

SN　74LS74　N

Prefix indicates manufacturer, IC type

Suffix indicates package type, temperature range

Core number

74　LS　74

Logic family

Number indentifies function of IC

Logic subfamily examples:

No letter = TTL
C = CMOS
H = High speed
L = Low power
LS = Low power/Schottky
S = Schottky

Fig. 10.26 *Identification marks on the digital ICs*

Core Number: Identifies the logic family and its functions. In 74 LS 51, the first two numbers indicate that the IC is a member of the 7400 series IC family. Last letters give the function of the IC. Letters inserted in the centre of the core number show the logic sub-family. Since TTL is the most common series, therefore, no letter is inserted in the centre of the core number. In case of other families, the following letters are used:

$$C = CMOS \qquad\qquad H = High\ Speed$$
$$L = Low\ Power$$
$$LS = Low\ Power\ Schottky$$
$$S = Schottky$$

The same numbered ICs in each family perform the same function and have the same pin numbers. They are, however, not inter-changeable because of differences in timing and power requirements.

A prefix to the core number identifies the manufacturer. For example, SN shows a device from Texas instruments.

A suffix to the core number indicates package type, temperature range, etc.

In some IC_s, marking is also provided for the year of manufacture and production batch. For example, 8234 indicates that the device was produced in 1982 in the 34th batch.

The manufacturer's logo (trademark of the manufacturer) is also printed along with other information about the IC. Further detailed information about the ICs can be obtained from several sources such as:

(a) Most of the IC manufacturers publish data books and product information data sheets. For most common series like 7400, you can get information from several sources. The information is provided on pin-outs, truth tables, etc.

(b) The IC Master, published regularly, serves as a reference book. This publication is usually available in a good technical library.

10.12 IC PIN-OUTS

The technical information on the IC includes pin connection diagram which shows the signals that are connected to each pin on the IC (Fig. 10.27).

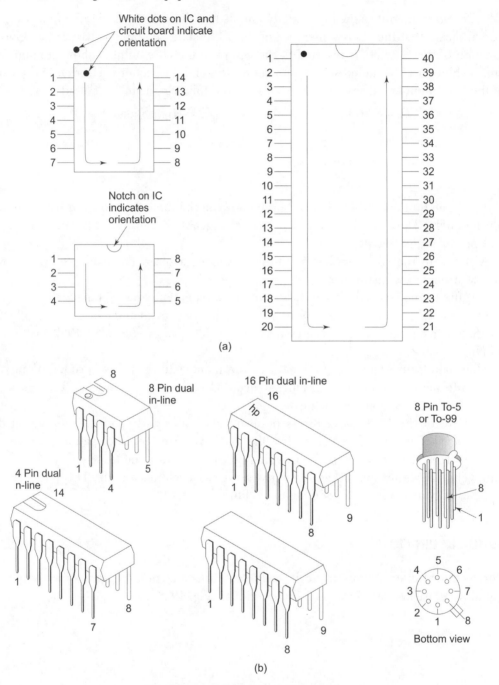

Fig. 10.27   *Standard IC pin numbering system:*
(a) Numbering System Scheme; and (b) Actual representation

One end of the IC is marked with a white dot or a notch on the plastic. Pin number 1 is always the upper left hand pin on the end of the IC that includes the notch.

The numbers run down the left side of the IC and up the right side.

In the pin connection diagram, the pin for supply voltage is indicated by V_{CC}. In most ICs, this voltage is $+5V_{dc}$.

The pin connection for ground is indicated by GND on the diagram.

In general, in digital ICs, the pin with the highest number is V_{CC} and the pin with half that number is GND. This is, however, not always true.

10.13   HANDLING ICs

ICs are delicate devices and can easily be damaged by rough and careless handling. The following precautions need to be observed while working with ICs:
  (a) Use the minimum possible amount of heat to solder or desolder connections.
  (b) Note the orientation of the IC before removing it. This should be done by drawing a sketch of the IC with surrounding parts and noting the position of the notch on the IC.
  (c) Always turn off the equipment before removing or replacing any IC.

10.13.1   Removing ICs from the Circuit Board

  (a) If the IC is mounted in a socket, it can be easily removed. Preferably use a tool as shown in Fig. 10.28.

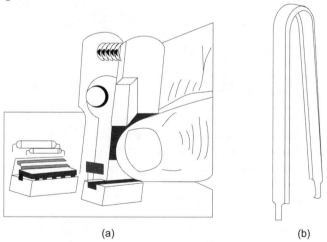

(a)   (b)

Fig. 10.28   *(a) DIP insertion–extraction clip; and*
                (b) IC Remover

(b) If the IC is to be discarded after its removal, simply clip each of the pins with a pair of cutting pliers and remove the body of the IC.

After removing the body of the IC, unsolder and remove each of the legs. Clean out each hold in the circuit board by heating and removing the melted solder with a wooden toothpick.

(c) Removing an IC that is to be used again needs more careful handling. In that case, work only on one pin at a time. Heat the pin until the solder at the base of the pin melts. Draw off the solder with a vacuum device. Pull out the pin with a pair of needle nose pliers while the solder is soft. Follow the same procedure with other pins.

If your technique is good, it is possible to re-use the IC.

10.13.2   Mounting IC in the Circuit Board

If the IC is to be installed in the socket, be sure the pins on the IC are straight. Some insertion tools include a device that straightens the pins.

As shown in Fig. 10.29, insert one row of pins first, then the other row. Be careful to place the IC with the notch in the right position.

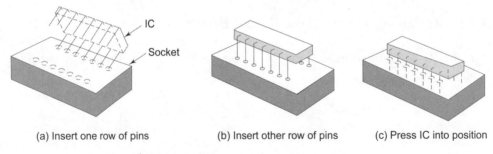

(a) Insert one row of pins   (b) Insert other row of pins   (c) Press IC into position

Fig. 10.29   *Mounting IC in a socket*

For installing the IC socket, the IC must not be in the socket. This eliminates the risk of heat damage to the IC during soldering. Once the socket is installed, clean the back of the circuit board with a flux remover. Any residual flux may be sticky and can attract dust. The dust can cause a slight leakage that can make the IC appear to be faulty.

10.13.3   Lifting an IC Pin

Many troubles can be isolated by lifting a pin or two from the circuit board, when the IC is mounted on a socket. In that case, bend one of its pins out of the way of the socket. To do this, use a pair of long nose pliers and bend the pin right at the point where it comes out of the package (Fig. 10.30). You will only need to bend it about 45° to clear the socket.

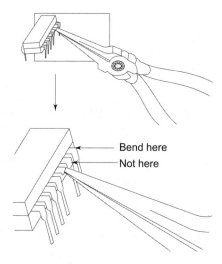

Fig. 10.30 *Correct use of a plier for bending a pin on the IC for lifting it from the socket*

On flip-flops that don't seem to be working right, you often need to lift more than one lead at a time. It will be necessary to tie some inputs, either HIGH or LOW, when they are lifted.

10.14 DIGITAL TROUBLESHOOTING METHODS

10.14.1 Typical Faults

With hundreds and thousands of semiconductor devices assembled on one small chip, the critical problem for the chip manufacturers is to get the voltages and signals in and out of such a tiny chip. Very thin wires are used as inputs and outputs to the chip. It is likely that the thermal stresses affect these tiny wires and after some time the bond may break away from the pad on the chip. This causes an *open*. Metal migration is another mechanism wherein metal particles begin to collect on parts of the chip. This can lead to pins being stuck at 1 or stuck at 0 irrespective of the input. High temperature, high voltage and power cycling can cause premature failure of the integrated circuits. Some failures take place due to poor assembly procedures and poor chip packaging.

Integrated circuits, when released in the market after quality checks are, however, fairly robust and reliable devices as much care is exercised during their manufacturing process. Therefore, before subjecting the IC to any test, it is advisable to check such faults in systems which could have been caused by:

(a) Dry joints;
(b) Breaks in PCB tracks (open circuited signal path);

(c) Shorts between tracks on the PCB; or

(d) Failure of discrete components external to the IC.

90% of all defects found in electronic equipment which may be associated with the digital ICs occur in the inter-connections between the chip and the supporting package. The typical failures include inputs or outputs shorted to ground, pins shorted to V_{CC} supply, pins shorted together, open pins and connections with intermittent defects.

The PCB can be easily examined with the help of a magnifying glass and any of the above faults rectified if so detected.

The faults in a gate output manifest themselves as follows:

(a) Stuck at '1', i.e. for TTL always above 2V irrespective of input states.

(b) Stuck at '0', i.e. for TTL always less than 0.8V irrespective of input states.

(c) High impedance state output, i.e. output not 0 or 1.

The causes of any such fault are possibly due to internal IC failure. The following kinds of internal failures may occur:

(a) An open lead which could be at either an input or an output pin; or

(b) A short circuit between an input terminal or an output terminal to V_{CC} or to the ground; or

(c) A short circuit between two pins, neither of which is at the V_{CC} or ground potential; or

(d) A defect in the internal circuitry of the IC.

Figure 10.31 shows possible fault conditions for a single gate. In Fig. 10.31(b), the output is stuck at '1' whereas with logic '1's on inputs, the output should be less than 0.8 V. The possible faults could be internal transistor open circuit or 0V line open circuits internally. It is assumed that the 0V line is externally good. In Fig. 10.31(a), the output is stuck at '0', whereas it should be logic '1'. The possible faults in this gate could be a short-circuited internal transistor, or V_{CC} line internally open.

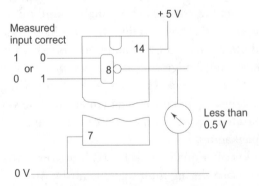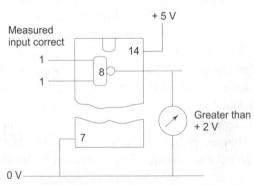

(a) Output stuck at 0, output should be logic 1
Possible faults: internal transistor short circuit or + 5V power line open either internally or externally

(b) Output stuck at 1 with logic 1s on inputs-output should be less than 0.8 V
Possible faults: internal transistor open circuit or 0 V line open circuit either internally or externally

Fig. 10.31 *Possible fault conditions in a single-gate*

It may be borne in mind that when a gate is used in a system, the fault diagnosis may be difficult because it may have its inputs and its own output may itself be driving several other gate inputs. Therefore, a fault may seem to be caused by a particular gate, but the same manifestations may result from a fault condition on the inputs of driving gates or the output of the driving gate. The fault diagnosis in such cases should be done systematically. For example, consider the circuit configuration shown in Fig. 10.32 in which the inputs of IC3 are driven from the outputs of IC2 and IC1. If the output of IC2 is open, it is floating and in TTL circuits, a floating input rises to approximately 1.5V and usually has the same effect on circuit operation as a 'high' logic level. This means that an open output bond in an IC will cause all inputs driven by that output to 'flat' to an incorrect level which is usually treated as a logic-high level by the inputs.

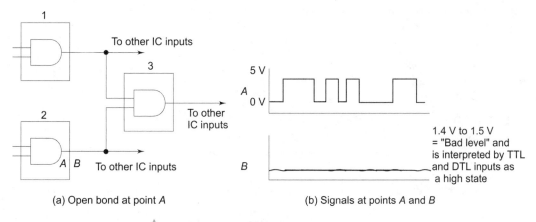

(a) Open bond at point *A*   (b) Signals at points *A* and *B*

Fig. 10.32   *Open output bond in an IC*

If there is an open input bond inside the IC3, the digital signal that drives the input will be unaffected and will be detectable at the input pin. However, the result will be detectable at the input pin. It will be as though the input were a static high level.

In case of the second type of fault to a short-circuit between an input and V_{CC} or ground, all signal lines connected to that input or output are held either high or low (high in the case of a short circuit to V_{CC} or low in case of a short circuit to the ground). Thus, fault usually causes normal signal activity at points beyond the short circuit to disappear and can be detected easily.

The third type of fault involving a short-circuit between two pins is shown in Fig. 10.33. This fault is not as quickly detected as a short circuit to V_{CC} or to the ground. In this circuit, whenever both outputs (of IC1 and IC2) tend to go high simultaneously or to go low simultaneously, the short-circuited pins will respond normally. However, when one output attempts to go low, the short circuit will be constrained to the low state.

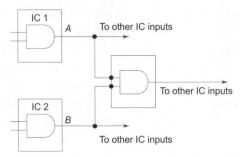

Fig. 10.33   *Integrated circuit with a short circuit between two pins*

A failure of internal or steering circuitry of an integrated circuit generally results in a 'stuck at' malfunction, which blocks the signal flow. Consequently it has catastrophic effect on circuit function.

While analysing fault conditions in a digital circuit, the following points may be kept in view:

(a) An open signal path in the circuit external to an IC has a result similar to that of an open output bond internal to the IC.

(b) A short circuit between a circuit connection and V_{CC} or ground is indistinguishable from a short circuit that is internal to the IC. Only a very close investigation of the circuit will permit determination of whether the failure is internal or external to the IC.

To summarize, Fig. 10.34 shows common faults encountered in digital ICs. These are:

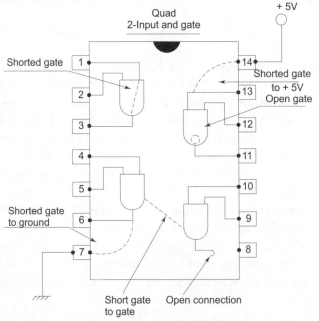

Fig. 10.34   *Common faults encountered in digital ICs*

- Shorted gate to $+V_{CC}$;
- Open gate (inside);
- Open connection to the pin;
- Short gate to gate;
- Shorted gate to ground; and
- Shorted gate (inside).

10.14.2 Testing Integrated Circuits with Pulse Generators

Three different types of tests can be performed on logic ICs. They are functional, DC and AC tests. Functional testing is the most commonly used test carried out in servicing, whereas DC and AC tests are more likely to be performed in production and research and development situations.

The functional test simply determines if the IC performs to its truth-table definitions. This test is carried out by employing a series of pulses from a pulse generator to step an IC through its various states. An oscilloscope or logic monitor can be hooked up to the IC's outputs so that those states can be monitored. The output of the pulse generator should be so set that the base line of the generator should be at 0 volts and the pulse amplitude should be set well above the minimum for logic 1. A functional test is adequate for most combinational logic ICs (gates, inverters, decoders, encoders, multiplexers, etc.).

Testing the DC parameters of an IC is usually required only if functional testing indicates that the IC is following its truth table, but the circuit is not still functioning. If replacing the suspected IC with another causes the circuit to operate properly, a failure to meet either the DC or AC voltage parameters is likely to be the cause of the problem.

Information on DC parameter measurement can usually be found in the manufacturer's IC data books. The commonly specified DC voltage parameters of a logic IC are as follows:

(a) The maximum DC voltage accepted by an input as a logic 0;
(b) The minimum DC voltage accepted by an input as a logic 1;
(c) The maximum DC output at logic 0 into full load, and the minimum logic 1 output voltage level under full load;
(d) Maximum power supply current; and
(e) Maximum output-current supplied by the IC into a short at logic 1.

All these tests can be made using a pulse generator with DC offset facility. A good rule to follow when setting the pulse generator parameters is to use the lowest possible repetition rate that will permit flicker-free oscilloscope observations. Usually, the pulse width is adjusted for approximately 50% duty cycle.

AC parameters need to be checked if the IC fails to operate in a relatively high frequency application, especially if used at or near its maximum rated frequency. The method and the specifications for AC parameter measurement are usually provided in the manufacturer's data handbook. Some of the important AC parameters are :

(a) Minimum and maximum input rise times;
(b) Minimum and maximum output rise time and fall times;
(c) Propagation delays through the IC; and
(d) Maximum repetition rates.

Depending on which logic family is used, a fairly complex pulse generator is required, along with a sophisticated dual-trace oscilloscope. A pulse generator with good rise-time specifications (5 to 10 ns), a repetition rate of 20 to 40 MHz and DC offset are the minimum requirements. Similarly, the oscilloscope must be able to measure rise-times of 10 to 40 ns if a complete series of AC tests are to be performed.

10.15   THE DIGITAL IC TROUBLESHOOTERS

Troubleshooting digital circuits poses some measurement problems not neatly solved by analog test equipment. The digital circuit typically has, as we have observed, only two narrow voltage ranges, high and low, that represent logic ones and zeros. A voltmeter or oscilloscope will easily give the voltage to more accuracy than is required, but won't translate these voltages into logic levels. Although the oscilloscope is sometimes indispensable for checking ringing, skews and so forth, in many instances a simple indication of logic levels will suffice.

Basically, the process of analysing faults in a digital circuit is by sequentially operating gates and ICs within the system and then comparing the resulting outputs with those which should be normally present. This can be done by applying suitable test signals and checking the resulting operations by suitable displays. There are several specially designed digital test and service aids which have been introduced over the years. The working principles and uses of some of these are described below.

10.15.1   Logic Clip

A logic clip provides field service users with a simple instrument which is so easy to use and so handy to carry. It clips on to any TTL or CMOS IC and permits the user observation of up to 16 pins of an IC at a glance. This feature is particularly useful with counters and shift registers or around any circuit with a truth table that needs checking. A logic clip is more convenient to use than analog meters in many digital applications, because it shows the state of each pin via an individual LED (Fig. 10.35). The operation is automatic and there are no adjustment switch settings or knobs to turn. Model 548A Logic clip from Hewlett Packard Co. USA is a multi-family clip which operates in TTL, CMOS and most other positive voltage logic families that employ power supply range of 4 to 18 volts DC. The maximum current drain is always less than 50 mA. The clip's buffered inputs draw less than 15 microamps from signal pins, insuring that circuit loading virtually never occurs. The clip has protection to 30V DC on all pins.

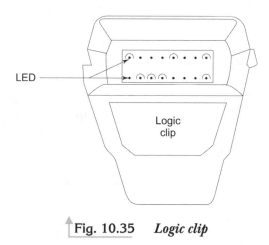

Fig. 10.35 *Logic clip*

The display on a logic clip shows logic high (lamp on), logic low (lamp off) and pulse activity (lamp dim; brightness dependent on duty cycle). If the system clock is replaced with a logic pulser (described later), sequential logic devices are slowly stepped through an entire cycle to verify the operation of shift registers, counters, flip-flops and adders.

10.15.2 Logic Probe (The Digital Screw Driver)

A logic probe (Fig. 10.36) is used as commonly by troubleshooters as the screw driver by mechanics. It simplifies troubleshooting by providing functional indications of in-circuit logic activity. For example, HP's 545A Logic Probe, uses a single lamp to indicate the various states possible on a digital signal path (high, low, single pulses, pulse trains, open circuit). The lamp indicator allows 360° viewing to clearly and quickly show the state of the circuit under test. The logic probe allows a TTL-CMOS- selectable operation. When switched to TTL position, it operates from 4.5 to 15 volt DC power supplies. In the CMOS position, logic threshold levels are variable from 3 to 18 volts. Overload protection is provided for ± 25 V DC for one minute.

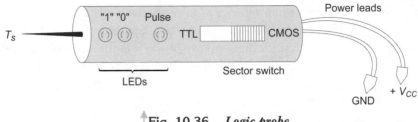

Fig. 10.36   *Logic probe*

An important feature of a logic probe is its ability to stretch pulses so that short, fast pulses are slowed down and lengthened at the display making them easy for the operator to see. For example, a 10 ns pulse is stretched to 100 ms so that the user can see it. This is accomplished by using the leading edge of a short pulse to trigger a circuit whose time delay is 100 ms. Single pulses cause the probe's lamp to flash once; pulse trains flash at ≤ Hz regardless of frequency, so you can easily detect the presence of activity. Figure 10.37 shows the response of digital logic probe to various input signals.

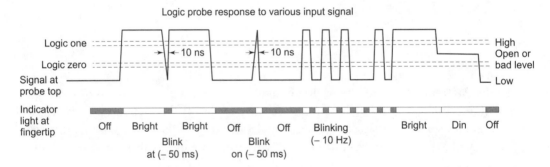

Fig. 10.37   *Logic probe response to various digital signals*

An auxiliary unit to logic probe is the pulse memory unit which is capable of capturing and displaying transient pulses, which are otherwise hard to see. When the probe tip detects such a pulse, it is stored in the memory and displayed until RESET is pressed. Use of the memory causes no change in the probe operation.

Some logic probes have two LED display indicators: a green LED for logic 0 and a red LED for logic 1. Such types of probes can be very versatile.

10.15.3   Logic Pulser

A logic pulser (Fig. 10.38) is used to stimulate digital circuits in-circuit and supplemented by a logic probe, it aids in testing for circuit response to easily check suspect gates, lines, busses and nodes. The pulse has a tristate output and until operated, its output remains in a high impedance state. This means that it pulses both HIGH and LOW; it also means that it has very high impedance when not operating, so it can be attached to a node and left there. There is no need to remove it when it is not pulsing.

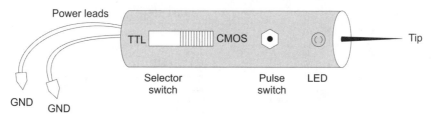

Fig. 10.38 *Logic pulser*

The logic pulser 546A from Hewlett Packard provides six different push button selectable output patterns according to the code shown in Fig. 10.39.

Operation		
Press and release code button 0		
Press and latch code button 0_		

Output modes		To output exactly 432 pulses		
o	Single pulse	1 100 Hz burst		
o_	100 Hz stream		oo_	98^1
				100
oo_	100 Hz burst			100
ooo_	10 Hz stream			100^2
				400
oooo_	10 Hz burst	2 10 Hz burst	oooo_	6^1
				10^2
oooo_	1 Hz stream			430
		3 Single pulse	o	1
		Single pulse	o	1
				432

Fig. 10.39 *Selectable output patterns in a logic pulser*

It means that the pulser provides versatile stimulus-response testing capability in both voltage and current applications for virtually any positive voltage logic family. While programming the pulser, the following may be kept in view:
  (a) The first pulse burst is outputted after subtracting the pulses produced when programming the output.
  (b) Release the pulse button during the final burst. The pulser will complete the burst, then shut off.

For example, to output exactly 321 pulses, proceed as follows.

(a) 100 Hz Burst 00

	98	–	(1st step)
	100		
	100	+	2(last step)
			300

(b) 10 Hz Burst 0000

	6	–	(1st step)
	10	+	4(last step)
			20

(c) Single pulse

	0	1
		321

The pulser is designed to drive high nodes low or low nodes high over a wide range of supply voltages automatically. Automatic pulse load sensing and width control effectively limit the amount of pulse energy that is delivered and ensure fast leading edges irrespective of the logic family under test.

The circuit in the pulser shuts it off faster for TTL than CMOS. This keeps the total energy low so as to eliminate any damage to the circuit being pulsed. Pulse height or amplitude is derived from the power supply that the pulser is connected to. For this reason, the pulser should always be powered from the circuit under test or power supply of the same voltage. The power supply requirements for the pulser are 3–18 V DC for CMOS and 4.5–5.5V DC for TTL. It draws less than 35 mA current and is protected against ± 25V when applied for one minute.

10.15.4 Logic Current Tracer

A logic current tracer primarily locates low-impedance faults in digital circuits by sniffing out current sources or sinks. Many troubleshooting problems such as wired-AND/OR configurations result in considerable waste of time in locating faults as several ICs may have to be removed before finding the bad one, and in the process the circuit board may be damaged. The use of a current tracer helps to pinpoint the one faulty point on a node, even on multi-layer boards. In addition, a current tracer locates those aggravating hair-line solder bridges that manage to pass unnoticed until a circuit is fired up.

A current tracer depends, for its action, on sensing the magnetic field generated by fast rise-time current pulses in the circuit (or, provided by a logic pulser), and display steps, single pulses, and pulse trains using a simple one light indicator. The tracer uses a shielded indicative pick up, and a wide band, high gain amplifier to provide the sensitivity needed to sense magnetic fields caused by current changes along PC board traces. The tracer has enough sensitivity (1 mA to 1 A) to follow traces in multi-layer boards, through the insulation in shorted cables, and for tracing faults on computer back planes and motherboards. Because it is not voltage-sensitive, the tracer operates on all logic families having current pulses exceeding 1 mA, and the repetition rates less than 10 MHz.

Prior to introduction of the current tracer, logic state indicators were limited to displaying voltage information such as a node was HIGH, LOW, open or pulsing. When a node is stuck, however, it may be trying to change state but is not able to cross threshold levels. The use of the current tracer adds the final bit of information necessary to pinpoint just such logic faults on bad nodes. For example, on a bad node, the tracer can verify that the driver is functioning and also show where the problem is by tracing current flow to the source or sink causing the node to be stuck.

Knowing both current and voltage information helps determine possible faults on a node. For example, when a node is active with current, but is unable to change state, the driver is delivering current pulses and the fault exists in the circuitry being driven. The tracer verifies this current activity, shows the current path and pinpoints the fault. Figure 10.40 shows how information about voltage and current can assist in the location of bad nodes.

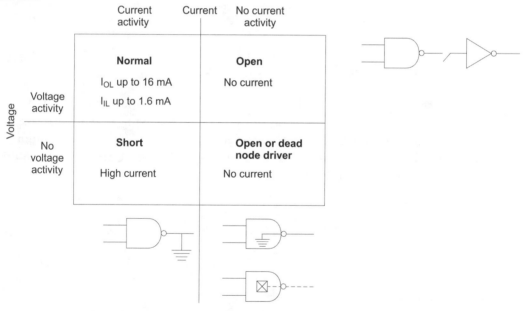

Fig. 10.40 *Current and voltage activity in failed ICs*

To use the tracer for location of the fault, the reference display is first set at the node driver output. The sensitivity control is then set to indicate the presence of AC current activity and then the circuit is traced to locate the main current path. The lamp will remain illuminated along the main path, the lamp extinguishes when the tip of the tracer is removed from the main current path. As you probe from point to point or follow traces, the lamp will change intensity, and when you find the fault, the tracer will most likely indicate the same brightness found at the reference point.

To illustrate how a solder bridge fault can be detected using current tracer, consider Fig. 10.41. A solder bridge between U1 and U2 causes both nodes to indicate functional logic failures. Tracing current flow in the circuit quickly shows the location and cause of the fault. In this case, the current under test has current activity.

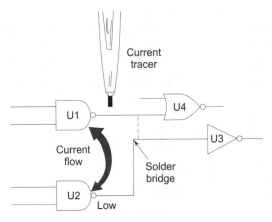

Fig. 10.41 *Solder bridge fault detection using a current tracer*

Figure 10.42 is an example which shows location of multiple input fault. Gate U 5 A is shorted to ground causing the node to be stuck LOW and sinking virtually all current from U 1 and other inputs. A current tracer quickly verifies this fault by providing a simple, clear, single lamp indicator of current activity on the node. Current pulses in this example, are provided by a logic pulser.

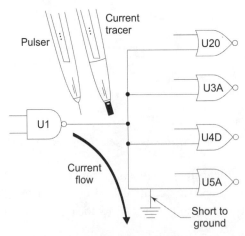

Fig. 10.42 *Multiple input fault detection using current tracer and logic pulser*

Digital Circuit Failure Modes: It has been observed from practical experience that ICs fail about three-fourths of the time by opening up at either the input or output. Failures of this kind can often be found using voltage-based methods such as logic probes, pulsers and clips. Repair to other types of failures (Fig. 10.43) can be conveniently carried out by using current tracers. For example, if there is little or no current flowing on the node, there is a likelihood of a dead driver (open output bond) or a lack of pulse activity in the circuit. In cases like this, use the logic probe and pulser to narrow down the symptoms of the fault, then use the pulser and tracer to investigate current flow on the node in question. It is worth noting that the current tracer is generally the last tool used after a fault has been located down to a specific node or set of nodes.

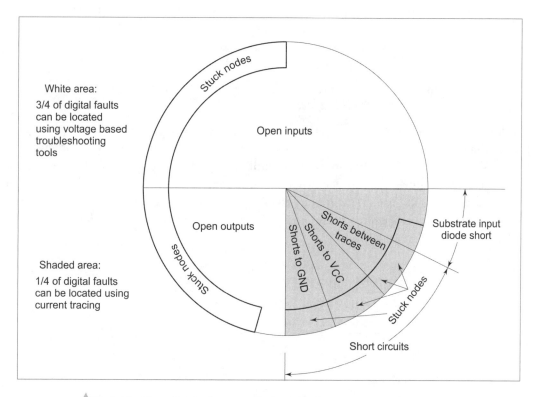

Fig. 10.43   *Digital circuit failure modes*

Stuck Node Caused by Dead Driver: Figure 10.44 illustrates a frequently occurring troubleshooting symptom. A node has been identified on which the voltage is stuck high or low. It is important to isolate as to whether the driver is dead or something like shorted input is clamping the node to a fixed value. Current tracer readily answers this question. If the driver is dead, the only current indicated by the tracer will be that caused by parasitic coupling from any nearby currents, and this will be much smaller than the normal current capability of the driver. On the other hand, if the driver is good, normal short circuit current will be present and can be traced to the circuit element clamping node.

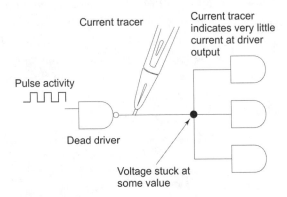

Fig. 10.44   *Stuck node caused by dead driver-use of current tracer for fault diagnosis*

V_{CC}-*to-Ground Short:* Locating V_{CC}-to-ground shorts is usually attempted by replacing all ICs on the board or all capacitors on the board. This is quite an exercise even though most people's experiences show that shorted decoupling capacitors account for the majority of supply to ground shorts and faulty capacitors cause most of the shorts. But if there are a lot of bypass capacitors, it usually won't pay to take them out one at a time to find the shorted one, so as to minimize damage to the board. By using a current tracer, such shorts can be easily traced out (Fig. 10.45).

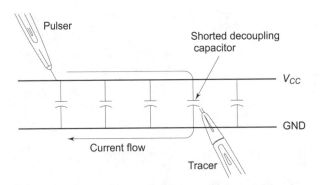

Fig. 10.45   *Use of current tracer for V_{CC}–ground faults*

To do so, disconnect the power supply and pulse the power supply terminal using the logic pulser with the supply return connected to the GND lead of the pulser. Even if the capacitors are connected between V_{CC} and ground, the current tracer will usually reveal the path carrying the greatest current (the shortened path).

Efficient use of the current tracer usually requires a larger familiarization period than does the operation of voltage-sensing instruments. Also, it requires some skill to avoid the cross problem, that is, if a small current is being traced in a conductor that is very close to another conductor carrying a much larger current, the sensor at the tip of the current tracer may respond to the current in the nearby trace. The operator can, however, learn to recognise interference or cross-talk with some experience.

10.15.5 Logic Comparator

A logic comparator works on the simple principle that a known good IC can act as a standard against which to measure the in-circuit performance of suspected faulty ICs. With this technique, TTL or DTL functional logic failures in terms of faulty node or IC can be detected much faster than old analog methods. The comparator performs this function by comparing the output responses of a REFERENCE IC against an in-circuit TEST IC and displaying subsequent errors in performance pin by pin. The IC output pin that does not correctly follow its inputs will produce an error indication, even when the error is a short-term (200 ns) dynamic fault. Figure 10.46 shows the comparator response to errors.

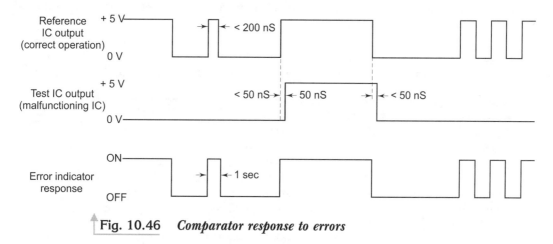

Fig. 10.46 *Comparator response to errors*

The comparator operation is quite simple; each input pin of the REFERENCE IC is connected in parallel with the IC in the circuit under test (Fig. 10.47) using a small circuit board. Each output pin of the REFERENCE IC is connected to an exclusive-OR to be compared with the output pin of the IC under test. Any difference in the operation is stretched and displayed as an error by the individual LED associated with the output pin. The comparator is accompanied with a switch programmable socket board that accepts 8, 14 or 16 pin ICs. To use it, the reference IC is placed in the socket, the 16 switches are programmed (inputs in, outputs out) and the testing is carried out. The socket board senses which of the input pins are supply and ground, and displays output errors greater than 300 ns.

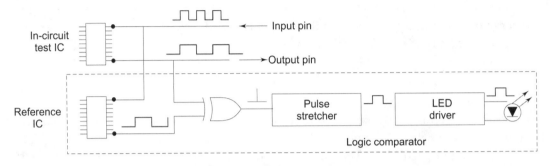

Fig. 10.47   *Typical input pin/output pin configuration*

The following chart (Table 10.1) lists out the capabilities of a comparator for various devices and logic families.

Under most conditions, the logic comparator does not effect the circuit operation due to its loading effect. However, when analog components such as register, capacitors or transistors are used to control-timing or to buffer signals, the timing or drive capability of these circuits may be adversely affected.

Table 10.1 ■ *Comparative Capability for Different IC Families*

IC family	Comparator capability
(a) Combinational Logic (AND, NAND, XOR, etc)	Excellent, this category covers majority of ICs in use.
(b) Sequential Logic (flip-flops)	Excellent, reference and test IC should be synchronized by a pulse on the 'Reset' input.
(c) Memories, shift registers	Excellent.
(d) One shots	Usually good, since reference and test IC share the RC timing components, circuit timing can be affected.
(e) Open collector and tri-state logic	Usually poor, when outputs are bussed together, a good gate is constrained to operate improperly and this will be indicated by the comparator.
(f) MOS devices	No, they require different power supplies exceeding the 7V input limit and will damage the comparator.
(g) Analog and linear ICs	No, the outputs are analog and cannot be tested by the comparator

10.16 SPECIAL CONSIDERATIONS FOR FAULT DIAGNOSIS IN DIGITAL CIRCUITS

The repair and servicing of digital equipment should be attempted only if the following are ensured:

(a) The service manual of the equipment must be available. The manual should contain circuits, layout diagrams, specifications of parts and their list. Study them carefully.

(b) The tools and test instruments specific to digital circuits should be available.

(c) Thoroughly understand the types of logic ICs employed in the equipment. The expected logic levels and specification for power supply voltages must be precisely known.

(d) Test signals must not be applied while the power is off.

(e) Power supply voltages should be checked at the actual IC pins, and not between board connections on the PCB. This would isolate the fault arising from an open circuit power line to the IC.

(f) An IC should never be removed or inserted while power is applied to avoid possible damage due to current surges.

(g) Large test probes must be avoided in digital circuitry. These can inadvertently short out IC pins and cause more faults.

(h) Carefully check temperature of components with your finger if there is no shock hazard. Faulty components tend to be hotter than good ones when operating.

(i) Probe pins on devices, not pins on sockets.

10.17 HANDLING PRECAUTIONS FOR ELECTRONIC DEVICES SUBJECT TO DAMAGE BY STATIC ELECTRICITY

(a) When testing static charge-sensitive devices, DC power should be on before, during and after application of test signals.

(b) Ensure that all pertinent voltages are switched off while boards or components are removed or inserted, whether hard-wired or plug-in.

(c) Avoid circumstances that are likely to produce static charges, such as wearing clothes of synthetic material, sitting on a plastic covered or rubber-footed stool, combing your hair or making extensive erasures. These circumstances are most significant when the air is dry.

(d) If any circuit boards or IC packages are to be transported or stored, enclose them in conductive envelopes and/or carriers.

(e) Handle IC packages without touching the pins.

(f) Ground yourself reliably, through a resistance to the work surface, for example, use a conductive strap or cable with a wrist cuff. The cuff should make electrical contact directly with your skin. The cuff should not be worn over clothing. Resistance between skin contact and work surface is typically in the range of 250 K Ω to 1 M Ω.

(g) Do not use those tools and items that can generate a static charge. Examples of such items are plunger type solder suckers.

(h) The work surface, typically a bench top, must be conductive and reliably connected to earth ground through a safety resistance of 250 KΩ to 500 MΩ .

(i) Ground the frame of any line-powered equipment, test instruments, lamps, drills, soldering irons, etc. directly to earth ground. The grounded equipment should have rubber feet or other means of insulation from the work surface. The equipment being serviced should be insulated while grounded through the power.

(j) Avoid placing tools or electrical parts on insulators, such as books, paper, rubber pads, plastic bags or trays, etc.

10.18   FUNCTION AND TESTING OF FLIP-FLOPS, COUNTERS AND REGISTERS

10.18.1   Flip-flops

A flip-flop is a two-state electronic device which can be either turned on or turned off when commanded to do so. It is a bi-stable logic element with one or more inputs and two complementary outputs. A flip-flop essentially remains in its last state until a specific input signal causes it to change state. Because of the ability of the flip-flop to store bits of information (1 or 0), it has memory characteristics. It is thus one of the most important basic building blocks in digital circuitry.

Nearly all flip-flops have two output levels (Fig. 10.48 (a)) designated as Q and $\overline{Q}$ on which the true state (Q) and the complement $(\overline{Q})$ of the stored function is available. The input terminals may receive either discrete level or pulse signals depending on the circuit. A flip-flop is known by several other names, the most common being bi-stable multi-vibrator, binary or latch.

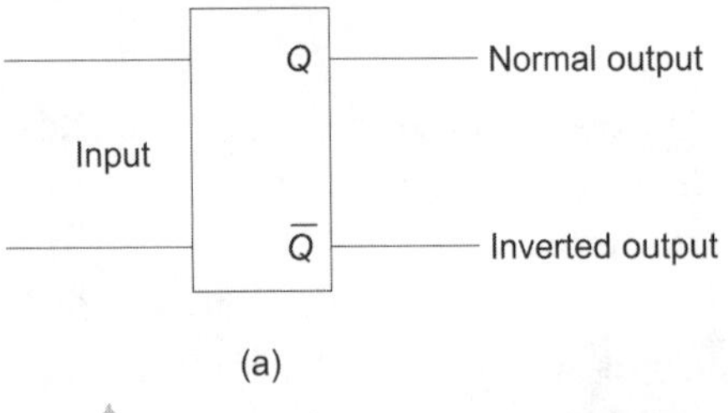

Outputs		Flip flop
1	0	State
High	Low	1
Low	High	0

(a)   (b)

Fig. 10.48   *(a) Basic flip-flop configuration; and*
(b) Truth table of a basic flip-flop circuit

The truth table (Fig. 10.48(b)) indicates the two possible output conditions for a flip-flop and the corresponding definition for the state. There are many forms of flip-flops, each of which has its specific features. However, all those various forms of flip-flops contain essentially the same bi-stable element, the RS flip-flop.

10.18.2 Reset-set Flip-flop or the R-S Flip-flop

A basic flip-flop can be constructed using two NAND gates (Fig. 10.49) cross-coupled to the inputs. When power is applied, opposite states will appear on the outputs of gates A and B. If the Q output of gate A is '1', this '1' will be applied to the input of gate B whose output ($\overline{Q}$) will then become '0'. When this '0' is applied to the input of gate A, a '1' will remain on the Q output of gate A. Thus, the gates are latched into a stable state. Since output Q is high, the flip-flop is in the high or '1' state. Any additional pulse at the SET input will have no effect on the output. However, when a pulse is applied to the RESET input, the output reverses or 'flips'. Further pulses at the RESET input have no effect on the outputs. Switching the inputs again causes the outputs to 'flop' back, to their original condition.

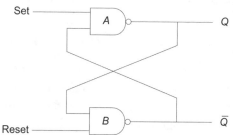

Fig. 10.49 *R–S flip-flop configuration*

The flip-flop is like a toggle switch, either in one position or the other; and once the change-over has been made, repeating the action has no further effect. The condition is stable, either way. Though R-S flip-flop in itself has limited applications, it is the basic building block of flip-flop chains in integrated circuit form. They normally utilize a clock input to synchronize the changes from one state to another.

Flip-flops are checked for correct performance using a logic probe and comparing the outputs with the truth table.

10.18.3 Clocked Flip-flops

The R-S flip-flop is commonly used when there are no possibilities of simultaneous set and re-set inputs. If both the inputs are simultaneously enabled with a low pulse, both outputs will go high for the duration of the pulse. Removing the enabling pulses will make the flip-flop to go over to an indeterminate state, i.e. it could latch in either the '1' or '0' condition. The stable state, which it will finally race to, depends upon the relative time delay of the two NOR gates used in the circuit.

The problem of races can be avoided by using a synchronous or a clocked flip-flop in which the input can only be applied in coincidence with a clock signal. It is convenient to enable or condition the flip-flop by applying the appropriate levels to the inputs first and then arrange for the flip-flop to change state on receipt of a pulse from another source. The pulse is called 'clock' which may be from an oscillator circuit. Thus, the clock signal is a signal that initiates action at regular spaced intervals. Figure 10.50 shows a typical clock signal. The operations in the system take place at the time when the transition occurs from 1 to 0 (falling edge) or 0 to 1 (rising edge).

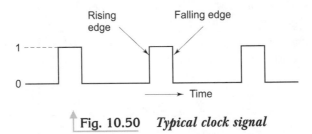

Fig. 10.50   *Typical clock signal*

Clocked flip-flops are designed to change states at the appropriate clock transition and will rest between successive clock pulses. Figure 10.51 shows the circuit configuration of a clocked R–S flip-flop. The flip-flop section (gates C and D) is identical to the R–S flip-flop. In addition, there is a circuit arrangement which applies clock pulse to either the SET or RESET input terminal of the flip-flop. If the SET input line is enabled with a high signal, gate A will be enabled when a high pulse is presented at the clock input. When gate A is enabled, it will provide a low SET signal to the R-S flip flop which will go to the '1' condition.

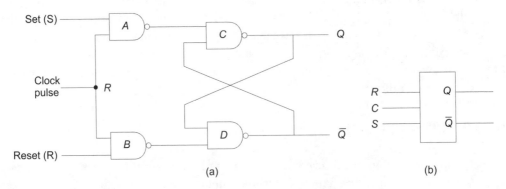

Fig. 10.51   *Clocked R–S flip-flop (a) Circuit configuration; and (b) Symbol*

If the SET input line has a low level and the RESET line is enabled with a high level, gate B will be enabled when a high clock pulse occurs. When gate B is enabled, a low RESET signal will be presented to the flip-flop and it will go to the '0' condition.

It may be noted that the flip-flop will only change state when a clock pulse is applied and this may be kept in mind while troubleshooting this circuit.

10.18.4 D Flip-flop

One way of avoiding the intermediate state found in the operation of the simple R–S flip-flop is to provide only one input which can be either high or low. This input is called the *D* input or 'data' input and the flip-flop thus constructed is called a *D* flip-flop. Figure 10.52 shows the circuit arrangement of a *D*-type flip-flop. A '1' or a '0' applied to this input is passed directly to one of the inputs of the flip-flop proper and inverted to the other input.

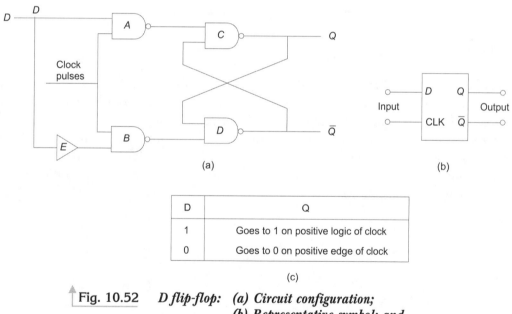

(a)

(b)

D	Q
1	Goes to 1 on positive logic of clock
0	Goes to 0 on positive edge of clock

(c)

Fig. 10.52 *D flip-flop: (a) Circuit configuration;*
(b) Representative symbol; and
(c) Truth table for D flip-flop

From Fig. 10.53 it is obvious that whatever information is present at the *D* input prior to and during the clock pulse is propagated to the *Q* output when the clock pulse is applied, while the inverse of that information appears at the $\bar{Q}$ output. The flip-flop is thus set in the '1' state if the *D* input is made '1' and in the '0' state if the *D* input is made 0. This type of circuit is known as edge-triggered *D* flip-flop and is one of the most commonly used in computers. In *D*-type latch, which is similar to the *D* flip-flop except that it can change states during the HIGH portion of the clock signal, i.e. as long as clock is HIGH, Q will follow the D input even if it changes when the clock goes LOW, Q will store (or latch) the last value it had and the D input has no further effect.

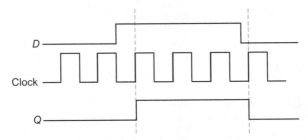

Fig. 10.53   *Positive edge triggering in a D flip-flop*

D-type flip-flop is checked by observing the activity on the CLK line. The flip-flop does not operate unless this line is clocked. The presence of signal at D and a pulse on the clock line should result in a signal at Q.

10.18.5   J-K Flip-flop

One of the most useful members of the flip-flop family is the J-K flip-flop. A unique feature of the J-K flip-flop is that it has no ambiguous state. It is the most widely used flip flop type in logic circuitry and is the ideal memory element to use.

A J-K flip-flop often has more than one J input and more than one K input. In this case, one J input and one K input are generally connected together for use as input for clock pulses. This kind of circuit arrangement is shown in Fig. 10.54. The operation of the circuit is controlled as follows:

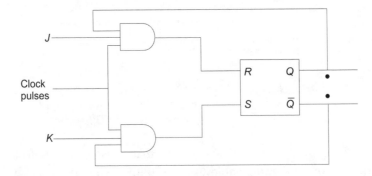

Fig. 10.54   *J–K flip-flop circuit arrangement*

(a) A clock pulse will not cause any changes in the state of the flip-flop if neither the J nor the K input is activated.

(b) If both J and K inputs are activated, the flip-flop will change state when the next clock pulse is received.

(c) The J and clock pulse inputs used together set the flip-flop in the set position, while the K and clock pulse inputs reset it.

It may be observed that propagation delay time prevents the J-K flip-flop from racing (toggling more than once during a positive clock edge). This is because the outputs change after the positive clock edge has struck. By then, new Q and $\bar{Q}$ signals are too late to coincide with positive spikes driving the AND gates. Commercially available J-K flips-flops also give the facility of synchronously setting the output to 1 (pre-set) or 0 (clear). DM 74733 has two J-K flip-flops with clear facility. Dual J-K with separate pre-set and clear facilities are available in DM 74766 and with common pre-set and clear in DM 7478.

It may be noted that the output of J-K flip-flop becomes stable only when the clock pulse goes zero. To avoid this problem, a special form of the J-K flip-flop known as the 'J-K master-slave flip-flop is used. As the name implies, it comprises two flip-flops—a master and a slave—being triggered at positive and negative edge of the clock respectively. This is advantageous because a sequence of such master-slave flip-flops could then be triggered simultaneously by a clock pulse derived from the same source, without ambiguity. Figure 10.55 shows master-slave J-K flip-flop.

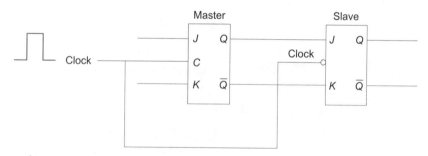

Fig. 10.55 *Circuit configuration of master-slave J–K flip-flop*

If J and C inputs of the J-K flip-flop are connected together, it becomes a Toggle or a T flip-flop whose output toggles whenever a '1' is applied at the input.

J-K flip-flop is tested in the same way as the R-S circuits, except for ensuring the presence of correct signals at the control lines.

Flip-flops are the basic blocks, using which, it is possible to build many other useful sequential circuits, some of which are described in the subsequent section.

10.18.6　Counters

A counter is a sequential circuit consisting of a series of flip-flops which go through a sequence of states on the application of pulses at its input. Counters are constructed out of T or J-K flip-flops.

In Fig. 10.56 it may be seen that three *J-K* flip-flops are combined to give a three-bit counter. One output gives the information at each stage, the other is to carry to the next stage. Let us assume that all outputs at Q are initially cleared to '0'. A pulse applied to the input of the first stage switches its counting output Q(A) to 1, while the other output $\overline{Q}$(A) becomes '0' and thus has no effect on the second state. The circuit has now counted the first pulse (binary 1). On the receipt of the second pulse, the count output Q(A) of the first stage switches back to '0' while the carry output $\overline{Q}$ (A) flips to '1'. This switches to second state (B) to count '1'. The circuit has now counted two, which in binary notation is 10. The third pulse switches stage one to count 1, Q(A), but the second stage is not switched. The count is 11, binary notation for three. The counting continues in this fashion until the circuit output is one-one-one, which as we know is 7 in binary notation.

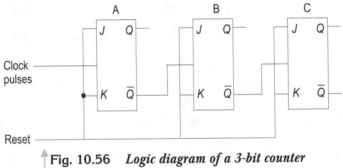

Fig. 10.56   *Logic diagram of a 3-bit counter*

The eighth count re-sets the circuit to zero. The pulse diagram of a three-bit binary counter is shown in Fig. 10.57(a). Figure 10.57(b) gives truth table of this counter.

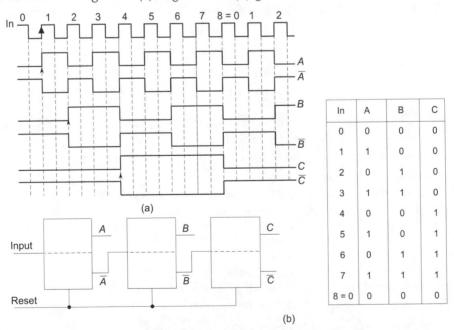

In	A	B	C
0	0	0	0
1	1	0	0
2	0	1	0
3	1	1	0
4	0	0	1
5	1	0	1
6	0	1	1
7	1	1	1
8 = 0	0	0	0

(a)

(b)

Fig. 10.57   *(a) Timing diagram of a 3-bit counter; and*
(b) Truth table of a 3-bit counter

Decimal counters are binary counters which employ four flip-flops and are so contracted that they count up to 9 (instead of 15) and re-set to 0000 on receipt of the tenth input pulse. By cascading more flip-flops to the chain, we can build a counter of any length. Eight flip-flops give an eight-bit counter, twelve flip-flops result in a twelve-bit counter, and so on.

These counters are capable of counting only upwards from zero to some maximum count and then re-set to zero. Counters have also been designed to count in either direction which are called up/down counters. These counters have an up/down input which is used to control the counting direction, i.e. one type of logic level causes the counter to count up from zero and the other logic level applied to it causes the counter to count down say from the 1111 to 0000 (in four-bit counters). In some up/down counters, two separate clock inputs are employed, one for counting up and the other for counting downwards.

A wide variety of counters are available commercially as standard integrated circuit packages. They avoid the necessity of constructing counters using individual flip-flops. Among the popular counters are DM 7492 which is a divide by 12 counter, and a binary counter. DM 74191, DM 74192 are synchronous up-down counters.

Counters and dividers can be quickly checked using logic probe or with an oscilloscope. The activity at each of the input and output pins is determined and with the help of the truth table, the problem is isolated.

The outputs checked with a logic probe should match one line of the truth table. Next, a pulse is sent to the output with the pulser and the outputs are checked again. Obviously, the outputs should correspond to the next line on the truth table; if not so, the chip is defective and need to be replaced. Some counters and dividers have 'enable' lines which must be made high or low before the device works.

10.18.7 Registers

A *register* is a group of memory elements employed to store binary information. Registers have an important place in digital computers, because the very operation of computers is based on transferring binary information from one register to another and carrying on certain operations before it is again transferred. The simplest register is a flip-flop.

In digital circuitry, a register usually consists of parallel latches. It can represent a number in the range from 0 to 2^n-1 where n is the number of latches in parallel. The register works under the control of a clock which signals when the register should record the input, the last latched data always appears at the output. Registers are internal to the microprocessor and are very important because of the rather lengthy process involved in accessing data in memory. For example, intermediate results can be temporarily kept in registers rather than returning them to main memory repeatedly. Hence, the number of programmable registers in the CPU is very important.

10.18.8   Shift Registers

A shift register is a group of serially connected flip-flops that is used for the temporary storage of information. They can also be used for shifting of the information stored in the register either one position to the right or left with each clock pulse. This is accomplished by gating the outputs of the flip-flops to the appropriate inputs for performing either a left or right shift. This shift direction is controlled by a mode input. When shifting information back into the register, it may be desired to replace the original information back into the register. This operation is achieved by employing an 'end-around-shift' feature.

A shift register can be built using series connected R–S, J–K or D type flip-flops. They are so connected that the output of each flip-flop becomes the input to the next flip-flop. As the register is clocked, the data is shifted one position to the left or right for each clock pulse. They are, however, constructed using a number of integrated flip-flops of TTL or CMOS families. The capacity of an integrated shift register ranges from 4 bits in the TTL family to 2048 bits in the CMOS family. CMOS shift-registers are usually only of the serial-in serial-out type because of the many stages involved. There are not enough pins for parallel-input or output connections.

Figure 10.58 shows a 4-bit shift register employing D flip-flops. This is a serial input shift register. Initially, a clear pluse (logic 0) is applied to the RESET which sets the outputs at Q_A, Q_B, Q_C and Q_D to 0. Next, the first data bit (D_1) is applied to the SERIAL INPUT. A pulse will appear at the leading edge of D_1. When the next data bit D_2 is applied to the input, $Q_A = D_2$ and $Q_B = D_1$. Continuing this process after four clock pulses, $Q_A = D_4$, $Q_B = Q_3$, $Q_C = D_2$ and $Q_D = D_1$.

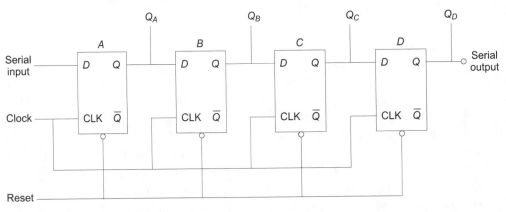

Fig. 10.58   *4-bit shift resister with serial entry and serial parallel output*

The circuit explained above pertains to serial-to-parallel converter. This arrangement is not practical in shift registers with a large number of bits. In such cases, serial output registers are employed.

Methods of Using Shift Registers: Shift registers can be used in any one of the following four ways:

(a) Serial-in/parallel-out;
(b) Parallel-in/serial-out;
(c) Serial-in/serial-out; or
(d) Parallel-in/parallel-out.

With a serial-in/parallel-out shift register (Fig. 10.59(a)) data is fed in the serial form and when the complete word is stored, the bits are read off simultaneously from the output of each state.

Figure 10.59(b) shows parallel-in/serial-out shift register in which the data is stored, after clearing all the stages, in each flip-flop. The data is then read out serially, i.e. one bit at a time under the control of the clock.

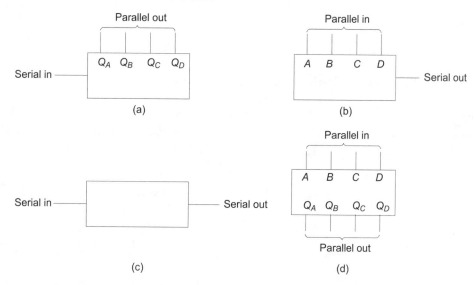

Fig. 10.59 *Methods of using a shift-register:*
(a) Serial in/parallel-out; *(b) Parallel-in serial-out;*
(c) serial-in/serial-out; and *(d) Parallel-in/parallel-out*

The serial-in/serial-out shift register (Fig. 10.59(c)) merely acts as a temporary delay circuit. The data is read out in the same order in which it has been stored.

Similarly parallel-in/parallel-out shift registers [Fig. 10.59(d)] act as a temporary storage.

The malfunctioning of a shift register can be mostly due to incorrect control signals. For example:

(a) If the clear line stuck, new data is not accepted by the chip.

(b) If the clock signal is missing, the outputs will not shift.

(c) Some shift registers need an 'enable' signal before the IC will accept data. Check for the presence of this signal before attempting to replace the IC.

10.18.9 Multiplexer

A multiplexer is a logic circuit which accepts several data inputs and outputs, but only one of them at a time. In essence, it behaves as a multi-position switch which operates under the control of SELECT or ADDRESS inputs. Figure 10.60 shows the representation of a digital multiplexer.

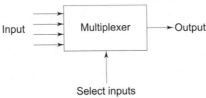

Fig. 10.60 *Symbolic representation of a multiplexer*

Multiplexers are available in the form of integrated circuits. For example, 74151A is an 8-input multiplexer which gives out complementary outputs. The layout of the pins of this multiplexer is shown in Fig. 10.61(a). Figure 10.61(b) shows the truth table of this multiplexer.

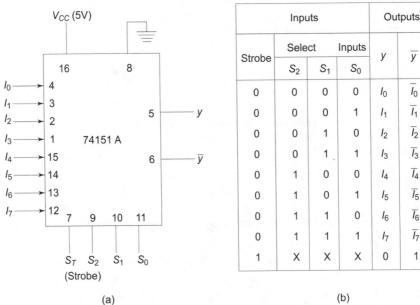

Strobe	Select			Outputs	
	S_2	S_1	S_0	y	$\bar{y}$
0	0	0	0	I_0	$\bar{I_0}$
0	0	0	1	I_1	$\bar{I_1}$
0	0	1	0	I_2	$\bar{I_2}$
0	0	1	1	I_3	$\bar{I_3}$
0	1	0	0	I_4	$\bar{I_4}$
0	1	0	1	I_5	$\bar{I_5}$
0	1	1	0	I_6	$\bar{I_6}$
0	1	1	1	I_7	$\bar{I_7}$
1	X	X	X	0	1

(a)   (b)

Fig. 10.61   *(a) Pin configuration of 74151A multiplexer; and*
(b) Truth table of this multiplexer

Other popular multiplexers are 74153 which is dual 4 : 1 multiplexer and 74150 which is 16 : 1 multiplexer.

10.18.10 De-multiplexer

A de-multiplexer performs the reverse operation of a multiplexer. It receives a single input and distributes it over several outputs. The SELECT input code determines to which output the data input will be transmitted. Figure 10.62 shows a schematic diagram of a de-multiplexer.

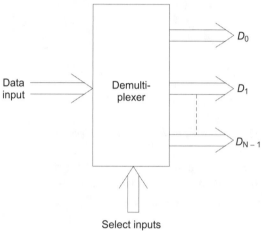

Fig. 10.62 *Schematic representation of a de-multiplexer*

A typical example of a de-multiplexer is 74155 which is a dual 1: 4 de-multiplexer. It converts 2-line inputs to 4-line outputs.

10.18.11 Encoders

In digital circuits including microcomputers, the data is handled in the binary form, whereas the most common language of communication is decimal numbers and alphabetic characters. Therefore, there is a need to devise interface circuits between the digital system and human operators. Several binary codes have been developed which carry out the function of code conversion. The process of generating binary codes is known as encoding.

Some of the commonly used codes are decimal-to-BCD encoder, octal-to-binary encoder and hexadecimal-to-binary encoder. 74147 IC is a priority encoder which can be used for decimal-to-BCD conversion. The block diagram of 74147 is given in Fig. 10.63. Similarly 74148 IC provides octal-to-binary encoding.

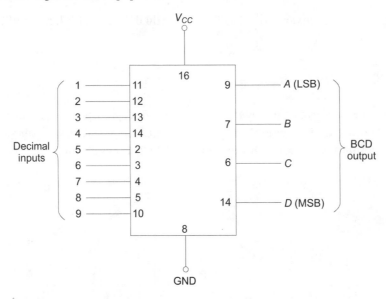

Fig. 10.63   *Block diagram of 74147 decimal-to-BCD encoder*

Hexadecimal code is a commonly used code in microcomputers especially when long binary words are handled. Hexadecimal-to-binary encoder can be realized by using two 74148 octal-to-binary encoders and a data selector.

10.18.12   Decoders

A decoder is a logic circuit which converts the input '*n*' bit binary code into an appropriate output signal to identify which of the possible 2^n combinations is present. The most commonly used decoder is BCD-to-decimal decoder which provides decoding from 4 line-to-10 line decoding function. The input is 4-bit binary information out of which only 10 BCD input codes are used. Correspondingly there are 10 output pins.

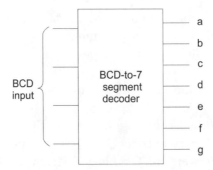

Fig. 10.64   *Block diagram of BCD-to-7 segment decoder*

In the digital display systems, the digits are displayed on 7-segment LEDs. Therefore, it is more convenient if the BCD code is decoded into 7-segments. Thus, BCD-to-7 segment decoder is the most popular display device used in digital systems. Figure 10.64 shows the block diagram of BCD-to-7 segment decoder along with 7-segement LED (Fig. 10.65) display unit. The decoder circuit has four input lines for BCD data and seven output lines to drive a 7-segment display. '*a*' through '*g*' output of decoder are to be connected to '*a*' through '*g*' inputs of the display respectively. The outputs of the decoder can be active-low or active-high and the 7 segments of the display may be 7 cathodes with a common anode or 7 anodes with a common cathode. The decoders normally include drivers in the chip itself. The typical examples of BCD-to-7 segment decoder driver ICs are 7446A and 7447A (active low, open collector) and 7449 (active-high, open collector).

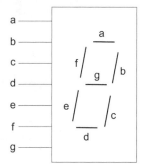

Fig. 10.65   *7-Segment LED display unit*

10.18.13   Tri-state Logic

Tri-state gates are designed so as to give output in three distinct states. Besides the normal two states of logic 1 (high) and logic 0 (low), a third state having a very high output impedance is available in tri-state gates.

Figure 10.66 shows two possible arrangements of tri-state buffers. When the control signal is high, the switch is off and no signals flow through the device. When the control line is low, the switch is on, and the input signals are passed through the output.

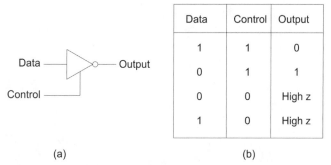

Data	Control	Output
1	1	0
0	1	1
0	0	High z
1	0	High z

(a)   (b)

Fig. 10.66   *TTL tristate invertor: (a) symbol; and (b) truth table*

Tri-state buffers are usually used at each outlet of a data and address bus in microprocessor-based circuits. This is one of the first ICs that a signal encounters as it comes into the computer. This makes these devices more vulnerable and causes a relatively high rate of failure.

Tri-state buffers can be checked with a logic probe. The circuit in which the buffer is used is turned on and the activity on the input, output and control lines is observed. The checks are made for the following:

(a) If the tri-state buffer is enabled, the same signal should appear on the input and output pins.

(b) A bad buffer will show a good input but no output. Also, a bad buffer may show an output stuck in a high or low state.

Tri-state buffers are usually mounted in groups of eight on a single chip. If one stage is faulty, the whole IC has to be replaced.

11

Troubleshooting Microprocessor-Based Systems

11.1 MICROPROCESSORS

11.1.1 Introduction

A microprocessor—also known as a CPU or Central Processing Unit—is a complete computation device that is fabricated on a single chip. A chip is basically an integrated circuit. Generally it is a small, thin piece of silicon onto which the transistors making up the microprocessor have been etched. A chip might be as large as an inch on a side and can contain as many as 10 million transistors. Simpler processors may contain only a few thousand transistors etched onto a chip.

The first microprocessor was the Intel 4004, introduced in 1971. The 4004 was not very powerful when compared with today's microprocessors. All it could do was add and subtract, and it could only do that 4-bits at a time. But it was amazing that everything was on one chip. Prior to the 4004, the computers were built either from a collection of chips or from discrete components. The development of 4004 was responsible for the first portable electronic calculator.

The first microprocessor to make it into a home computer was the Intel 8080, a complete 8-bit computer on one chip introduced in 1974. The first microprocessor to make a real revolution in the market was the Intel 8088, introduced in 1979 and incorporated into the IBM PC (International Business Machines—Personal Computer) which appeared in the market around 1982. If you are familiar with the PC developments and its history, then you must be knowing about the PC market moving from 8088 to the 80286, to the 80386, to the 80486, to the Pentium, to Pentium II, Pentium III and the now Pentium IV. All of these microprocessors are made by Intel and are improvements on the basic design of the 8088. The new Pentiums can execute any piece of code that ran on the original 8088, but the Pentium II runs about 3000 times faster than 8088.

11.1.2 The Microcomputer

A digital computer making use of a microprocessor as the central processing unit (CPU) is called a microcomputer. Because of its low cost, small size and tremendous capabilities, the range of microcomputers applications is so large at present that they are limited only by the imagination of the user. The microcomputer applications include those for industrial control, hobby and personal computers, office automation products, medical equipment, scientific instruments and process involving data acquisition and analysis.

A simplified block diagram of a microcomputer is shown in Fig. 11.1. It consists of the following modules:

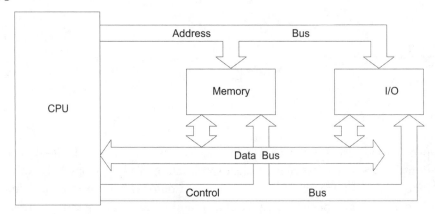

Fig. 11.1 *Block diagram of a microcomputer*

- *CPU:* Contains the central processing unit (CPU), system timing and interface circuitry to memory and I/O devices;
- *Memory:* Contains memory for programme storage and data storage; and
- *Input/Output (I/O) Ports:* Contains circuitry that allows communications with devices outside of the system, such as keyboard and display unit.

The following three buses that inter-connect the blocks:

(a) *Data Bus:* This is a bi-directional bus that allows data flow between CPU and memory or I/O.
(b) *Address Bus:* This is a uni-directional bus that transmits an address that identifies a particular memory location or I/O device.
(c) *Control Bus:* This is a uni-directional bus that carries the signals indicating the type of activity in current process. The type of activities are: memory read, memory write, I/O read, I/O write and interrupt acknowledge.

The memory stores the data to be manipulated by the CPU, as well as the programme that directs that manipulation. A programme is a group of logically related instructions. The

CPU reads each instruction from memory in a logically determined sequence, and uses it to initiate processing actions. The CPU can rapidly access any data stored in memory. Also, the CPU can address one or more input ports, added to receive information from external equipment, and input the data contained therein.

The computer also requires one or more output ports that permit the CPU to communicate the result of its processing to the attached equipment. Like input ports, output ports are addressable, the input and output ports permit the processor to communicate with the outside world.

The CPU unifies the system by controlling the functions performed by the other components. The CPU must be able to fetch instructions from memory, decode their binary contents and execute them. It must also be able to refer the memory and I/O ports as necessary in the execution of instructions. In addition, the CPU should be able to recognize the response to certain external control signals, such as INTERRUPT requests.

Hardware alone does not make a microcomputer. Before any microcomputer, and for that matter any computer, can be put to work, it must be given a set of instructions, known as a programme. It is the programme that states the procedure the computer is to follow in solving the programme at hand. By changing the programme, the same hardware can perform many different functions.

11.1.3   Inside a Microprocessor

The microprocessor is the most important component of a microcomputer. It executes a collection of machine instructions that tell the processor what to do. Based on the instructions, a microprocessor does three basic things:

- Using its Arithmetic/Logic Unit (ALU), a microprocessor can perform mathematical operations like addition, subtraction, multiplication and division.
- A microprocessor can move data from one memory location to another.
- A microprocessor can make decisions and jump to a new set of instructions based on these decisions.

Figure 11.2 shows a simplified diagram of a microprocessor capable of performing the above functions. This microprocessor has:

- an address bus (that may be 8-, 16- or 32-bits wide) that sends an address to memory;
- a data bus (that may be 8-, 16- or 32-bits wide) that can send data to memory or receive data from memory;
- An RD (read) and WR (Write) line to tell the memory whether it wants to set or get the addressed location;
- a clock line that lets a clock pulse sequence the processor; and
- a re-set line that re-sets the programme counter to zero (or whatever) and re-starts execution.

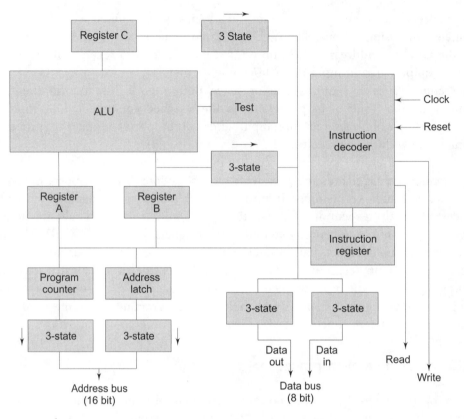

Fig. 11.2 *Simplified block diagram of a microprocessor*

Let us assume that both the address and data buses are 8-bit wide. Thus, the components of this simple microprocessor are:

- Registers A, B and C are simply latches made out of flip-flops.
- The 'Address' latch is just like registers A, B and C.
- The programme counter is a latch with the extra ability to increment by 1 when told to do so, and also to re-set to zero when told to do so.
- The ALU can be as simple as an 8-bit adder or it might be able to add, subtract, multiply and divide 8-bit valves.
- The 'Test' register is a special latch that can hold values and compare two numbers to determine if they are equal or if one is greater than the other etc. It stores these values in flip-flops and then the instructions decoder can use the values to make decisions. The test register can also normally hold a carry bit from the last stage of the adder.

- The '3-State' boxes represent tri-state buffers. A tri-state buffer can pass a 1, a 0 or it can essentially disconnect its output. A tri-state buffer allows multiple outputs to connect to a wire, but only one of them to actually drive a 1 or a 0 onto the line.
- The instruction register and instruction decoder are responsible for controlling all the other components.

Most practical microprocessors use 40 or more pins, and even keeping to that number requires that the eight data pins be used for both reading and writing. Figure 11.3 shows a representation of a typical CPU and the different types of packages in which they are available.

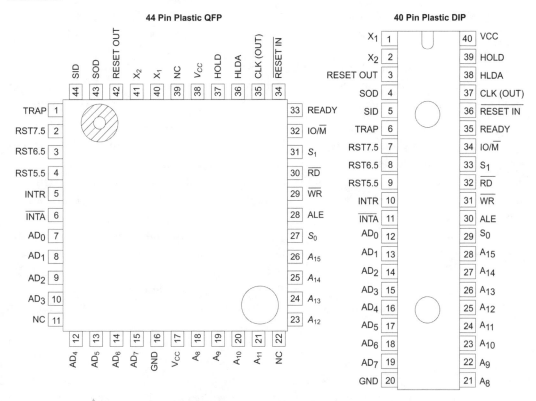

Fig. 11.3 *Typical CPU Pin-configuration (Intel 80C85) and different packages of 80C85*

As already mentioned above, the function of the microprocessor is to manipulate data in accordance with the instructions stored in the memory. For this, the microprocessor transfers data and internal state information via an 8-bit bi-directional 3-state data bus. Memory and peripheral device addresses are transmitted over a separate 16-bit 3-state address bus. Timing and control outputs are given out for synchronization and control inputs like re-set, hold and ready and interrupt are used to perform specific functions.

Data Lines   Data lines are used to transfer instructions or data from memory or data from input device to the microprocessor. Data lines are also used to transfer results from microprocessor to memory or output devices. The number of data lines are related to the word length of the computer. Normally, an *n*-bit processor has *n* data lines. In an 8-bit microprocessor, one bus of eight lines can be used to carry all the data signals, whether for reading or writing. To ensure that this operation is correctly carried out, a read/write control is provided in the microprocessor package. A bus which carries signals in both directions is called a bi-directional bus and the data bus is the only one which is bi-directional.

The processor's internal data bus is isolated from the external data bus by an 8-bit bi-directional 3-state buffer. In the output mode, the internal bus content is loaded into an 8-bit latch that, in turn, drives the data bus output buffers. The output buffers are switched-off during input or non-transfer operations.

During the output mode, data from the external data bus is transferred to the internal bus. The internal bus is pre-charged at the beginning of each internal state, except for the transfer state.

Address Lines   These are employed to address the memory to fetch instructions or data from it. These are also used to address and connect I/O devices to the microprocessor. The number of address lines determine the size of the memory a particular microprocessor can handle or the maximum number of I/O devices that can be connected to it.

A typical microprocessor has sixteen pins as address outputs (Fig. 11.3) which are labelled as A_0 to A_{15}. They are used to select particular locations in the memory. For example, the first signal on the address lines will comprise of sixteen logic 0, address 0. As the microprocessor starts operating, A_0 will change to 1 and when the read/write control signal is used for reading, a byte will be fetched from that particular memory location corresponding to this address. After this, the address will step at the end of the first instruction and call for the next instruction.

The sixteen address lines can carry up to 2^{16} (65, 536) bytes of information which is equivalent to a 64K memory. Microprocessors in general will, however, need much smaller memory for small machines. Thus, there will always be unused address lines which can be connected to video display unit or output signals to other circuits under the control of the programme.

Multiplexing is a useful technique employed with microprocessors for obtaining additional address and data lines. For example, by multiplexing, it is possible to handle 16-bit data signals, through an 8-line data bus. Similarly, the address lines and data lines are also multiplexed and we have address and data available on the same physical lines at different instants of time. However, the result of any time-multiplexing is slowing down the system.

In the context of microprocessors, the following terms must be understood:

- **Microns** is the width, in microns, of the smallest wire on the chip. For comparison, a human hair is 100 microns thick. As the feature size on the chip goes down, the number of transistors rises. Modern microprocessors operate on 0.25 micron technology.
- **Clock speed** is the maximum rate that the chip can be clocked. The clock speed of a modern microprocessor is nearing 2 GHz.
- **Data Width** is the width of the ALU. An 8-bit ALU can add/subtract/multiply/etc. two 8-bit numbers, while a 32-bit ALU can manipulate 32-bit numbers. An 8-bit ALU would have to execute four instructions to add two 32-bit numbers, while a 32-bit ALU can do it in one instruction. In many cases the external data bus is of the same width as the ALU, but not always. The 8088 had a 16-bit ALU and an 8-bit bus, while the modern Pentiums fetch data 64 bits at a time for their 32-bit ALUs.
- **MIPS** stands for Millions of Instructions Per Second, and is a rough measure of the performance of a CPU. The MIPS value of modern processors is around 1000 as compared to 1 MIPS for 80286 processor.

11.1.4 8085 Microprocessor

One of the most popular microprocessors used in electronic equipment is the Intel 8085. The 8085 is an 8-bit parallel, central processor unit contained on a single 40-pin (DIP) or 44-pin (QUAD) LSI chip (Fig. 11.3). Figure 11.4 gives the functional blocks within the 8085. It transfers data and internal state information over an 8-bit, bi-directional data bus. Memory and device addresses are transmitted over a multiplexed 16-bit address bus, capable of addressing as many as 65, 536 memory locations. Six timing and control outputs (SYNC, DBIN, WAIT, WR, HLDA and INTE) emanate from the 8085. Instructions for the 8085 are located in memory, from where they are fetched and executed sequentially. There are over 70 separate instructions possible, though many are similar, the only difference being in various CPU internal registers specified.

The 8085 contains the following:
- (a) Instruction register;
- (b) Program counter;
- (c) Memory address register;
- (d) Stack pointer;
- (e) Arithmetic and logic unit;
- (f) Bi-directional, 3-state data bus buffer; and
- (g) Other assorted registers and logic elements.

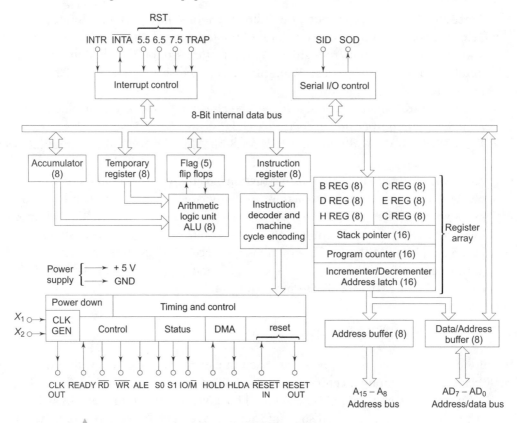

Fig. 11.4   *Functional blocks within the Intel 80C85*

The instruction register contains the 8-bit instruction code. The programme counter maintains the 16-bit memory address of the current programme instruction and is incremented automatically during every instruction fetch. The memory address register, referred to as the H,L register pair, is made up of two 8-bit registers and is used to address memory during the read and write memory instruction. The stack pointer maintains the address of the next available main programme instruction while an interrupt sub-routine is being executed. The arithmetic and logic unit (ALU) performs the arithmetic and logical operations. The remaining elements internal to the microprocessor perform the input and output operations over the data bus.

The data bus isolates the processor's internal bus from the external data bus (D0–D7). In the output mode, the contents of the internal bus are loaded into a latch that drives the data bus output buffers. During the input mode, data from the external data bus is transferred to the internal bus.

The processor has five interrupt inputs. These interrupts are arranged in a fixed priority that determines which interrupt is to be recognized if more than one is pending. The interrupts are INTR, RST 5.5 RST 6.5, RST 7.5, and TRAP. The TRAP interrupt is useful for catastrophic events such as power failure or bus error. The manufacturer data-sheets may be referred to in order to get details of the functioning of various controls and functions available on the microprocessor.

The modern CMOS technology-based microprocessors operate from single +3 to +6 volt power supply, have a power-down mode with on-chip clock generator. Timing for the 8085 is developed from an external clock source or with a crystal. The crystal should be of a minimum frequency of 6 MHz for a clock frequency of 3 MHz.

Basic CPU Operation: Operation of the CPU is divided into time periods called 'cycles' and 'states'. There are two types of cycles: instruction cycles and machine cycles.

Instruction Cycle: An instruction cycle includes both the fetching of the instruction from memory and the execution of the instruction. Each instruction can be either one, two, or three 8-bit bytes in length. Multiple byte instructions are stored in successive memory locations.

Machine Cycle: A machine cycle is required each time an I/O port or the memory is accessed. Each instruction cycle can contain from one to five machine cycles. There are ten different types of machine cycles possible, including:
  (a) Instruction Fetch;
  (b) Memory Read;
  (c) Memory Write;
  (d) Stack Read;
  (e) Stack Write;
  (f) Input;
  (g) Output;
  (h) Interrupt Acknowledge;
  (i) Halt Acknowledge; and
  (j) Interrupt Acknowledge while in Halt.

11.1.5 Microprocessor Instructions

A microprocessor executes a collection of machine instructions that tell the processor what to do. Even the simplest microprocessor will have a fairly large set of instructions that it can perform. The collection of instructions is implemented as bit patterns, each one of which has a different meaning when loaded into the instruction register. Humans are not particularly good at remembering bit patterns, so a set of short words are defined to represent the different bit patterns. This collection of words is called the **assembly language** of the processor. An **assembler** can translate the words into their bit pattern very easily, and then the output of the assembler is placed in memory for the microprocessor to execute. Every instruction can be broken down as a set of sequenced operations that manipulate the components of the microprocessor in the proper order. Some instructions, like ADD instruction, might take two or three clock cycles. Others might take five or six clock cycles.

11.2 SEMICONDUCTOR MEMORIES

Memory is technically any form of electronics storage. A digital memory is an array of binary storage elements arranged in a manner that it can be externally accessed. The memory array is organized as a set of memory words. Each word consists of a number of single bit storage elements called *memory cells*. The word length of a memory word is typically one, four or eight memory cells. Therefore, 1 bit, 4 bits or 8 bits (byte) of information can be stored by the memory word respectively. The memory capacity is the product of the number of memory words and the number of memory cells in each word. It is measured in bits and frequently expressed in kilobits where 1 kilobit $= 2^{10} = 1024$.

All microprocessor-based systems make use of two types of memory—ROM (Read Only Memory) and RAM (Random Access Memory).

11.2.1 Random Access Memory (RAM)

Random access memory (RAM) is used in a microprocessing system to store variable information. The CPU (central processing unit) under programme control can read or change the contents of a RAM location as desired. RAMs comprise a generic category that encompasses all memory devices in which the contents of any address can be accessed at random in essentially the same time as any other address.

There are two types of RAMs: static and dynamic. Both dynamic and static MOS random access memories are popular; the dynamic ones for their high circuit density per chip and low fabrication costs and the static RAMs for single power supply operation, lack of refresh requirements and low power dissipation. Bi-polar RAMs are used for very high speed scratch-pad memories.

Dynamic RAM: In the dynamic RAM, information is stored as electrical charge on the gate capacitance of MOS transistors. Since these capacitors are not perfect, the charge will leak away and the information is likely to be lost with time if the charge is not periodically refreshed. This can be done in several ways and depends upon the type of device in use.

Static RAM: Static RAM does not need to be refreshed, as the memory cells are bi-stable and similar in design to conventional flip-flops. In general, a static RAM consumes more power than its dynamic counterpart. However, it requires less support circuitry. Also, there are no problems of synchronizing the memory refresh cycles with normal CPU read and write operations.

When the information is stored in the memory, it is written into the memory. When information is retrieved from a semiconductor memory, it is read from the memory. These are the only two functions that are done to static memories. Writing information into a memory is done in a write 'cycle'. Reading information from a memory is done in a read cycle. The term 'cycle' means a fixed period of time required to perform the function of writing into or reading from a memory. In fact, the electrical data or information is stored as a level of DC voltage. One DC voltage level corresponds to a '1' stored in the memory. A different DC voltage level corresponds to a '0' being stored in the memory.

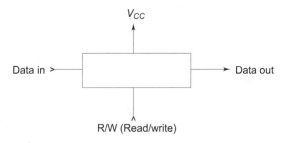

Fig. 11.5 *Block diagram of one-bit memory*

In a semiconductor memory (Fig. 11.5), data is entered on an input pin on the physical device labelled 'data in'. Data being read from a memory, are read on a device output pin labelled 'data out'. Therefore, one bit memory device will have four major physical connections: power input (V_{CC}), Data input (D_1), Data output (D_0) and read cycle or write cycle (R/W). If we want 16 bits of information storage, four address pins will be required ($2^4 = 16$) in the memory device (Fig. 11.6). Under those conditions, no particular sequence will be needed to read or write information in the memory.

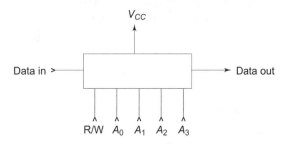

Fig. 11.6 *Block diagram of a 16-bit memory*

Figure 11.7 shows pin configuration, logic symbol and block diagram of a typical 1 K × 1 bit static random access memory, 2125A from intel. It is packaged in 16-pin dual in-line package and operates on single +5V supply. This is directly TTL compatible in all respects, inputs, outputs. It has three state data output and can be used for memory expansion through chip select ($\overline{CS}$) enable input. Besides 10 address input lines for addressing all the 1024 words, it also has control to choose either READ or WRITE operation. When chip enable (CE) is high, the D_{out} is a high impedance state.

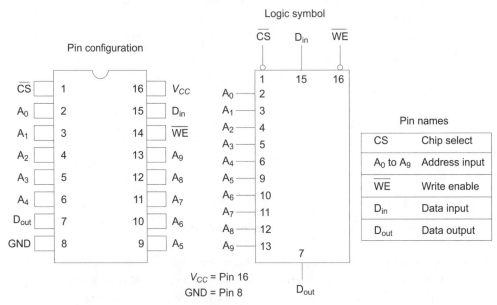

Fig. 11.7 *Pin configuration of Intel 2125A*

11.2.2   Read Only Memory (ROM)

In a microprocessor-based system, ROMs are normally used to hold the programme of instructions and data constants such as look-up tables. Unlike the RAM, the ROM is non-volatile, i.e. the contents of the memory are not lost when the power supply is removed. Data stored in these chips is either unchangeable or requires a special operation to change. This means that removing the power source from the chip will not cause it to lose any data.

There are five basic ROM types. These are
- ROM;
- PROM;
- EPROM;
- EEPROM; and
- Flash Memory.

(a) *Mask Programmed ROMs:* These are programmed by the manufacturer to the user's requirements. ROM chips contain a grid of columns and rows and use a diode to connect the lines if the value is a (1). If the value is (0) then the lines are not connected at all. This type of ROM is only used if fairly large number of units are required, because the cost of preparation of creating the bit pattern on the chip is quite high. The contents of these ROMs cannot be altered after manufacture. Once the chip is made, the actual chip can cost very little money. They use very little power, are extremely reliable and, in the case of most small electronic devices, contain all the necessary programming to control the device.

(b) *Programmable Read Only Memory (PROM):* This is programmed by the user. Selectively fusing (open-circuiting) the metal or poly-silicon links in each memory cell sets that cell to a fixed state. The process is irreversible. In one form of PROM, the information is stored as a charge in a MOSFET cell. Blank PROM chips can be coded with anyone along with their programmer.

PROM chips have a grid of columns and rows just as ordinary ROMs do. The difference is that every intersection of a column and row in a PROM chip has a fuse connecting them. A charge sent through a column will pass through the fuse in a cell to a grounded row indicating a value of "1". Since all the cells have fuse, the initial, or blank, state of a PROM chip is all "1"s. To change the value of a cell, to "0", the programmer is used, which sends a specific amount of current to the cell. The higher voltage breaks the connection between the column and the row by burning out the fuse. The process is known as **burning the PROM**.

The contents of a PROM can be erased by flooding the chip with ultraviolet radiation. Following this process, a fresh pattern can be entered. PROMs are used in the microprocessor-based systems during the system development phase and in the production system when the total production run is not high enough to justify the use of mask-programmed ROMs.

PROMs can only be programmed once. They are more fragile than ROMs. A jolt of static electricity can easily cause damage to the chip. But blank PROMs are inexpensive and are great for prototyping the data for a ROM before committing to the costly ROM fabrication process.

(c) *Erasable Programmable Read Only Memories (EPROM):* These devices provide the facility of re-writing the chips several times. EPROMs are configured using an EPROM programmer that provides voltage at specified levels, depending upon the type of EPROM used.

For erasing the chips of its previous contents, an EPROM requires a special tool that emits a certain frequency of ultra-violet (UV) light. Because the UV light will not penetrate most plastics or glasses, each EPROM chip has a quartz window on top of the chip. The EPROM is kept very close to the eraser's light source, within an inch or two, to work properly. An EPROM eraser is not selective, it will erase the entire EPROM. The EPROM must be removed from the device it is in and placed under the UV light of the EPROM eraser for several minutes. An EPROM that is left under UV light too long can become over-erased. In such a case, the chip cannot be programmed.

(d) *Electrically Erasable Programmable Read Only Memories (EEPROM) or Read-Mostly Memories (RMM):* These are designed such that the contents of these memories can be altered electrically. However, this is a fairly slow process. It often requires voltages and circuit techniques that are not commonly found in normal logic circuitry.

EEPROMs remove the following drawbacks of EPROMs :
- The chip does not have to be removed to be rewritten;
- The entire chip does not have to be completely erased to change a specific portion of it; and
- Changing the contents does not require additional dedicated equipment.

(e) *Flash Memory:* This is a type of EEPROM that uses *in-circuit wiring* to erase by applying an electrical field to the entire chip or pre-determined sections of it called *blocks.* Flash memory works much faster than traditional EEPROM because it writes data in chunks, usually 512 bytes in size, instead of a byte at a time.

Figure 11.8 shows a typical symbol of ROM, for storing 1024 8-bit words. This is also called a 1K × 8 ROM where 1K represents 1024. Similarly, a 2048 × 8 can be written as a 2 K × 8 and so on. Since 1 K ROM stores 1024 different words, it needs 10 address inputs ($2^{10} = 1024$). The word size is 8-bits, so there are eight output lines. The memory chip is enabled or disabled through the chip select (CS) input. ROMs do not provide for data input or read/write control because they do not normally have write operation. Some ROMs do provide for special input facilities for initially writing the data into the ROM which is generally shown on the symbol.

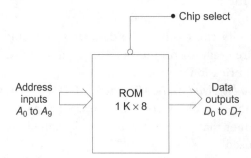

Fig. 11.8   *Typical symbol of ROM*

The Intel 2716 is a 16,384 (2K × 8) bit ultra-violet erasable and electrically programmable read only memory. Figure 11.9 shows pin configuration and block diagram of the 2716 EPROM.

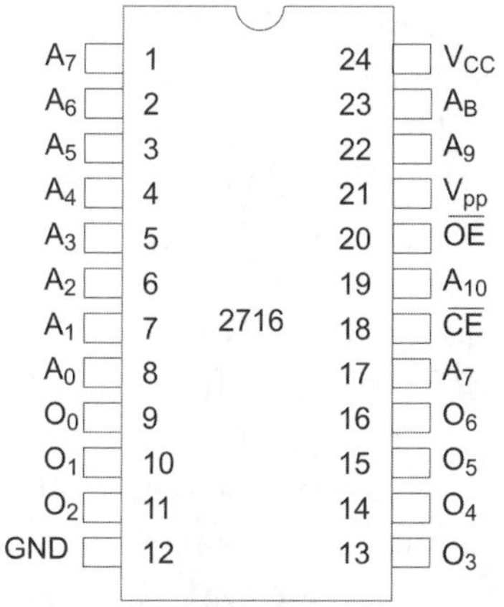

```
        A7  [   1         24   ]  VCC
        A6  [   2         23   ]  AB
        A5  [   3         22   ]  A9
        A4  [   4         21   ]  Vpp
        A3  [   5         20   ]  OE
        A2  [   6         19   ]  A10
        A1  [   7   2716  18   ]  CE
        A0  [   8         17   ]  A7
        O0  [   9         16   ]  O6
        O1  [  10         15   ]  O5
        O2  [  11         14   ]  O4
       GND  [  12         13   ]  O3
```

Pin names	
$A_0 - A_{10}$	Address
$\overline{CE}$	Chip enable
$\overline{OE}$	Output enable
$O_0 - O_7$	Outputs

Fig. 11.9 *Pin configuration of Intel 2716 EPROM*

The 2716 has a stand-by mode which reduces the active power dissipation by 75%. The 2716 is placed in the stand-by mode by applying a TTL high signal to the $\overline{CE}$ input. When in stand-by mode, the outputs are in a high impedance state, independent of the $\overline{OE}$ input.

The 2716 provides two line control functions which allow for the low power dissipation and assurance that output bus contention will not occur. To use these control lines, $\overline{CE}$ (pin 18) is decoded and used as the primary device selecting function while $\overline{OE}$ (pin 20) be made a common connection to all devices in the array and connected to the READ line from the system control bus. Thus the output pins are active only when data are desired from a particular memory device.

11.3 MICROCONTROLLERS

A microcontroller is a small inexpensive computer usually employed for controlling devices based on the input which is sensed from the real world. Most electronic devices that we use today have a microcontroller in them in some form or another. They are easy to use with some sensors and output devices and they can communicate with desk top (personal) computers. A microcontroller is basically a single chip microcomputer provided in a single integrated circuit package, which contains a CPU, clock circuitry, ROM (flash), RAM, serial port, timer/counter and I/O circuitry. As such, it does not require a host of associated chips for its operation, as conventional microprocessors do.

Most of the microcontrollers come in 40-pin DIP packages; the pin-out consists of essentially up to 32 I/O lines with the remainder being used for power, re-set interrupt and timing. The instruction set of a single chip microcomputer generally bears a close resemblance to that of the microprocessor family to which the microcontroller belongs.

Microcontrollers offer several advantages over conventional multi-chip systems. There is a cost advantage as extra chips and printed circuit board and connectors required to support multi-chip systems are eliminated. These are essential requirements when designing small control systems or similar items such as high volume consumer products, requiring relatively simple and cheap computer controllers.

Microcontrollers have been available for a long time. Intel introduced 8048 (MCS-48, an 8-bit microcontroller in 1976). It has developed upwards and its new version is 8051 family, which is a high performance 40-pin DIP package. Microcontrollers functionality has been tremendously increased in recent years. Today, one gets microcontrollers which are stand alone for application in data acquisition systems and control. They have A/D converters on the chip which enable their direct use in instrumentation.

One of the popular microcontrollers available in the CMOS technology is 89C51, which has 4K bytes of flash programmable and erasable read-only memory (EEPROM). Its instruction set and pin-out are compatible with the industry standard MCS-51. Flash allows the programme memory to be re-programmed in-system or by conventional non-volatile memory programmer. This microcontroller is available in both DIP and QUAD packages.

The procedure for troubleshooting microcontroller-based equipment follows the same logical steps as are used for other digital systems and microprocessors.

11.4 MICROPROCESSOR-BASED SYSTEMS

Every microprocessor based system is composed of various functional sections (Fig. 11.10), which are generally as follows:

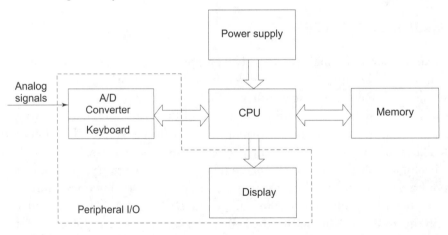

Fig. 11.10   *Various functional blocks in a microprocessor-based system*

- *CPU Section:* comprising microprocessor, clock and re-set circuitry and local bus buffers and latches;
- *Memory Section:* comprising of read/write and Read Only Memories;
- *Peripheral I/O Section:* which can be further sub-divided depending upon a specific application, of which some kind of keyboard and display are important; and
- *Power Supply.*

It is useful to consult the block diagram, which is normally available in the service manual. If it is not available, you may draw one on paper. The good starting point for the microprocessor part of a microprocessor-based product is as shown in Fig. 11.11. The block diagram lets you identify the major product components.

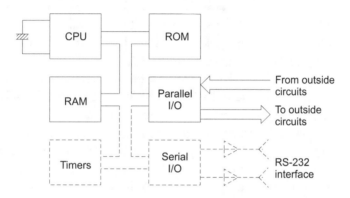

Fig. 11.11 *The basic block diagram for the microprocessor system part of any microprocessor-based product*

11.5 TROUBLESHOOTING TECHNIQUES

The failure of a microprocessor-based system can be due to a hardware or a software problem. It is preferable to test the system for a software bug before taking the system apart for failure analysis. It is also important to have any or all of the manufacturer's documentation about the system. This documentation may include operation manuals, programming manuals, instruction set cards, logic/schematic diagram, troubleshooting and testing manuals and equipment assembly information. The following three step approach will be useful to troubleshoot microprocessor-based systems:

- *Understand the system and the problem,* which involves knowing or being able to find:
  - control signals produced by a microprocessor;
  - functions of the peripheral devices to which the control signals are sent; and
  - logic levels that must appear at specific points to accomplish a particular function.

- Have a *test plan* for localizing problems and eliminating unaffected sections; and
- Have proper *test equipment* to implement the test plan.

Failures generally occur in circuits that are used or stressed the most. The most frequently used devices include the RAM and ROM memory chips, the microprocessor and the input/output devices. The microprocessor is a highly reliable device and does not fail very often. Most commonly the problem comes in the power supply section, the problem may be a harmless loose power connector to a grievous short circuit. Most often, a problem in power supply causes failure of discrete components like transistors, diodes, resistors and capacitors, etc.

The digital ICs (TTL or MOS) which connect the microprocessor and the peripheral devices fail far more often as compared to the discrete components. The main cause is thermal stress. These devices heat up when the system is on and cool down if the system is off. This hot-cold-hot effect can cause a break in the connection of a wire leading from inside the chip to a pin, producing an open circuit. In such a case, the chip may work intermittently or simply refuse to work at all. An output can be stuck at 1 or bestuck at 0.

Hardware problems can normally be identified using specialized troubleshooting tools such as logic clip, logic probe, logic pulser, current tracer, logic analyser, signature analyser, in-circuit emulator and specialized microprocessor troubleshooters.

Most microprocessor-based products are susceptible to transient electrical noise much more than in other electronic equipment. Transient electrical noise can come from voltage spikes on power lines or other communication cables or directly from static discharge to the product. The electrical transients may cause a program to generate incorrect data or to crush. Similarly, if the transient reaches the processor, it may charge the programme centre values. As the microprocessor depends upon exact register and memory values, even a single bit error in a programme counter can usually result in an immediate crash.

11.5.1 Testing Microprocessors

Being a part of the microprocessor system, microprocessors interact with several other sub-systems. With all these connections, the CPU needs to be checked in a very careful and systematic manner. The troubleshooting procedure consists of checking important signals on CPU which could give a fairly good idea about the working of a CPU. The following are some of the important signals which must be checked.

Clock Signal: A faulty clock signal can completely upset the working of the computer. It may stop working or may show intermittent behaviour. In some microprocessors, the clock generator is built separately and it is carried to the CPU at pins usually marked on the pin diagram as CLK or $\phi 1$, $\phi 2$. In most modern CPUs, the clock generator is built within the *chip* and it only requires an external crystal. The clock pulse can be checked with a logic probe or oscilloscope. If the clock circuit is faulty, check the capacitor or the crystal which usually go bad.

Re-set System: The operation of the re-set key on the computer makes the computer go back to a known starting point. Check for 're-set' signal on the pin of the CPU. Normally, the pin should go low when the 're-set' key is pressed and then return to high. The presence of the 're-set' signal can be checked with a logic probe. If no activity is observed, check for a fault in the lines that carry the 're-set' signal.

Address and Data Lines: Usually there could be two problems:
(1) A short circuit or any other problem with one of the data or address lines on the printed circuit board; or
(2) A problematic IC served by these lines.

A quick check of the data and address lines can be made by using a logic probe. Each line should show activity which should be observed carefully and analysed for the following troubles:
(1) If the activity appears to be stopped in a steady high, it suggests that the line is shortened to the + 5V DC power supply, or
(2) In case the activity shows a steady low, it is an indication that the line is shortened to ground.

All the address and data lines should be checked in this way. Address and data lines are associated with tri-state buffers. These drivers ensure that correct devices are connected to the bus at the correct moments. If there is a failure in any one of the buffers, correct signals will not reach the CPU.

Supposing a problem has been located in a certain line which does not show short or break and the tri-state buffers associated with this line are normal, there is the possibility of a problem in the other ICs associated with the line.

As we observe from Fig. 11.4, both the address and data lines run from the CPU to RAMs and ROMs. Check these ICs as per the procedure given for these devices.

If everything is found to be OK replace the CPU.

11.5.2 Testing Random Access Memory (RAM)

For testing a RAM in a system, proceed as follows:
(a) Touch the surface of the RAM. A bad RAM will often feel warm to the touch.
(b) Using a logic probe, observe the activity on the address and data lines:
  (i) The address lines should show constant activity.
  (ii) Observe for data flow into and out of the chip over the data in and data out lines.

If these signals appear to be normal, it is quite possible that the chip is not storing and reproducing information accurately. This is generally tested with diagnostic software, similar to the procedure described for ROM in the next section. The software writes a pattern into each memory space in each RAM and it verifies those spaces and sees if the correct information is stored. In case of a problem, the software will tell which chip is bad.

In case of dynamic RAMs, if the refresh signal is delayed or absent, the RAMs may become unreliable. When testing a dynamic RAM in a circuit, find out which line handles the refresh function and check for the correct signal on the line.

Most semiconductor random access memory is volatile, i.e. the stored information is lost when the power supply is removed. The problem is avoided by using a battery-maintained power supply.

11.5.3   Testing Read-Only Memory (ROM)

(a) Testing of ROMs is usually difficult without proper test equipment. Of course, with the help of a logic probe, it is possible to determine whether the chip is working or not. The power supply is checked first and the activity is observed on the input, output and control lines of ROM.

(b) In most of the equipment, the chips are mounted on IC sockets. They can be easily removed and replaced to isolate a faulty chip. The replacement action should be carried out step by step to see if the equipment has begun to operate normally. However, ROM replacement procedure does not work if there is a problem in the monitor (ROM), i.e. the chip that gets the equipment started.

(c) The quickest way to test a ROM in a computer is to use a diagnostic software program which is usually provided if the ROM is used in a microcomputer. The diagnostic program verifies the data that is supposed to be stored in each ROM located and compares this with the data from a ROM that is known to be good. Any discrepancy is thus located. To run the diagnostic program, the computer must be operating, which is not possible with some ROM problems.

11.6 DATA CONVERTERS

In the world of electronic equipment, most circuitry is concerned with acquisition, amplification and processing of signals which are available in an analog form. However, the advent of low cost and highly programmable microprocessors and compact digital computers has made possible the digital manipulation of analog signals after they have been converted into a digital form. On the other hand, the display systems are often of analog type, requiring the digital output to be converted into an analog form. The class of devices for converting analog signals to digital form and digital signals to analog form are called Data Converters. They are obviously of two types:

(1) *Analog-to-Digital Converter (ADC)* is a device which has an analog signal as its input and a digital representation of that input as its output.

(2) *Digital-to-Analog Converter (DAC)* is a device which receives a digital input and outputs an analog quantity.

Figure 11.12 shows the principle of data converters. It may be seen that the output of a DAC is purely analog, but it is quantized and made up of discrete steps. This is because the input to DAC is digital. There can only be a finite number of input states, and there can be only a finite number of output states. The size of the discrete steps depends on the resolution and conversion rate of the system.

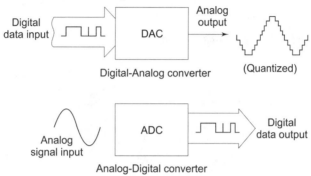

Fig. 11.12 *Principle of data converter*

The *resolution* of an ADC is defined as the smallest analog change that can be discriminated. From a digital point of view, resolution implies the number of discrete steps into which the full analog input signal range of the converter is divided. It is usually expressed as a number of bits (binary) or a number of digits. For example, in a 12-bit ADC, if the input voltage range is ±5V, the smallest analog change that can be resolved is $10 / 4096$ (2^{12}) = 2.44 mV dc.

Conversion Time is the time elapsed between application of a convert command and the availability of data at its output.

The working of a data converter is usually checked by using an oscilloscope and observing the analog/digital waveform at the input/output terminals.

11.7 DATA ACQUISITION SYSTEMS

The analog voltage signals must be converted into digital form in order to communicate with a microprocessor. Converting analog signals to digital signals is the essence of data acquisition, whose functional block diagram is shown in Fig. 11.13 whereas Fig. 11.14 gives various sub-systems in a microprocessor based data acquisition system. It basically consists of a multiplexer, buffer (differential amplifier), sample and hold amplifier, A/D converter and controlled logic. These components operate under the control of interface logic that automatically maintains the correct order of events.

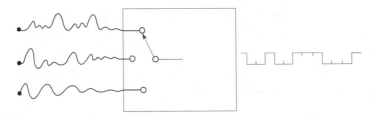

Fig. 11.13 *Data acquisition system*

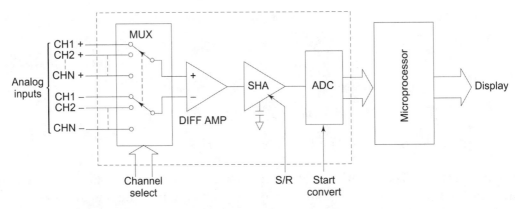

Fig. 11.14 *Microprocessor-based data acquisition system*

Multiplexer: The function of the multiplexer is to select an analog input channel and connect it to the differential amplifier. Although the number of channels is usually 8 or 16, the channel capacity can be easily expanded to accommodate larger number of inputs with the addition of the necessary external hardware.

Differential (Instrumentation) Amplifier: This amplifier conditions the selected input signal to a suitable level for application to the analog/digital converter. It provides impedance buffering, signal gain and common mode rejection.

Sample and Hold Amplifier: The analog/digital converter requires a finite time for the conversion process, during which time the analog signal will still be changing according to its frequency components. It is therefore necessary to sample the amplitude of the input signal and hold this value on the input to the A/D converter during the conversion process, thereby providing an essentially constant voltage to the A/D converter throughout its conversion cycle.

A/D Converter: This block carries out the process of the analog-to-digital conversion. The resolution of A/D converter is usually 8, 10 or 12 bit.

Control Logic: This provides the necessary interface between the microprocessor bus and the elements of the acquisition unit in providing the necessary timing control. It helps to ensure that the correct analog signal is selected and sampled at the correct time, initiates the A/D conversion process and signals to the microprocessor on completion of conversion.

11.8 SPECIAL METHODS OF TROUBLESHOOTING LSI-BASED SYSTEMS

Digital systems making use of LSI (Large Scale Integrated Circuits) components like microprocessors, ROMs and complex interface devices are usually bus-oriented machines. Unlike random logic circuits, LSI-based machines defy fault analysis unless adequately supported by documentation and extensive use of circuit isolation techniques.

Several approaches have been tried out to find failing components on digital circuits boards. They mostly use stimulus response testing techniques, all of which have had some shortcomings. Some simply do not test a set of realistic input conditions, while others perform well at detecting logical errors and stuck nodes but fail to detect timing-related problems. In view of this, test systems capable of detecting one-half to two-thirds of all possible errors occurring in a circuit have been considered quite good. These systems tend to be complex, for factory based use only, as they are computer-driven requiring program support, software packages and specific hardware interfaces for each type of device or board to be tested. Therefore, field troubleshooting beyond the logic probe capability to detect stuck nodes is not very common, apart from the exchange of complete boards.

11.8.1 Problems of Troubleshooting LSI-based Systems

The testing of LSI based digital circuits and systems presents special problems for the following reasons:

- **Complexity:** Chip complexity has reached a state that a single microprocessor may contain more gates than a board containing 100–200 MSI devices. The board on which a microprocessor is placed may contain several other complex LSI chips. In a stimulus-response testing technique, such complexity makes the task of generating effective task sequence very difficult.
- **Speed:** Boards containing LSI chips typically exhibit greater susceptibility to timing-related parameters. Many such devices function only under dynamic conditions and cannot therefore be tested under static conditions. This implies that LSI board testers must work at rates corresponding to the sub 1 μs cycle level.
- **Fault Isolation:** The presence of bi-directional busses and specified input-output (I/O) devices in the LSI boards can make the diagnosis of faults to the component level a difficult task. It may be possible to isolate a faulty node, but the problem of

determining which of the several ICs connected to the bad node is generating the fault is important. For a bi-directional bus structure, the input to one device becomes an output for another requiring resolving a feedback loop system.

- **Visibility:** With the increased chip complexity, visibility of functionality has decreased. There are simply too many functions, too many states and too much memory to test the entire spectrum of circuit functions.
- **Long Data Streams:** LSI-based systems are characterized by long digital data streams rather than with repetitive waveforms so that conventional test equipments like oscilloscopes and multi-meters are no longer helpful. Further, more circuit designs have already evolved from LSI devices to VLSI to UHSI (ultra high scale integration), making troubleshooting a challenging job.

11.8.2 Troubleshooting Alternatives for LSI Devices

Board Exchange: Board exchange is the method of choice for field servicing of LSI-based systems. It minimizes downtimes and results in economies of scale through centralized board repair. It enables the field personnel to repair a wide range of products with minimum training. However, board exchange systems are normally supported by expensive board test systems and huge idle inventories.

Logic Analysis: Logic analysis is an excellent aid for troubleshooting complex systems but in a field test environment, the logic analysis is not really useful, requiring as it does, rightly trained personnel. Also, finding faulty components in a microprocessor or bus-structured product using a logic analyser requires a detailed knowledge of circuits under test and can be very time-consuming.

Signature Analysis: In order to carry out component level field repair, it is required to compress the long data streams present in a system running at its normal operating speed, into a concise, easy to interpret and meaningful readout. In fact, the readout need not tell any more than whether or not a particular node is operating correctly.

Signature analysis is an effective technique to test a board with a microprocessor and other peripheral chips. Signature analysis is a synchronous process whereby activity (response) at an electrical node, referenced to a clock signal, is monitored for a particular stimulus condition during a measurement time period. The complex data stream pattern at the node being tested is fed into a 16-bit shift register whose output is then decoded to give a 4-digit signature. Correct signatures for a particular circuit are determined empirically from a known good product. Testing is performed by probing inter-dependent nodes to determine the functional origin of bad signatures. If a bad signature is found, the signatures of lower nodes are checked until a component can be located with good signatures on the input but bad signatures on the output.

Isolation of faults to the components level on circuit boards which employ feedback connections is dependent upon hardware and software capabilities for breaking the feedback paths. The effective signature analysis troubleshooting of circuits with feedback can be obtained by breaking the feedback path at appropriate points and providing the correct stimulus.

How would a service engineer use signature analysis if a product failed?

The pre-requisite is that the signature analysis method is designed into the product by the following:

(a) Instructions on the schematic or in the service manual show how to switch the product into the diagnostic mode and how to connect the signature analyser to the device under test. Each node on the schematic is marked with a signature (Fig. 11.15). With the aid of the schematic, read the output signatures of the device under test. If they are bad, trace back to a point in the circuit where a good signature appears at the input side of a component and a bad on at the output side. This is called 'backtracking'.

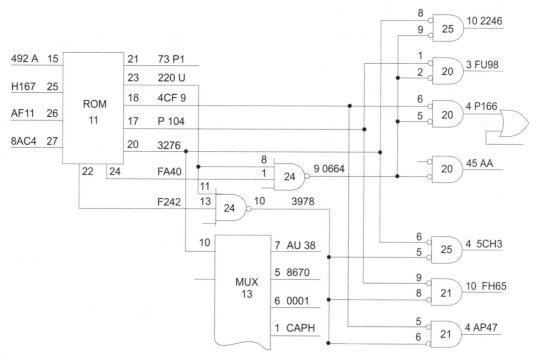

Fig. 11.15 *An example of an annotated schematic, showing correct signatures at various circuit nodes*

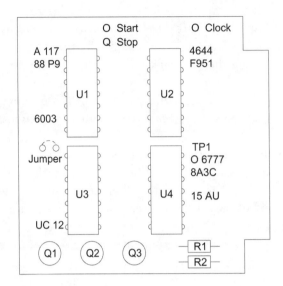

Fig. 11.16   *Proper signatures at various test points on the circuit board*

(b) Another method is to print the signatures onto the printed circuit board itself, with arrows indicating the signal flow. Yet another one is to print a test template that is attached to the component side of the circuit board when the service is required (Fig. 11.16). Holes in the appropriate locations, signatures and other instructions printed on the template guide the service person to the faulty node.

12

Rework and Repair of Surface Mount Assemblies

12.1 SURFACE MOUNT TECHNOLOGY

The conventional through-hole mounting technology used for printed circuit assemblies is being increasingly superseded by surface mount technology. Instead of inserting leaded components, through the holes, special miniaturized components are directly attached and soldered to the printed circuit board. The surface-mounted components and their packing are particularly suitable for automatic assembly. The advantages of surface mounting are rationalized production, reduced board size and increased reliability.

Figure 12.1 shows the conventional through-hole technology in which the components are placed on one PCB side (component side) and soldered on the other side. In surface mount technology, the components can be assembled on both sides of the board (Fig. 12.2). The components are attached to the PCB by solder paste or non-conductive glue and then soldered.

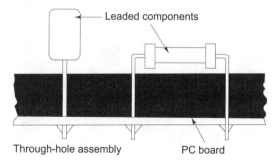

Fig. 12.1 *Conventional through-hole technology—components are placed on one side and soldered on the other*

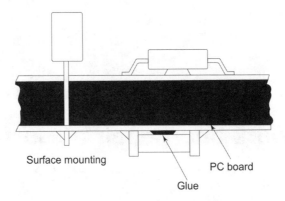

Surface mounting   PC board

Glue

↑Fig. 12.2   *Surface-mount technology—components are mounted on both the sides*

In addition, there are hybrid circuits consisting of thick and thin film circuits, which are basically leadless components. They are reflow-soldered onto the ceramic or glass substrate in addition to the components already integrated on the substrate. The mounting of the hybrid circuits is shown in Fig. 12.3.

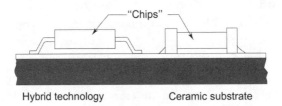

"Chips"

Hybrid technology   Ceramic substrate

↑Fig. 12.3   *Hybrid technology—consisting of components integrated on the substrate and soldered chips*

Since not all component types are available as surface mount version, a combination of leaded and surface mounted components is generally encountered in the manufactured electronic equipment.

The prime motivation for introducing surface mount technology (SMT) is density increase and board area reduction due to continuous demand and the market trend for miniaturization in electronic assembly, particularly in portable products. The advantages of this technology are:

- *Board Area Reduction:*
    - smaller products requiring smaller PCBs; and
    - components mounted on both sides.
- *Better Product:*
    - more functions in less space; and
    - improved electric and mechanical performance.

- *Lower Manufacturing Cost:*
  - SMT components are cheaper;
  - Fewer components as multi-function ICs occupy less space than conventional components;
  - Production equipment is universal and highly automated;
  - Capital, people and space can be reduced by up to 50%; and
  - Component handling is simplified and less chance of damage and wastage of componets.

12.2 SURFACE MOUNT DEVICES

The abbreviation SMD for 'Surface Mount Device' is the most common designation for the components used in surface mount technology. SMDs are designed with soldering pads or short leads and are much smaller than comparable leaded components. In contrast to conventional components, the leads of which must be inserted into holes, SMDs are directly attached to the surface of the PCB and then soldered.

Resistors, ceramic capacitors and discrete semiconductors represent 80% of the total available SMDs. Normally, in the SMDs, the cubic shape prevails over cylindrical versions, as the latter can only have two pins, thus being exclusively suitable for resistors, capacitors and diodes. If the development of a special SMD package is not possible for electric or economic reasons, the DIP package can be converted into a surface mountable version by bending the leads.

Lead Styles: SMDs are constructed with different types of lead styles. The commonly used lead styles are shown in Fig. 12.4.

Type	Drawing	Components
Gull-wing		SOIC QEP TSOP
J-lead	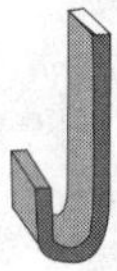	PLCC SOJ
Ball		BGA Chip Scale Flip Chip (Bump)
Metallized Terminations	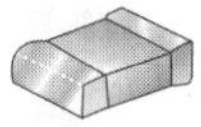	Capacitors Resistors Ferrites

Fig. 12.4 *(a) SMD lead styles*

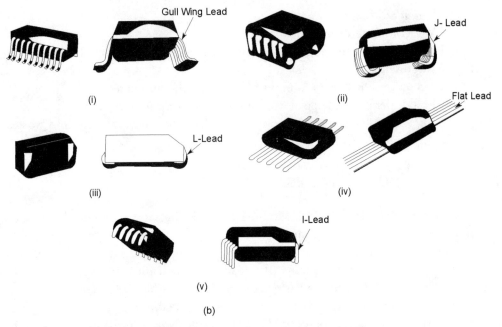

(i)

Gull Wing Lead

(ii)

J- Lead

(iii)

L-Lead

(iv)

Flat Lead

(v)

I-Lead

(b)

Fig. 12.4   *(b) Shapes of components with different lead styles*
  (i) Gull wing lead: metal lead that bends down and away
  (ii) J-lead: metal lead that bends down and underneath a component
    in the shape of letter J
  (iii) L-lead: inward formed underneath a component
  (iv) Flat lead: protrudes directly out from the body of the component
  (v) I-lead: a through-hole lead cut short for surface mounting

Lead Pitch: The lead pitch in an SMD is measured from the centre-to-centre of leads, and is not the air gap between the leads as shown in Fig. 12.5.

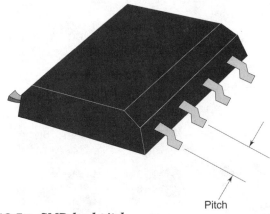

Pitch

Fig. 12.5   *SMD lead pitch*

Component Packaging: Automated assembling of printed circuit boards is carried out by pick and place machines. For this purpose, proper packaging is required to protect the components, particularly the SMDs, from damage during transport. The various packaging methods to provide proper feeder to receive the components are trays, tubes, tape and reel and bulk feed cassettes. They are shown in Fig. 12.6.

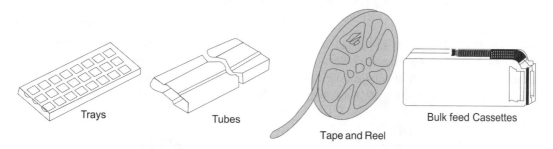

Trays   Tubes   Tape and Reel   Bulk feed Cassettes

Fig. 12.6   *SMD packaging techniques*

Chip Size Codes: The size of chip components are defined by a 4-digit size code. The thickness of the component is not defined in the size code. Figure 12.7 illustrates the examples of specifying the size of the components. Size code may be stated in inches or in the metric system. For example, the size code of ceramic capacitors and resistors is usually stated in inches, while that of tantalum capacitors is stated in metrics.

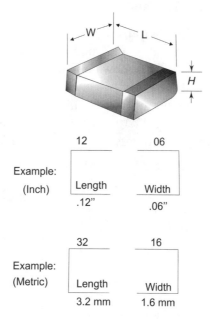

Example: (Inch)

12	06
Length .12"	Width .06"

Example: (Metric)

32	16
Length 3.2 mm	Width 1.6 mm

Fig. 12.7   *Chip size codes*

12.3   SURFACE MOUNTING SEMICONDUCTOR PACKAGES

As the technology of surface mounting has developed, a range of packaging types has emerged. The following are the commonly used packages in semiconductor SMDs:

12.3.1 SOIC (Small Outline Integrated Circuit)

This is a plastic package, available in 6, 8, 10, 14, and 16 versions with a body width of 4 mm, and in 16, 20, 24, and 28 pin versions with a wider body of 7.6 mm. The leads are on standard 1.27 mm centres and are formed outwards so that the tips of the leads lie in contact with the PCB (Fig. 12.8). In addition, there are also two Very Small Outline (VSO) packages with 40 and 56 leads at a pitch of 0.762 mm.

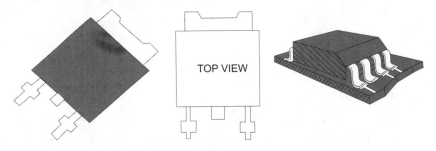

Fig. 12.8   *Small Outline Integrated Circuits (SOIC)*

12.3.2 SOT (Small Outline Transistor)

These packages are used for discrete transistors and diodes. The most common packages are the SOT–23 and the SOT–89 (now renamed as TO-236 and TO-243 respectively). The construction of a typical SOT–23 is shown in Fig. 12.9 with the standard dimension of the package. The package has three leads, two along one edge and a third in the centre of the opposite edge. Semiconductors on larger chips (up to about 1.5 mm square) are packaged in the SOT-89 format (Fig. 12.10). Its three leads are all along the same edge of the package but the centre one extends across the bottom to improve the thermal conductivity.

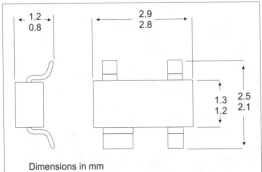

Fig. 12.9   *SOT-23 package*

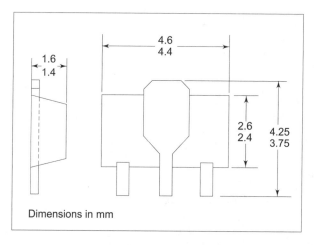

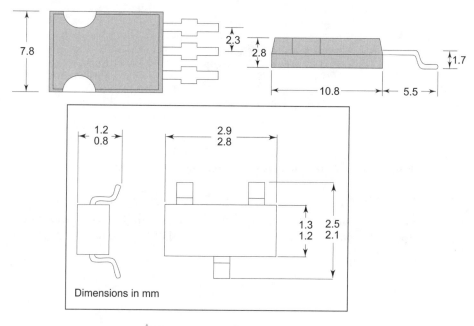

Fig. 12.10 *SOT-89 package*

For high power applications, SOT–194 as shown in Fig. 12.11 has been developed which can allow power dissipation of upto 4 watts when used with a suitable heat dissipating inter-connection substrate. An outline of a four-lead package, SOT–143 is shown in Fig. 12.12.

This is used for dual gate devices.

Fig. 12.11 *SOT-194 package*

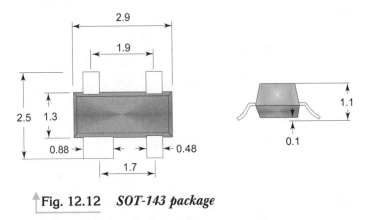

Fig. 12.12   *SOT-143 package*

12.3.3 Cylindrical Diode Packages

The two most popular packages developed specially for diodes in the cylindrical shape are:

- *SOD (Small Outline Diode)* package is specifically designed for small diode chips, limited to a power dissipation of 250 mW. A typical example of this type of device is SOD-80 whose dimensions are shown in Fig. 12.13.

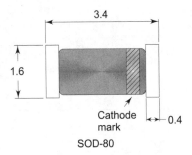

SOD-80

Fig. 12.13   *Small outline diode package*

- *MELF (Metal Electrode Face Bonded)* package in larger cylindrical encapsulation is used when more power handling capability is required. The typical dimensions of MELF diode are shown in Fig. 12.14. The SOD-80 package is also sometimes referred to as MiniMELF.

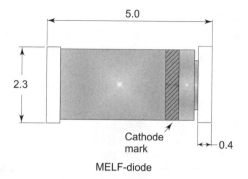

MELF-diode

Fig. 12.14 *High power diode package*

12.3.4 LCCC (Leadless Ceramic Chip Carriers)

The term 'chip carriers' refers to those IC packages that are square or nearly square with their terminations brought out on all four sides. LCCCs are those devices which do not carry any leads and excess packaging material. These packages are suitable for direct soldering or for attachment by sockets with added leads. In these devices, the IC chip is bonded to a ceramic base and connections are brought out with wires to solderable contact pads as shown in Fig. 12.15.

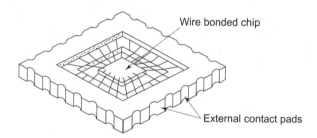

Fig. 12.15 *Leadless ceramic chip carrier (LCCC)*

LCCCs are commonly available in 18, 20, 28, 32, 44, 52, 68, 84, 100, 124 and 156 termination versions. The component height is typically 1.5–2.0 mm. The pitch between the terminations is always 1.27 mm.

12.3.5 PLCC (Plastic Leaded Chip Carriers)

They are available in a wide range in the same sizes and formats as the LCCCs. The leads of these devices also have a pitch of 1.27 mm. The majority of PLCCs are available with 'J' leads that are folded underneath the package. This is shown in Fig. 12.16. Since the 'J' leads are tucked under the device, and are not protruding, they present difficulties during the inspection and testing of circuits.

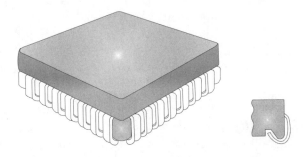

Fig. 12.16   *J-leaded PLCC package*

12.3.6   Flatpacks and Quad Packs

The flatpack package has its lead frame co-planar to the body of the package. The original flatpack had leads on two sides of the body, but presently it is available with leads on all four sides, and is accordingly called a 'quad pack'. The quad packs are usually high lead–count plastic packages in the range of 40 to 200. The size of the package remains the same, but the pitch of the leads varies with the lead count. For example, the pitch is 1.0 mm on packages up to 64 leads, 0.8 mm on the 80 lead and 0.65 mm on the 100 pin package. The shape of the package is shown in Fig. 12.17.

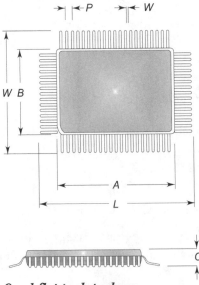

Fig. 12.17   *Quad flat-pack package*

12.3.7 LGA (Land Grid Arrays)

In these devices, the pins emanate from an array on the under side of the package (Fig. 12.18) rather than its periphery. The surface mounting version of the leaded grid array is the land grid array, whereby the pins are substituted by an array of solderable pads on the base. They are available in various sizes, pad sizes and pad densities to meet the requirements of different lead-out arrangements.

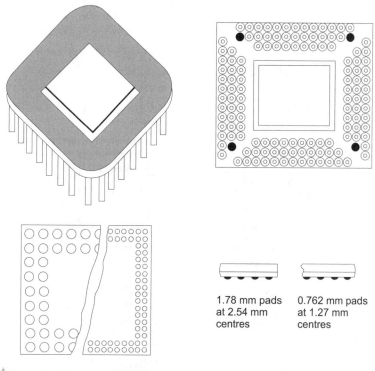

1.78 mm pads at 2.54 mm centres

0.762 mm pads at 1.27 mm centres

Fig. 12.18   *Land Grid Array (LGA)*

12.3.8   BGA (Ball Grid Arrays)

A BGA is actually any type of IC package that routes from the die and connects to the PCB via solder bumps. The package provides solutions for high performance, high pin count applications, with pin counts now nearing 1000, with 300 to 600 being standard. A typical BGA package consists of a BT laminate substrate with two metal layers and through-hole vias. The IC die is mounted to the top of the substrate, and is enclosed in a plastic mould (Fig. 12.19). Ball pitches are decreasing from 1.27 mm to 1.0 mm and as low as 0.5 mm for new chip scale packaging. For high reliability applications, several vendors offer a ball grid array package with a ceramic substrate. Multi-chip packages are usually BGAs or QFPs containing 2 to 4 die.

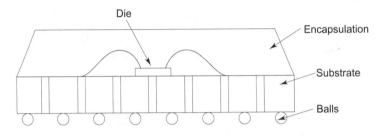

Fig. 12.19   *Plastic base grid array package*

Different ball patterns available on BGA are full grid, peripheral, stagger and thermal via (Fig. 12.20). Acronyms for other 'Grid Arrays' are:

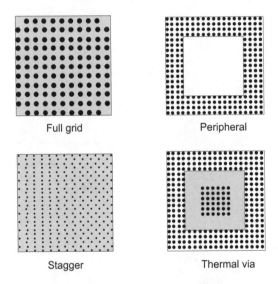

Fig. 12.20   *Different pattern types of ball grid arrays*

CBGA	– Ceramic Ball Grid Arrays, for high temperature requirements.
fBGA	– Flex BGA, uses a flex polymide substrate.
SBGA	– Super BGA, has metal heat spreader on top.
PBGA	– Plastic BGA, industry standard BGA.
LGA	– Land Grid Array, pads without the balls.
CGA	– Column Grid Array, solder columns instead of balls.

CSP	– Chip Scale Package, Fine-pitch BGA. Package is max 120% > chip size.
μ*BGA*	– Chip Scale Package, trademark of Tessera.
Flip Chip	– Die with solder bumps, very small.

12.3.9 COB (Chip-on-Board)

With chip-on board, a bare (unpackage) semiconductor is attached with epoxy directly to a PCB, wire bonded and then encapsulated with polymeric materials. It offers high packaging density and fast signal speed by means of wire bonding the chip directly onto the board.

12.4 PACKAGING OF PASSIVE COMPONENTS AS SMD

In order to utilize the full potential of surface mount technology, it is desirable that the components other than semiconductors also be surface mounted. These components include resistors, capacitors, inductors, etc. The capacitors and resistors are available in cubic dimensions and are often referred to as 'chips'. The most common chip components are resistors, capacitors and diodes. However, every kind of two terminal devices can be had in the chip form. The most commonly available package is in the rectangular form and has its solderable terminations only on the end face, or on the top and bottom faces as well as the end or on the sides in addition. Figure 12.21 shows typical examples of different types of metallizations on the chip component.

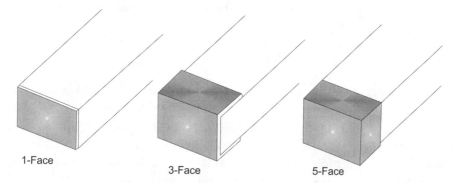

1-Face   3-Face   5-Face

Fig. 12.21 *Difference between 1-face, 3-face and 5-face metallization on chip components*

Typical packages of different types of capacitors, resistors and inductors are shown in Fig. 12.22.

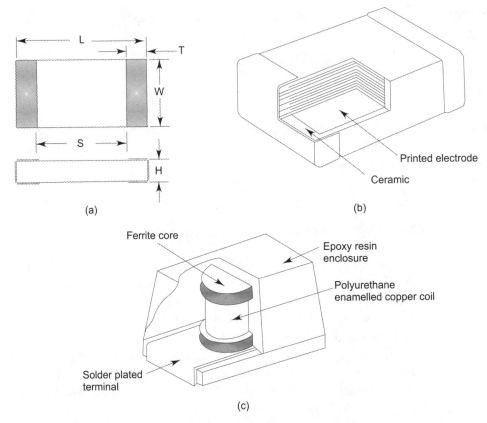

Fig. 12.22 *Surface mount: (a) resistors; (b) capacitor; and (c) inductors*

12.5 REPAIRING SURFACE MOUNTED PCBs

In the next few years, it will be hard to find a manufacturer who will build anything that will not contain surface mount devices. Having the proper tools to do the job and the knowledge to use the tools that you choose will thus be important to your survival in the electronics repair business. Surface mount technology has been around for a long time, but the tools for removal and replacement of surface mount devices have been slow to be accepted by the service and repair personnel and organizations. Possibly, it was because of the high cost of such tools. The technicians looked for innovative ways to use the tools at hand to do the work, sometimes even ruining the whole board by not using the right tools. The justification for investment on the repair and rework stations comes from the fact that the high priced equipment certainly deserves to be treated in a better way so that you are in a position to perform the job worthy of your knowledge and experience.

As the assembly of electronic components moved away from the use of single-sided printed circuit boards towards double-sided boards with plated-through holes, component removal became somewhat easier. Also, the associated damage occurring to PCBs during rework/repair became less. However, the SMT printed circuit boards are, as far as rework is concerned, essentially single-sided PCBs. Therefore, with the increasing use of surface mount components, we are seeing an increasing number of damaged pads and tracks due to inappropriate or careless component removal practices. It has been observed that these damages are mostly due to the inadequate training of the operators/repair workers in properly understanding and handling the SMT PCBs.

Reworking faulty SMT assemblies usually requires component removal and replacement. Occasionally, the replacement of damaged PCB pads and tracks also becomes necessary because of poor reworking practice. The methods of removing a faulty multi-lead surface mount component are discussed below.

12.5.1   Cut All Leads

Cutting all leads is the simplest method to remove a faulty component. It is recommended if other methods are not practical. The technique is to carefully cut through each leg in turn and take off the device. Each joint is then melted with a fine tip, temperature controlled soldering iron and remaining IC leg are removed with tweezers. After allowing a cool down period, excess solder can be removed with a desoldering braid.

The advantage of this method is that it is cheap and can be carried out in the field, as it does not require any special tool. The disadvantage is that it damages the component and there is a possibility of damaging the PCB substrate and copper pads. Also, soldering the replacement component in position using a soldering iron requires processing one lead at a time, a difficult if not impossible task with fine pitch multi-lead devices.

12.5.2   Heating Methods

There are two basic heating methods for reworking PCBs whose components include SMDs: conductive and convective. Conductive rework involves a heated tool that contacts the solder joints to effect reflow. The convective approach employs heated gas or air to melt the tin lead alloy.

Conductive Method

Soldering tools fitted with tips designed to heat all the component's leads are available. They rely on electrodes coming into contact with the component legs and holding them flat to the copper pads on the PCB. The more sophisticated rework stations employ a precisely controlled pulse of current which passes through the electrode heating them to solder reflow temperature very quickly (approximate three seconds). This melts the solder on the joints and a built-in vacuum pick-up will lift the component from the surface. The technique enables all the leads to cool down rapidly after the soldering operation and so allows the leads to be held in position while the solder solidifies.

This method has several advantages. It is very fast and repeatable and there is no heating of the component body. It is very good for replacement as the electrodes will hold the legs flat to the pads during solder reflow, while the alignment and positioning is ensured with a microscope. The disadvantage is that it is expensive and machines are dedicated solely to gull wing (QFPs) and TAB components.

Dual Heater with Vacuum Pick-up

This is a special tool for handling larger component removals. Dual heating brings the larger tips up to the required temperature quickly and the built-in vacuum pick permits single-handed removals once the reflow is established. The tool facilitates their removal of all conventional flat packs as well as several BGAs.

The *thermal tweezer*, with dual heaters and a squeezing action, can remove a variety of parts ranging from small chips to large PLCCs and leadless packages. The tweezer action permits the tips to contact the solder joints, thus ensuring high heat delivery, but at the lowest possible temperature.

Convective Method (Hot Gas Soldering)

Most production and rework stations use a hot gas or hot air as the heat transfer medium. With a single point nozzle, small parts such as chips, transistors, SOICs and flat packs can be removed. The hot gas is swept over the leads until full reflow is achieved, after which the part is lifted with a tweezer. Although removal times are longer than with a conductive tool, one tool and nozzle shape handles several applications. With longer components, a component-specific nozzle is fitted to the hand-piece and brought around the part to remove almost any two- or few-sided SMDs. The provision of vacuum provides component lift-off after reflow. A rework station that uses infra-red radiation to reflow the solder joints is also available. Ancillary features frequently include a vacuum pick-up mechanism for removing the faulty device and magnification systems, sometimes with video display unit (VDU), to aid observation of the work in progress.

12.5.3   Removal and Replacement of Surface Mount Devices

The following steps should be taken to remove a component using a hot gas machine:

- Apply a small amount of liquid flux to all joints.

- Choose the correct head to suit the component.
- With the PCB in place, activate the gas flow to reflow the solder on every joint (use microscope / VDU to check).
- If the component has been bonded with an adhesive, rotate the head to shear the bond.
- Remove the component with the vacuum pick-up and allow the PCB to cool.
- Remove any remaining solder by the use of fine desoldering braid.
- Allow a further cool down period.
- Inspect the pads to ensure that they are not damaged.

For replacement of the component, the following procedure is followed:

- The new component should be carefully inspected to ensure that the legs are not bent or distorted. Ideally the legs of the device will slope down from the body by 1–2 degrees. This will allow the legs to flatten onto the pads when the component is placed onto the PCB.
- A thin film of flux is lightly applied to the pads.
- The component is then placed into the head of the hot gas machine and carefully lined up to the PCB. The fingers on the SolderQuick tape will help to align the component.
- Before the gas flow is initiated, the component should be lifted away from the board surface until the legs are just clear of the pads.
- The gas glow should then be applied. The gas will heat up the legs and the solder on the pads.
- When the solder flows, the component should be carefully brought down onto the board, ensuring that the legs of the component are sitting between the fingers of tape and hence are central over the pads.
- Allow gas flow to continue for a few seconds to ensure that the solder flows correctly around each leg.
- When the solder flows correctly, switch off the gas and allow the board to cool at least for one minute to avoid disturbing the joints before removing the PCB from the machine.
- After removing the PCB, carefully remove the SolderQuick tape and clean all excess flux from the joints.
- Inspect all joints with an X10 magnifier to ensure correct reflow.
- Clean the PCB with iso-propyl alcohol in the aerosol form to ensure penetration of the solvent under the component to wash out any flux. The area can then be brushed to remove all traces of flux.

The most common damage on surface mount boards is lifted pads on quad flat pack (QFP) layouts. The most probable reason for this is that operators have difficulty in knowing when the solder joints on all four sides of the device package are molten. In order to repair such damage, the following method is suggested:

(i) Remove the damaged pad/track and clean the immediate area on the board.

(ii) Select appropriate replacement track/pad (These are available from a number of suppliers).

(iii) Solder the replacement pad/track to the undamaged track on the board. Figure 12.23 shows the replacement pad and portion of track together with the undamaged track to which it will be joined. The replacement track is cut so that it overlaps the undamaged track and the two parts are soldered together (Fig. 12.24).

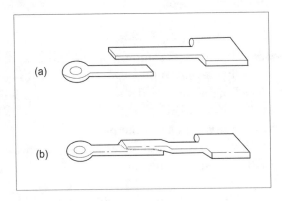

(a)

(b)

Fig. 12.23   *Working repair*

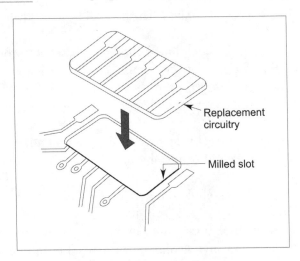

Replacement
circuitry

Milled slot

Fig. 12.24   *Serviceable repair*

(iv) Using an adhesive capable of withstanding high temperature, glue the new pad/track to the PCB substrate. Clamp together until the adhesive has cured.

(v) Solder the replacement component in place.

(vi) Clean off flux residue.

(vii) Re-apply any solder resist that has been removed.

Removing a surface mount component can be compared to 'steaming a stamp off of an envelope'. It is actually done by simultaneously melting the solder around a component's joints and then picking the component off of the PCB. The substrate is then cleaned and a new component is soldered back onto the circuit board. The best way to apply heat to the solder and component leads is a method of choice. There are conductive tools, convective tools, single-point, multi-point, tinable and non-tinable tips which can be used for this purpose.

On a through-hole assembly, a defective component is desoldered before the component can be removed and replaced. It is ensured that all of the solder is removed from the hole, and then cleaned, before a new component is placed on the board. On a surface mount board, it is unsoldered. The difference between the two is that on the through-hole board, molten solder is removed or sucked away from the lead and plated-through-hole by vacuum. With the use of hot air tool or solder pot, all leads on a through-hole can be reflowed simultaneously, allowing the component to be removed. On the surface mount board, all the device leads must be heated simultaneously and the component must be lifted off the board before it can re-solidify. If all leads are not heated concurrently and the device is pulled off before all the solder has been melted, the footprint on the board may be damaged. This can affect the co-planarity of the new component when it is placed on the PCB.

As more and more manufacturers include fine pitch technology in their surface mount designs, the rework process becomes even more complex. As board pitch becomes finer, boards become more sensitive to component misalignment and PCB heat damage. Reworking fine pitch boards usually requires some sort of a vision system. As lead count becomes finer and finer, it is essential to develop a vision system that allows for simultaneous viewing of the PCB and component. Therefore, optical devices should be used for the placement of fine pitch components to ensure proper alignment. Considering all these factors, an ideal rework station would include:

- A vision system that can be used when placing and soldering the component;
- A placement tool that will allow for movement which is smaller than the smallest pitch being used on the board;
- A heating method that can control the heating process and can heat the board and the component in a manner that approximates the method used in the original production; it must be able to apply uniform heat without de-laminating the board or damaging the component during removal or replacement; and
- It should be simple to use, both by an operator and engineer, without much training.

12.6 REWORK STATIONS

Today's printed circuit boards, with BGAs, DCAs, CSPs and fine pitch SMDs require a level of precision and performance that cannot be ensured with hand-held tools. Adding to the difficulties of rework are area array components. Because the bumps are on the bottom of the chip, inter-connections with the pads are not easily aligned and inspected, and voids, bridges and other defects can go unnoticed until functional testing discovers them. Also, with manual desoldering, using soldering iron and a wick control is required over several parameters such as tip temperature, dwell time at each pad, applied pressure, affected area, contact area and location. On the other hand, vacuum desoldering tools require control over vacuum flow, distance from pad, hot air flow (if applicable) temperature, source pressure, etc. Most of these parameters are directly related to the operator skill and may result in over-heating and damage to pads, traces and solder marks. An automated work station, which eliminates dependance on technician skills offers a practical way to secure consistent quality and cost-effectiveness in rework operations.

There are many different types of rework equipment in the market today. One typical example of a rework station is that of Model SD-3000 from M/s Howard Electronic Instruments, USA. It is a microprocessor-controlled equipment using a single nozzle blowing out hot air which traces along the soldered points of the SMD. The equipment is suitable for any size and shape of QFP, SOP, PLCC, PGA, BGA, etc. to remove and/or reflow (solder). It is able to handle all SMDs without changing the nozzle head. A built-in timer helps to prevent damage to the PCB and the nearby parts caused by over-heating.

The various controls provided on this equipment are shown in Fig. 12.25 and are as follows:

- *X-Axis:* This knob is used to adjust the nozzle width of the component to be re-flowed. It is also used as the inside adjustment for BGA/PGA removal.
- *Y-Axis:* This knob is used to adjust the nozzle length of the component to be re-flowed. It is also used as the outside adjustment for BGA/PGA removal.
- *Z-Axis:* The Z-Axis control adjusts the height of the nozzle above the solder points to be re-flowed.
- *Nozzle:* The nozzle is adjusted by the X-, Y- and Z-Axis knobs to whirl around the solder points of the component to be removed. Holes in the nozzle allow the operator to visually inspect the temperature of the heater according to the colour of the heater coils.
- *Timer:* The timer is used to set the time required to reach the solder melt temperature after the start button is pushed. At the end of the time cycle, the unit automatically goes into its cool down cycle and shuts off after reaching its cool down temperature.
- *Temperature:* This control is used to control the temperature of the heater in all modes, at the discretion of the operator.

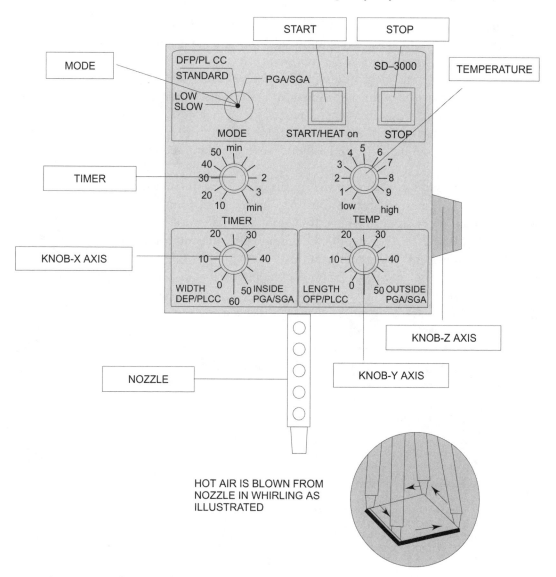

⇑Fig. 12.25 *Controls on typical rework station*

- *Mode:* The mode switch provides facilities for the removal of BGA/PGA, QFP/ PLCC standard packages. The mode also has high air flow (12 liters/min.) and high temperature.

In the LOW mode, lower rate of air flow (6 litre/min.) can be obtained. The mode is used for heavily populated boards so that small chips won't accidentally be blown from the board during re-flow.

In addition, SLOW mode is available which is used for replacing the QFP/PLCC packages after old solder has been removed from pads and new solder paste has been applied to the new component to be reflowed.

- *Start:* Pushing the start button the first time starts the nozzle rotating to allow adjustment of the width and length of the nozzle. After adjustment is complete, pushing the start button a second time starts the air flow, heat and timer.
- *Stop:* Pushing the stop button at any time will stop the heat and raise the nozzle approximately one half inch to allow vacuum picking the component from the PC board.

The other facilities available on the rework station are the mechanism for holding the PCB, applying vacuum to pick up the IC to be removed from the board and providing ease of sliding the PCB for alignment of component under the hot air nozzle.

While reworking on QFP or PLCC components, the following will assist in getting faster reflow times and lower temperatures:

- Keep nozzle height at 1 or 2 mm above the board at all times. This might require a fixture to hold the board and heater head/nozzle assembly.
- Use as high an air flow rate as possible without over-heating peripheral solder joints.
- Use flux if desired.

With these steps, the technician should be able to develop his own rework process using connective tools and to understand the effects on assemblies.

13

Typical Examples of Troubleshooting

13.1 POWER SUPPLY CIRCUITS

All electronic equipment and systems need a power unit for operation. Although in some equipment, the power may be derived from a battery, but more often, it is obtained from the single phase AC power mains. Since most of the modern solid state equipment operates on low DC voltages, the purpose of the power supply is to convert the 230V, 50 Hz AC mains supply into a form necessary for operating the internal circuitry of the equipment, which is usually a regulated DC voltage. Since all parts of an equipment need power to run, *a failure in the power supply can often lead to a complete failure of the equipment.*

Besides DC regulated power supply, electronic circuits also sometimes contain converters and inverters. An inverter is a power unit that converts DC power into AC power output. The DC source is usually a battery. The frequency of the AC output may be 50 Hz or higher like 400 Hz. A converter is basically an inverter followed by rectification, i.e. it converts DC into DC. A typical example of a DC to DC converter is that of generating high voltage, say 2 kV, from a 9-volt battery supply.

Most of the modern equipment may need one or more of the voltages in three ranges, ± 5 V for operation of digital and logic circuits, ± 12V or ± 15V for operation of linear integrated circuits and transistors, higher voltages for operation of specialised parts of the circuit, say 24V for operation of certain types of solenoids, 100 volts for driving deflection coils on a video monitor and 2kV for EHT (extra high tension) supply in an oscilloscope.

A typical line connected power supply performs the following functions:

- *Voltage conversion*—changing the 230V AC line voltage into one or more other voltages as determined by application;
- *Rectification*—converting the AC into DC;
- *Filtering*—smoothening the ripple of the rectified voltage(s);
- *Regulation*—making the output voltage(s) independent of line and load variations; and
- *Isolation*—separating the supply outputs from any direct connection.

Figure 13.1 shows a typical power supply arrangement in an electronic equipment. Practically in all power supply circuits, a rectifier is used to convert alternating voltage into uni-directional voltage, which is followed by a filter circuit to smooth out the pulsating DC. The rectifier circuit commonly used is a diode bridge, and the filter circuit mostly comprises only a large capacitor. For each output voltage, a separate transformer secondary winding is employed. This needs a separate regulating circuit for each of the outputs.

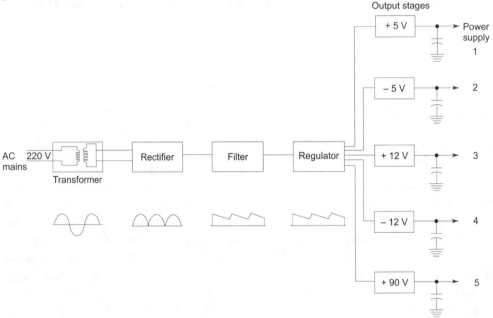

Fig. 13.1   *Typical power supply arrangement in an electronic equipment*

13.1.1   Types of Regulators

Two techniques are used to provide regulated DC voltages:
  (a) Linear series regulator; and
  (b) Switched-mode power supply.

13.1.1.1   The Linear Power Supply

A linear power supply essentially, operates on line frequency and has the following blocks as shown in Fig. 13.2:
  (a) Isolation and step down power transformer;
  (b) Rectifier and filter;

(c) Series pass element; and

(d) Feedback and control.

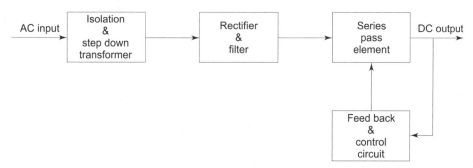

Fig. 13.2 *Block diagram of a linear power supply*

The isolation transformer provides the necessary isolation of the power supply circuit from the input mains AC, and steps down the voltage to the required level. This secondary voltage is then rectified and filtered to give unregulated DC which is fed to the series pass element (active device such as a transistor) operating in the active region.

By sampling a portion of the output voltage and comparing it to a fixed reference voltage, the series pass element is used in the form of a variable resistor to control and regulate the output voltage. Since the series-pass transistor operates in the active region, it dissipates a lot of power as heat, necessitating the use of heavy heat sinks and resulting in low efficiency between 30 and 50%.

13.1.1.2   Linear Regulators

A typical example of a linear regulator is shown in Fig. 13.3. Basically, it is a high gain control circuit that continuously monitors the DC output voltage and automatically corrects the output so that it remains constant irrespective of the changes in the load current and input AC voltage. In the circuit shown, the regulated output voltage is 10V from a 15V unregulated supply. The control element in the circuit is formed by the Darlington pair (Q_2 and Q_3). The base current for the Q_3 is provided by Q_2 which gets adjusted by the error amplifier Q_1. The emitter of Q_1 is held at the zener voltage (5.6V) and under normal conditions, the base voltage of Q_1 will be 0.6V higher than its emitter voltage. In case of a drop of voltage across the output, a portion of this drop appears across the base of Q_1 resulting in a decrease in bias of Q_2 which then tends to increase the base/emitter voltage. This will raise the collector voltage of Q_1, increasing forward bias of Q_2 and Q_3 which then tends to increase or correct the output voltage, restoring it to the set value.

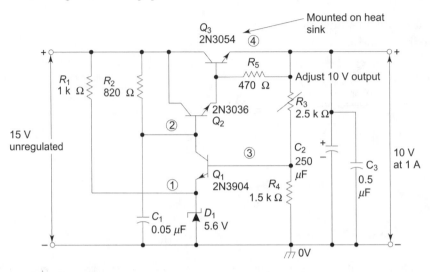

Fig. 13.3   *Linear regulator circuit*

Several features need to be incorporated in the series regulator circuits to protect them from overload currents. The two schemes that are usually employed for this purpose are discussed below.

Fold Back Current Limiting: This makes the power supply to switch, to give almost zero output voltage if the value of the load current is exceeded beyond its specified rating. In this technique, a current monitoring resistor is placed in the power supply return line, and the voltage developed across it is used to switch on a SCR (Fig. 13.4). In case of an overload, the SCR is triggered on, the voltage across it falls to approximately 0.9V, which is insufficient to forward bias the series transistor, with the result that zero volts appear across the output. Once triggered, the SCR will remain on and until the fault is removed, after switching off the power supply, the output will remain zero.

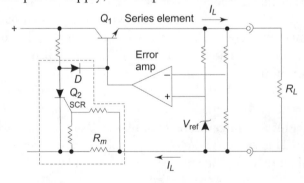

Fig. 13.4   *Foldback current limiting circuit*

Over-voltage Protection: All digital circuits must be protected from over-voltages. For example, TTL ICs will be damaged if they receive 7V on their V_{CC} terminal instead of 5V.

To achieve over-voltage protection, a zener is used (Fig. 13.5) to sense the voltage across the power supply's output terminals. In case of a rise in DC voltage, the zener conducts, turning on the SCR. The voltage at Q_1 collector falls rapidly to zero and the fuse blows. A circuit such as this is called *Crowbar Circuit*, which limits the output to the pre-set value.

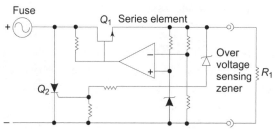

Fig. 13.5 *Over-voltage protection circuit*

Linear power supplies have a tight regulation band, low output noise and ripple. There are obvious disadvantages of low efficiency, bulky and heavy heat sinks, large power transformer, cooling fan, etc. All this makes them almost unfit for today's compact electronic systems.

13.1.1.3 IC Regulators

Monolithic IC regulators are used in most of the modern power supplies. These regulators have simplified the design and troubleshooting process. The most common IC regulator in use has been the uA 723, though several other ICs are also available. The pin configuration of this IC is shown in Fig. 13.6. This IC is available in a metal can type with 10 leads and 14-pin DIP encapsulation. This IC contains, internally, a reference source, error amplifier, series transistor and a current limiting resistor. The maximum output current and capability of the IC is 150 mA. However, the output current of the regulator can be increased by using an external power transistor in the output circuit.

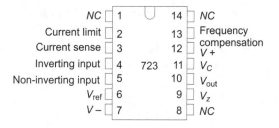

Fig. 13.6 *μA 723-pin configuration*

Two points must be remembered when troubleshooting power supplies based on μA 723 regulator:

(1) The input voltage must always be at least 3V higher than the expected output voltage.

(2) A low value capacitor must be present between the frequency compensation pin to the inverting input to avoid oscillations at high frequencies.

13.1.1.4 Three-pin Voltage Regulators

These are three-pin devices in which the input is applied between the centre leg and the input terminal and the output is taken between the output terminal and the centre leg. They are available for fixed voltages like 5V, 9V, 12V, etc. Most of the modern equipment makes use of these regulators.

Voltage regulators can be conveniently checked in-circuit using a multi-meter as shown in Fig. 13.7. The following steps are followed:

(1) With equipment turned on, check the input voltage to the regulator. It should be at least one volt higher than the specified output voltage.

(2) Check the output voltage, which should be nearly equal to the specified output voltage.

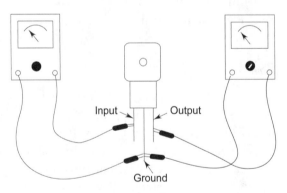

Fig. 13.7 *Testing 3-pin voltage regulators*

If the input voltage is present, but the output voltage is not correct, check for proper ground connection of the centre leg of the regulator.

Some voltage regulators are provided with short-circuit protection. In case of overload or short-circuit, the regulator cuts off the output. To check this condition, the output of the regulator is disconnected from the circuit to be driven and the output voltage is measured. If the output still does not return to the normal value, the regulator is faulty and should be replaced.

13.1.2 Power Supply Troubleshooting

(a) The equipment is connected to the AC mains by a cable and a plug. Ensure that all the three wires in the mains cable are intact and connected to the respective pins in the plug.

(b) Check the presence of the mains voltage in the socket to which the equipment is intended to be connected.

(c) Most electric equipment uses fuse to protect the power supply. The fuse is usually placed at the back of the equipment enclosed in a black cap. If the equipment is not working, disconnect it from the mains supply, unscrew this cap and check the fuse with a VOM.

(d) In troubleshooting power supplies, the following steps are followed:

 (1) Shut off the equipment.

 (2) Disconnect each of the power supply branches from the circuit in turn to determine if the problem is in the circuit fed by the power supply or in the particular power supply itself.

 (3) Turn on the power again and check. If the power supply returns to normal, the circuit it is feeding is bad, but if there is still a problem, the particular power supply itself is bad.

 (4) If tests suggest that the power supply is bad, the problem is most likely caused by a bad transistor, IC or diode. The component is probably shorted.

 (5) Even though a problem in the power supply itself has been spotted out, there could still be a problem in the circuit it feeds, maybe a faulty IC or a short circuit in printed circuit traces. Inspect carefully the circuit board for obvious problems. Look for a solder bridge or cracked circuit foil for broken connections.

   If the fault cannot be located, there is a chance that a component is probably shorted to ground. Because of this, the power supply would fail to supply enough current. This would necessitate a detailed troubleshooting procedure for isolating a faulty component in the circuit.

   In case the transistors or ICs are mounted on sockets, it would be convenient to try the substitution method for isolating the problem. In equipment where these components are soldered on the board, this option is not available. In such cases, try substitution only if all other means of diagnosing the fault have failed.

(e) A capacitor is usually used to filter out AC and provide a smooth DC output voltage. If this capacitor fails, some excessive AC will be present in the DC voltage as ripple, which can show up as different faults. The ripple is even interpreted in some circuits as digital signals and results in intermittent problems.

The presence of a ripple is checked with an oscilloscope. At each power supply branch, you should see a fairly smooth, clear output. If the ripple is found to be present (Fig. 13.8), the filter capacitor should be checked and replaced if faulty. In some cases, excessive ripple is also observed if the regulating circuit has some problem.

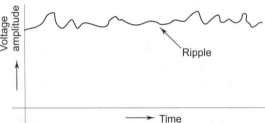

Fig. 13.8   *Noise (ripple) in the power supply*

(f) Sometimes, a power supply may be found to give high or low output voltage. Some circuits provide for adjustment of this voltage to allow you to correct the same. The adjustment is made by turning a 'trimmer pot' with a screw driver. However, voltage IC regulators do not need such type of adjustment provision. If the output voltage from these regulators is not correct, check the regulator.

13.1.3   Switched Mode Power Supplies (SMPS)

Switched mode power supply is also called switching power supply or sometimes chopper controlled power supply. It uses high frequency switching devices such as bi-polar junction transistors (BJTs), MOSFETS, insulated gate bi-polar transistors (IGBTs) or thyristors (SCRs or Triacs) to take directly rectified line voltage and convert it to a pulsed waveform. Most small SMPS use BJTs or MOSFETS whereas IGBTs are generally found in large systems. SCRs or Triacs are used where their advantages of latching in the on state and high power capability are required.

Switching mode power supply regulators are preferred over their linear counterparts for the following benefits:

- *Size and Weight:* Since the transformer and the filter run at a high frequency (in the range of 20 KHz to 1 MHz or more), they can be much smaller and lighter than the big bulky components needed for 50 Hz operation. The power density for SMPS compared to linear power supplies may easily exceed 20 : 1.
- *Efficiency:* Since the switching devices are ideally fully 'on' or fully 'off', there is relatively little power lost. So the efficiency can be much higher for SMPS than for linear power supplies, especially near the full load. Efficiencies of SMPS may exceed even 85%, whereas it is 50–60% for typical linear supplies.

With the advent of the laptop computer, cellular phones and other such portable devices, the importance of optimizing power utilization has increased dramatically. Many ICs are now commercially available for controlling and implementing SMPS with relatively few external components.

Switch mode supplies had been in common use in military and avionic equipment long before they found their way into consumer electronics. However, now-a-days, all TVs, monitors, PCs, most laptop and camcorder power packs, many printers, FAX machines and VCRs, and even certain audio equipment like portable CD players, use this technology to reduce cost, weight and size.

13.1.3.1   *Functional Block Diagram*

The functional block diagram of a typical SMPS is shown in Fig. 13.9. The four major blocks are:

(a) The input rectifier and filter section;
(b) High frequency inverter section;
(c) Output rectifier and filter section; and
(d) Feedback and control circuit.

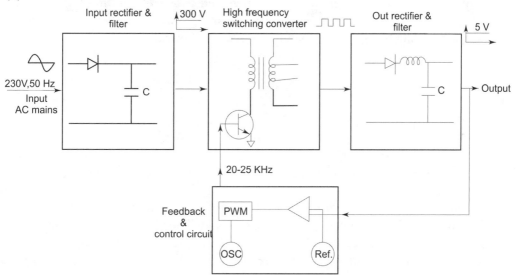

Fig. 13.9   *Functional block diagram of a switched mode power supply*

The switched mode power supply, as can be seen from the block diagram, is a relatively complex circuit. The AC mains is rectified and filtered. The high voltage DC is then fed to the high frequency inverter. The operating frequency range is from 20 kHz to 1 MHz.

The high frequency square wave thus generated is stepped down by the high frequency transformer and then rectified and filtered to produce the required DC output. The output is sensed, compared with a reference and pulse width modulated to get the desired regulation by the control circuit. The regulation of the output voltage is achieved by varying the duty cycle of the square wave. As the load is removed or input increases, the slight rise in the output voltage will signal the control circuit to deliver shorter pulses to the inverter. Conversely as the load is increased or the input voltage decreased, wider pulses will be fed to the inverter. The voltage control circuit can be incorporated either in the primary or secondary side of the transformer.

In the inverter, since the switch, i.e. the transistor is either 'on' or 'off', it dissipates very little energy resulting in a very high overall efficiency in the range of 70 to 80%. Also, the power transformer size is reduced because of high frequency operation. The hold-up time is greater for switched mode power supplies because it is easier to store energy in high voltage capacitors (200–400V) than in lower voltage (20–40V) filter capacitors common to linear power supplies. This is due to the fact that the physical size of the capacitor is dependent on its CV (capacitor X voltage) product while the energy stored is proportional to CV^2. Hence, the switched mode power supply offers an irresistible combination of high efficiency, smaller size, wider input voltage range and good hold-up time.

The input section of an SMPS has some special circuit components which require consideration. A block diagram of the input circuit is shown in Fig. 13.10.

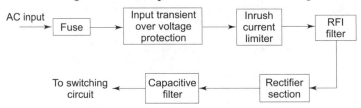

Fig. 13.10   *Input section of a typical switch mode power supply circuit.*

The fuse provides safety from over-current and voltage conditions. In addition to the fuse, flameproof or fusable resistors are also employed. They look like power resistors but are coloured blue or grey or may be rectangular ceramic blocks. They serve as a very important safety function: they cannot catch fire when over-heated and will open rather than changing value to provide an overload protection function.

This is followed by the input transient voltage protection circuit. High voltage spikes caused by inductive switching or natural causes like electrical storms or lightning, though of short duration, sometimes carry energy that could prove fatal to the different devices operating on AC mains, unless they are successfully suppressed.

The most common suppression device in use is the MOV (metal oxide varistor). MOV is used across the AC line input. Varistors are symmetrical non-linear voltage dependent resistors. At normal operating voltages, varistors are nearly open circuited. Upon the application of a high voltage pulse, such as lightning charge, they conduct a large current, thereby absorbing the pulse energy in the bulk of the material with only a relatively small increase in voltage, thus protecting the circuit. MOVs look like brightly coloured plastic coated disk capacitors but are not marked with capacitance. There will usually be either one MOV between the hot and neutral lines or three across hot, neutral and safety ground. If they are visibly damaged in any way, just remove and replace. Test with an ohmmeter. The resistance should be nearly infinite for a good MOV.

An off-the-line switching power supply may develop high peak inrush currents during turn-on because of the filter capacitors, which at turn-on, present very low impedance to the AC lines, generally only their ESR (effective series resistance).

Different methods can be used to limit to inrush current, like resistor-triac arrangement or a negative temperature coefficient thermistor. An arrangement with the thermistor is shown in Fig. 13.11. Thermistors offer high resistance when the power supply is switched 'on', thus limiting the inrush current. As the filter capacitors begin to charge, the current starts to flow through the thermistor, heating it up. Because of their negative temperature coefficient, the thermistor resistance drops. Thermistors should be properly chosen so that their resistance at steady-state load current is minimum, thus not affecting the overall efficiency of the power supply. Thermistors may be one or two in series with the AC input. They often look like flat black disk capacitors.

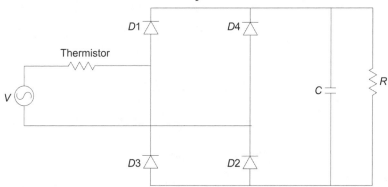

Fig. 13.11 *Use of a thermistor in the input circuit to limit the inrush current*

The most common method of noise suppression at switching power supply AC mains is the utilization of an LC filter for differential and common-mode RFI (radio frequency interference) suppression as shown in Fig. 13.12. Normally a coupled inductor is inserted in series with each AC line, while capacitors are placed between lines (called x-capacitors) and between each line and the ground conductor (called y-capacitors). The resonant frequency of the input filter is lower than the working frequency of the power supply. Resistance R is a discharge resistance for the x-capacitors.

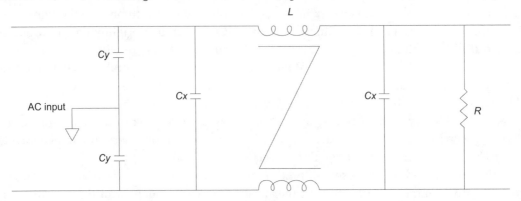

Fig. 13.12 *RF noise suppression circuit*

13.1.3.2 Quick Checks on a Faulty SMPS

If the SMPS is not working, identify the following four sections on the printed circuit board as per Fig. 13.13 and undertake the tests as indicated below:

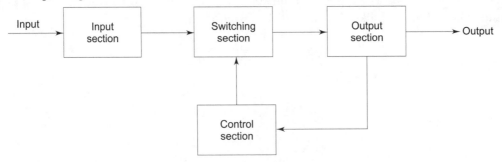

Fig. 13.13 *Various sub-systems of an SMPS*

Input Section: The main function of this section is to convert input 230V AC to high voltage DC.
> If SMPS is not working (dead) then:
* Check input fuse or fusible resistor.
* Check continuity across RFI coil.
* Check input rectifier diodes.
* Check filter capacitor.

> If AC mains fuse blows again and again:
>   • Check MOV (Metal Oxide Varistor) for short circuit.
>   • Check rectifier diodes.

Switching Section: This section is the heart of the SMPS. It converts high voltage DC into high frequency pulse train (AC) with help of switching devices (transistor/MOSFETs). Once the high frequency pulse train is applied to the primary of the high frequency transformer, the stepped down AC voltage will appear across the secondary terminals of the transformer.

> If the fuse blows:
>   • Check switching device for short circuit.
> If input section is working properly, but SMPS is not giving the output:
>   • Check switching devices.
>   • Check base resistors of the switching device.
>   • Check freewheeling diode connected across the emitter to the collector of the switching device.

Output Section: This section converts stepped down AC voltage into DC voltage which is delivered to the load.

> If the output voltage is zero or SMPS is generating some noise:
>   • Check output section rectifier diodes.
>   • Check output filter capacitors.
>   • Check continuity across output section inductor for dry soldering.

Control Section: The main function of this section is to generate high frequency pulses (20–25 KHz) to drive the switching devices and to provide protection of SMPS against input voltage variation and output short circuit.

> If SMPS is not working after testing of all the above-mentioned sections:
>   • Check supply voltage (V_{CC}) to PWM (Pulse width modulator) IC.
>   • Check if V_{CC} is not below 8V.
>   • Check capacitor across V_{CC} pin.
>   • Check if IC is getting heated up.
> If V_{cc} is in range of 12 to 20V:
>   • Check PWM IC output voltage (it should be between 2 to 10V).
>   • Check driver transistors.
> If SMPS is working but after connecting the load, SMPS switches off after some time:
>   • Check input filter capacitors (220 μF/400V).

13.1.3.3 Common Problems with SMPS

The commonly encountered problems, which account for 95% or more of the SMPS ailments, are detailed in Table 13.1.

Table 13.1 ■ *Common Ailments of SMPS*

Symptoms	Cause
(i) Supply dead, fuse blown	– Shorted switch mode power transistor and other semiconductors – Open fusible resistors or other bad parts
(ii) Supply dead, fuse not blown	– Bad start-up circuit (open start up resistor). – Open fusible resistors (due to shorted semiconductors) – Bad controller components

(iii)   – Outputs out of tolerance
     – Excessive ripple at line frequency (50 Hz)
     – Excessive ripple at twice the line frequency (100 Hz)
	– Dried-up main filter capacitor(s) on affected outputs
	    rectified AC input

(iv) – Outputs out of tolerance – Excessive ripple at the switching Frequency	– Dried-up or leaky capacitors on affected outputs
(v) Audible whine with low voltage on one or more inputs.	– Shorted semiconductors, – Faulty regulator circuitry – Faulty overvoltage sensing circuit or SCR – Faulty controller
(vi) – Periodic power cycling – Blinking power light	– Shorted semiconductors – Faulty over-voltage or over-current sensing components. – Bad controller

Notes : 1. The actual cause of failure may be power surge / burn-out / lightning strikes, random failure or primary side electrolytic capacitor open or with reduced capacity. Test them before powering up the repaired unit.

     2. In all cases, bad solder connections are a possibility as well. There are usually large components in these supplies and soldering to their pins may not always be perfect.

     3. An excessive load can also result in most of these symptoms or may be the original cause of the failure.

When testing SMPS, each branch of supply will shut itself off if its load is disconnected. For this reason, a load resistor must be connected to each branch during testing (Fig. 13.14). The value of the load resistor to be connected would depend upon the current rating of the supply. Further, an isolation transformer and an auto-transformer must be placed in the AC line during tests. The isolation transformer is necessary for protection and an auto-transformer for varying the AC voltage.

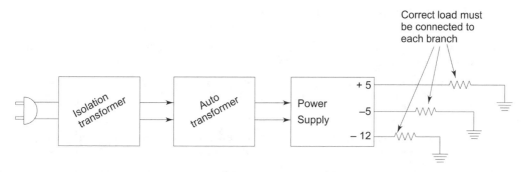

Fig. 13.14 *Testing switched-mode power supply*

If these equipments are not available, the troubleshooting of switched mode power supplies should be limited to visual inspection for burnt or damaged parts.

Many times it is felt that repairing a power supply may not always be an economical option. It is becoming very common for service centres to simply replace the entire power supply board or module even if the problem lies with a capacitor costing only a rupee or two. It may simply not pay for them to take the bench time to diagnose the fault down to the component level. Many problems with switch mode power supplies are, however, easy to find and easy and inexpensive to fix. This may not be true for with all the cases, but surprisingly it is so for most of them. So, do make a try rather than rejecting the power supply board as an uneconomical effort. Repairing the board may be a professionally satisfying exercise.

13.1.3.4 *Fans*

Many small SMPS do not have any in-built fan-but expect there to be a fan elsewhere in the equipment which provides air over the power supply. Most computer power supplies do have a fan inside and these are high failure items. A bad fan or even clogged air filters can result in over-heating and outright failure or could result in increased stress on components and reduced life expectancy. Thus, periodic maintenance of fans and associated cooling assembly are highly recommended.

The most common problem with fans is dry/gummed up/worn bearings. A quick test to check whether the problem is with the bearing or with the power, is to give the fan a spin. If it continues to rotate at least a couple of seconds, the bearings are probably good. If it stops instantly, the bearings are jammed or gummed up. In such a case, replace the fan. It is really not worth attempting to dis-assemble and oil the bearings.

Fan motors do go bad but this is much less common than bad bearings. With modern brushless DC motors, one phase could be defective resulting in sluggish operation and/ or failure to start if stopped in just the wrong position. In some sophisticated equipment, the fan speed is adjusted on the basis of temperature sensing. In the event of a bad temperature sensor, the fan may not be getting its full supply voltage and working efficiently.

13.1.4   High Voltage DC Power Supplies

There are numerous applications that require well-regulated high voltage DC supplies (1kV to 30kV). Examples of the use of such supplies are in cathode ray oscilloscopes, photo-multiplier tubes, X-ray image intensifiers, camera tubes, insulation testers, etc. Although these supplies can be made by using an AC power line transformer with a secondary wound with the necessary number of turns, the units tend to be bulky. Therefore, in most of these equipments, a preferred method to produce high voltage DC is to use a DC to DC convertor (often unregulated). In this technique, a low voltage DC is switched across the primary winding of a ferrite core transformer. The switching frequency could be 10 kHz to 25 kHz. The transformer secondary has a large number of turns, and it produces high voltage AC. Voltage multiplier rectifiers then convert this secondary AC into a high value DC.

Voltage multiplier circuits (Fig. 13.15) can provide only a very limited amount of current. Also, the regulator is poor because of the high values of output impedance which increases approximately with the number of stages used. Voltage doubler, tripler and quadrupler circuits are quite common.

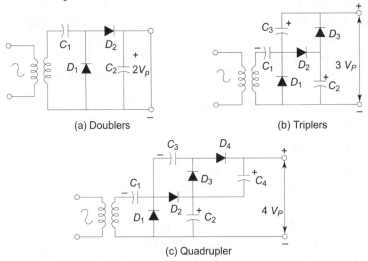

Fig. 13.15   *Voltage multiplier circuits*

High voltages should be measured very carefully. Some moving coil meters enable you to measure voltages up to 3kV. But they require about 50 μA current for full scale deflection which may be quite high for some high voltage supplies. Electrostatic voltmeters are preferred for measuring high voltages up to 50 kV with negligible loading on the supply voltage. The electrostatic meter has a non-linear scale, which makes it difficult to measure lower voltages accurately.

13.2 OSCILLOSCOPE

The basic blocks making an oscilloscope are shown in Fig. 13.16. It has three distinct sub-systems:

(a) Vertical deflection circuits, consisting of vertical pre-amplifier and vertical voltage amplifier;

(b) Horizontal deflection circuit, comprising time-base circuit and horizontal voltage amplifier; and

(c) CRT circuit-high voltage circuit for focus, intensity and other CRT controls.

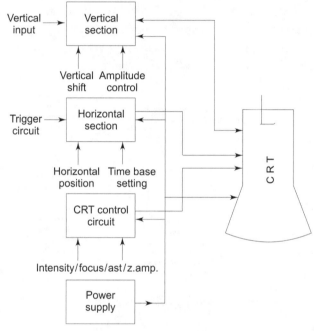

↑Fig. 13.16 *Subsystems in an oscilloscope*

While troubleshooting the oscilloscope, it is most important to isolate the sub-system in which the fault is suspected, and then to make efforts to locate the defective component. In many modern instruments, horizontal and vertical units come as plug-in units thus making it convenient to carry out independent tests on the three sub-systems.

Some preliminary steps should be taken to identify a visually apparent problem. After removing the cover, inspect the instrument thoroughly and carefully examine for the following faults:

• Loosely fixed printed circuit boards;

• Damaged printed circuit boards (physically broken or burnt);

• Badly soldered or loose components;

• Discoloured or burnt components caused by over-heating;

• Loose connector or cables;

- Examine connections to the front panel parts, control potentiometers and switches for poor soldering or disconnection;
- Examine socketed components (transistors and ICs) for proper fixation; and
- Identify the various functional printed circuit boards.

After following the above procedure, if there is no obvious fault detected, follow the step-by-step troubleshooting procedure as indicated in the service manual of the instrument or as per the general procedure explained in the following section.

13.2.1   Fault Diagnosis Chart

Figure 13.17 provides a guide for locating a defective circuit in an oscilloscope. This chart does not include checks for all possible defects, but could be helpful if followed for troubleshooting. Start from the top of the chart and perform the given checks until a step is found which does not produce the indicated results.

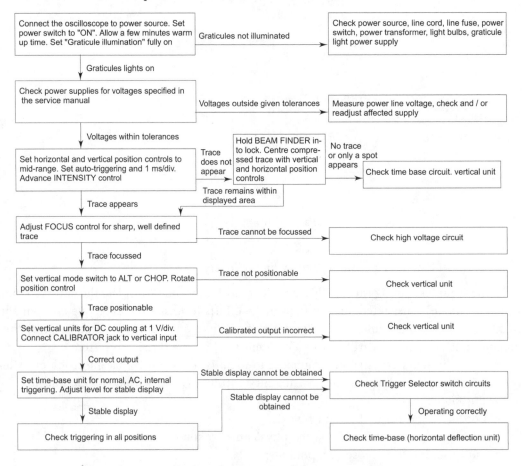

Fig. 13.17   *Fault diagnosis chart for an oscilloscope*

13.2.2 Cathode Ray Tube Replacement

Handle a CRT with care. Protective clothing and safety glasses should be worn. Avoid striking on any object which might cause it to crack or implode.

When storing a CRT, place it in a protective carton or set it face down in a protected location on a smooth surface with a soft mat under the face plate to protect it from scratches.

13.3 ELECTROCARDIOGRAPH (ECG MACHINE)

The electrocardiograph is an instrument which records the electrical activity of the heart. Figure 13.18 shows a block diagram of this machine. The potentials picked up by the patient's electrodes are taken to the lead selector switch. In the lead selector, the electrodes are selected two by two according to the lead programme. By means of a capacitive coupling, the signal is connected to a long-tail pair differential pre-amplifier. The pre-amplifier is usually a three- or four- stage differential amplifier having a sufficiently large negative current feedback, from the end stage to the first stage, which gives it a stabilizing effect.

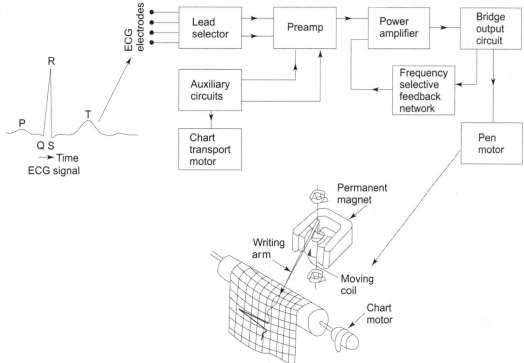

Fig. 13.18 *Block diagram of an ECG machine*

The amplified output signal is picked up single-ended and is given to the power amplifier. The power amplifier is generally of the push-pull differential type. The base of one input transistor of this amplifier is driven by the feedback signal resulting from the pen position and connected via a frequency selective network. The output of the power amplifier is single-ended and is fed to the pen motor which deflects the writing arm on the paper.

A direct writing recorder is usually adequate since the ECG signal of interest has limited bandwidth (0.02 Hz to 100 Hz). The Frequency selective network is an R–C network which produces necessary damping of the pen motor and is pre-set by the manufacturer.

The auxiliary circuits provide 1 mV calibration signal and automatic blocking of the amplifier during a change in the position of the lead switch. It may include a speed control circuit for the chart drive motor.

A 'standby' mode of operation is generally provided on the machine. In this mode, the stylus moves in response to input signals, but the paper is stationary. This mode allows the operator to adjust the gain and baseline position controls without wasting paper.

Electrocardiograms are almost invariably recorded on graph paper which is 50 mm wide. For routine work, the paper recording speed is 25 mm/s. Most ECG recorders use a thermal writing method in which the writing is done by a heated stylus on specially treated paper which turns black when it is heated. Some ECG recorders use pens and pressurized ink to make the recording.

13.3.1   Controls on an ECG Machine

Figure 13.19 shows the following important controls on an ECG machine panel.

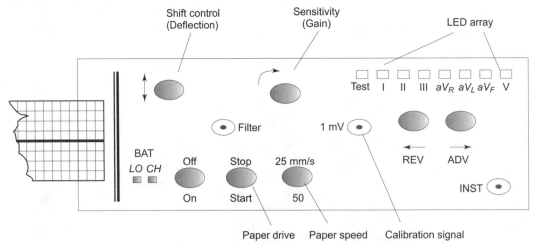

Fig. 13.19   *Controls on an ECG machine front panel (Model BPL 108 T)*

- *ON/OFF Switch*—light or flag, indicating when the equipment is 'ON';
- *Paper Speed*—25 mm/s, 50 mm/s;
- 1 mV signal—calibration signal;
- *Lead Selector/Test/Calibrate*—I, II, III, AVR, AVL, AVF, V lead (may be available at the back or on the side of the machine);
- *Sensitivity (Gain) Control*—continuous, in some machines, with three position switch for 5 mm/mV (½) 10 mm/mV (1) and 20 mm/mV (2);
- *Heat Control*—controls stylus heat (may or may not be available on the front panel);
- *Observe/Record/Stop*—three position switch;
- *Shift Control (Trace Position)*—adjusts the stylus up or down;
- *INST*—used to give zero signal during lead changeover;
- *ADV/REV*—advances or reverses the lead selection;
- *Filter*—Muscle tremor filter; and
- *LED Array*—Indicates the current lead selection.

13.3.2 Troubleshooting ECG Machines

ECG machines are heavily stressed machines in the hospitals. Therefore, internal electrical or mechanical faults occur occasionally in these machines. In most cases, the malfunction is an operator error or can be corrected by a simple adjustment or repair. It is thus essential to quickly establish whether the problem is INSIDE or OUTSIDE the instrument.

The following fault diagnosis charts (Fig. 13.20) help to isolate the problem in a non-functional ECG machine. Proceed as follows for making quick checks:
(a) Set the lead selector to CALIBRATE/TEST;
(b) Insert the patient lead plug into the socket; and
(c) Switch on the instrument, and allow about one minute to warm up.

13.3.3 Stylus Adjustments

Thermal writing stylus has three adjustments which affect the quality of tracing, heat, pressure and tilt adjustment. The heat adjustment is always a front panel control where as the position of pressure and tilt adjustments are shown in Fig. 13.21.

Fault Diagnosis Table 'A'

Fault	Possible cause	Action
1. Stylus remains stationary in centre	(i) Batteries low or wrongly fitted	(i) If low, recharge; check functioning of 'Batteries low' warning lamp, change the control board if faulty.
	(ii) Driver amplifier faulty	(ii) Check voltages; if incorrect, replace driver board.
	(iii) Galvanometer faulty	(iii) Change the galvanometer
2. Stylus remains hard on one side	(i) Battery low	(i) Change the battery
	(ii) Driver amplifier faulty	(ii) Check voltages; if incorrect, replace amplifier driver board.
	(iii) Preamplifier faulty	(iii) Check voltages; if incorrect, replace preamplifier board.
	(iv) Galvanometer faulty	(iv) Replace galvanometer.
3. Stylus insufficiently deflected or stylus not deflected fully on either side of the chart	(i) Stylus moves more on one side than on the other	(i) Batteries low
	(ii) Stylus does not move fully to both edges of the paper	(ii) Output stage faulty, change the driver amplifier. Proceed to perform further tests.

Fault Diagnosis Table 'B'

Fault	Possible cause	Action
1. Chart paper does not move	(i) Paper jammed	(i) Slide out the cover to release the paper and gently pull the paper. If the chart paper does not come out freely, look for mechanical fault.
	(ii) Check voltages on the recording block (across the motor).	(ii) If voltages are correct, replace the motor.
	(iii) Faulty motor	(iii) Replace motor and gear box assembly.
2. No black line on chart paper.	(i) Broken stylus, inspect visually.	(i) Replace if necessary
	(ii) Examine for wiring open circuited, check voltage on the stylus heat oscillator.	(ii) If oscillator is faulty, check oscillator board
3. Calibration signal of less amplitude or absent	(i) Defective standard cell (check with a precision voltmeter).	(i) Replace if necessary

(contd)

Fault	Possible cause	Action
	(ii) Faulty CAL switch	(ii) Correct for broken connections or replace switch.
	(iii) Faulty preamplifier board	(iii) Replace the preamplifier board.
4. Baseline drift.	(iv) Faulty lead selector switch Run motor for 3 or 4 seconds. Stop for about 1 second. Run motor again for 3 to 4 seconds.	(iv) Change control board.
	(i) If second trace starts where first trace finished, and drift continues as before	(i) Preamplifier is faulty. Replace preamplifier board.
	(ii) If second trace starts where first trace started and second trace is a copy of the first trace	(ii) Control board is faulty, replace control board.

Fault Diagnosis Table 'C'

Fault	Possible cause	Action
1. Presence of excessive 50 Hz	(i) Open circuited lead.	(i) Repair or charge patlent lead.
	(ii) Faulty lead selector switch.	(ii) Change the switch.
2. No signal	(i) Faulty lead selector switch or faulty control board.	(i) Change the control board or Selector switch.

Fault Diagnosis Table 'D'

Fault	Possible cause	Action
1. Ripple on base line.	(i) External ac source faulty	(i) Check external ac source
	(ii) Faulty motor circuit	(ii) Change faulty item.
	(iii) Faulty galvanometer (rarely)	(iii) Replace galvanometer.
2. Noise on base line (random fluctuations).	(i) Preamplifier faulty	(i) Replace preamplifier board or Faulty component.
	(ii) Loose patient plug connection.	(ii) Inspect and rectify
3. Drifting baseline	(i) If present on all leads	(i) Refer to Table 'B'
	(ii) If present on leads AVR, AVL, AVF only, control board Faculty.	(ii) Replace control boards

(contd)

Fault	Possible cause	Action
	(iii) If present on V leads.	(iii) Check the chest electrode-for connection and cleanliness.
4. Poor linearity.	(i) Faulty drive amplifier	(i) Replace with a tested circuit board Proceed to further tests.

Fault Diagnosis Table 'E'

Fault	Possible cause	Action
1. No patient signal on one or more lead positions-but calibrating signal present	(i) Wrong application of electrodes. (ii) Faulty patient lead (iii) Faulty patient lead selector switch	(i) Check correct application of electrodes. (ii) Replace patient lead with a new one. (iii) Check control board.
2. Stylus vibrates violently on any one or more lead positions.	(i) Faulty patient lead (ii) Faulty patient lead selector	(i) Change the patient lead. (ii) Check control board

Fault Diagnosis Table 'F'
Problems relating to lead selection

Fault	Possible cause	Action
1. One or more LED indicator does not glow but correct signal available on stylus	(i) Corresponding LED faulty or (ii) Connector wire to LED faulty	(i) Replace LED or repair connector
2. Either advance or reverse is not possible.	(i) Corresponding push-button faulty (ii) Trigger pulse not available (iii) Its connector wire faulty/not getting voltage.	(i) Replace push button, corresponding bias circuit or wire.
3. Lead selection not possible (i.e. neither advance nor reverse possible)	(i) Source for providing trigger faulty (ii) Faulty counter (iii) Multiplexing circuit faulty (iv) Connector to switches faulty/losse.	(i) Replace corresponding component

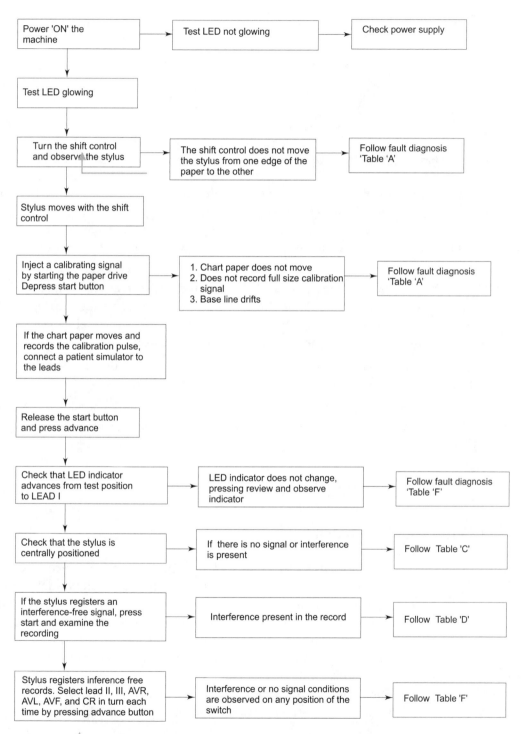

Fig. 13.20 *Fault diagnosis tree for an ECG machine*

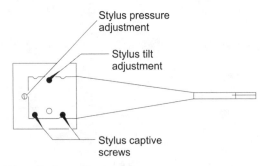

Fig. 13.21   *Stylus adjustment*

The heat and pressure adjustments determine the density of the recorded trace. The tilt adjustment is used to precisely set the edge definition of the recorded trace. These adjustments are factory-set and would normally not require field adjustment. However, if the stylus is replaced, make the adjustments as follows, when the machine is running at 25 mm/s.

Stylus Pressure: If a pressure gauge is available, set the pressure as recommended by adjusting the pressure adjustment screw. Ensure that the stylus is resting on the chart paper. If the pressure gauge is not available, the stylus pressure should be set for optimum recorded display of square wave when the CAL 1 mV button is pressed.

Stylus Tilt: Stylus tilt is adjusted correctly when both edges of the recorded tracing are clear and sharp, with no uneven spots.

Stylus Heat: Always start by setting the front panel heat stylus fully counter-clockwise, then slowly advance control clockwise until the recorded trace is sharply defined, but does not flood the paper.

13.4 CORDLESS TELEPHONES

13.4.1 Principle of Operation

A telephone wherein the handset is linked to the base unit not through an inter-connecting wire but by radio channel is called the cordless telephone. Cordless telephones make use of the narrow band frequency modulation for communication between the handset and the base unit. Basically, a cordless telephone contains all the circuits and functions of a conventional telephone, plus a two-way radio communication link. The two-way radio link must handle not only voice communications, but also some method of dialling, ringing and controlling the hook switch.

Cordless telephones are made up of two separate units (Fig. 13.22): a base unit and a handset (portable) unit. Communications between the base and handset units are full duplex: this means that both the base to handset link and the portable to base link can operate simultaneously. Both the base and handset units have a transmitter and a receiver each. The base unit is connected directly to the telephone line and serves as a link between the handset unit and the telephone line.

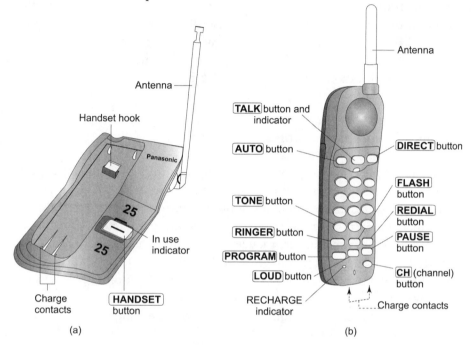

Antenna

Handset hook

Antenna

Panasonic

25

25

In use indicator

Charge contacts

HANDSET button

(a)

TALK button and indicator

AUTO button

TONE button

RINGER button

PROGRAM button

LOUD button

RECHARGE indicator

Antenna

DIRECT button

FLASH button

REDIAL button

PAUSE button

CH (channel) button

Charge contacts

(b)

Fig. 13.22 *Cordless telephone (a) Base unit (b) Handset*

Cordless telephone base unit to handset links use a carrier signal in 46 MHz band and a telescopic antenna. The base unit contains a power supply that is used to power the transmitter and receiver and also to charge the batteries of the handset unit. The handset to base unit link uses a carrier signal in the 49 MHz frequency band. There are 10 pairs of carrier frequencies available for cordless telephone operation. Each base to handset channel has a specific corresponding handset channel.

False ringing is reduced by using several specific ring frequencies for the base to handset link. In that way, two cordless phones with overlapping range and operating on the same channel, but with different ring frequencies, will not cause ringing of the neighbouring unit. Similarly, the use of several specific guard tones (pilot tones) for the handset to base link reduces the unauthorized use of a telephone line. The base will not respond to a handset unit unless the proper guard tone is sent. That prevents capture of a base unit by a nearby handset unit on the same channel, but with a different guardtone. Earlier cordless telephones used only guard tone and ring frequencies to help prevent unauthorized capturing of telephone lines and false ringing. New cordless telephones use digital coding. With digital coding, ringing, dialling and disconnection are usually caused by a frequency shift of the carrier frequency.

It is to be noted that the human ear does not respond equally to all audio frequencies and the intelligence of the human voice is conveyed between 300Hz to 3400Hz. So generally a bandwidth of 3.1kHz is sufficient for the transmission of audio intelligence and it is this audio bandwidth that is made use of in telephones. Cordless telephone manufacturers have to follow certain standards in the design of cordless telephones, for which the specifications are framed by the regulatory authorities. Cordless telephones must meet such specifications before they can be permitted to connect to the national telephone network.

13.4.2 Block Diagram of Cordless Telephone

While trying to understand the circuit details of a cordless telephone, a preliminary knowledge of the analog communication is a must, as the study of cordless telephones involves the establishment of a communication link through frequency modulation. The modulation technique corrects variation in one of the parameters (frequency, phase, amplitude) of the carrier wave in accordance with the amplitude variation of the signal to be transmitted. In frequency modulation, it is frequency that is varied. Frequency modulation has a definite advantage over amplitude modulation in the sense of being more noise-immune.

In a cordless telephone, each unit (handset and base unit) has a transmitter and receiver. The transmitting frequency of the base unit is the receiving frequency of the handset and the transmitting frequency of the handset is the receiving frequency of the base unit and vice versa. It follows logically that each unit has a mutually compatible transmitting and receiving path.

The block diagram of a cordless telephone is shown in Figs 13.23 and 13.24.

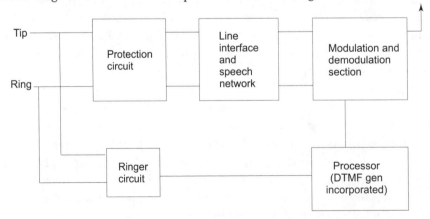

Fig. 13.23 *Block diagram of the Base Unit of the Cordless telephone*

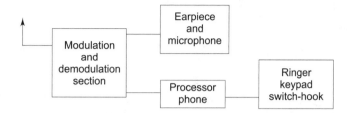

Fig. 13.24 *Block diagram of the handset unit of the cordless telephone*

13.4.2.1 Base Unit

The functions of the various blocks are detailed below:

- *Protection Circuit:* It provides protection from over-voltages resulting from lightning, switching transients, etc. It also provides polarity as well as short circuit protection.
- *Line Interface and Speech Network:* This network used to ensure 2–4 wires conversion and to allow full duplex (both way communication) operation. It also provides impedance matching and line balancing to allow sidetone and to prevent howling.
- *Modulation and Demodulation Section:* The function of this section is to transform the base band signal to RF signal for transmission purposes and vice versa.
- *Ringer Circuit:* This amplifies the incoming ring signal from the phone line.
- *Processor Section:* This co-ordinates the data transmission in the telephone and also acts as a DTMF (dual tone multi frequency) generator.

13.4.2.2 Handset Unit

- *Modulation and Demodulation Section:* The function of this section is to transform the base band signal to RF signal for transmission purposes and vice versa.
- *Processor Section:* This co-ordinates the data transmission in the telephone and also acts as a DTMF generator.
- *Earpiece:* This is the receiver part of the telephone.
- *Microphone:* This is the transmitter part of the telephone.
- *Ringer:* It produces the output signal from the ringer amplifier.
- *Keypad:* It enables tone dialling.
- *Switch Hook:* It establishes the connection between the telephone circuit and the telephone line.

13.4.3 Detailed Functional Block Diagram

13.4.3.1 Ringing of Cordless Telephone (Incoming Call)

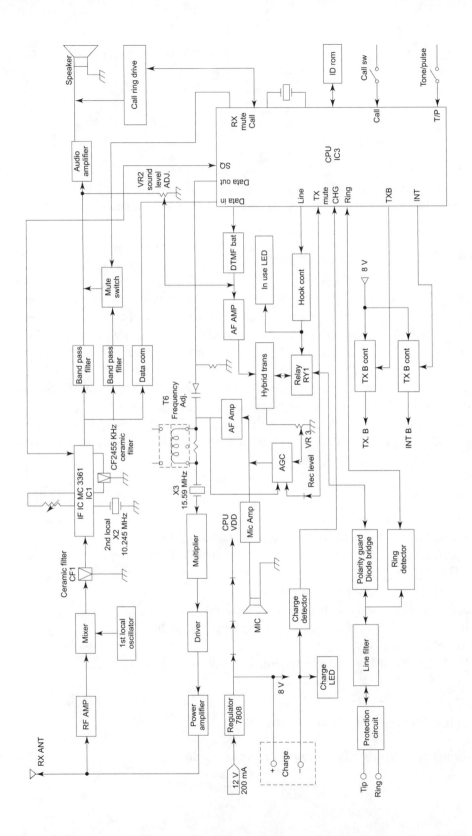

Fig. 13.25 *Functional block diagram of the base unit*

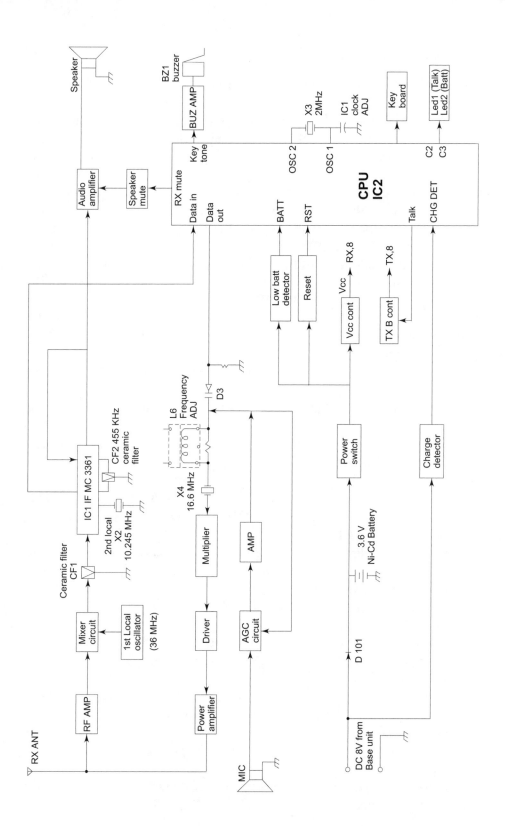

▲ Fig. 13.26   *Functional block diagram of the handset unit*

Referring to Fig. 13.25 of the base unit, the following are the major functional blocks in a cordless telephone:

- *Tip and Ring:* Telephone line from the exchange at –48V and ground is connected to Tip and Ring (T & R) of base unit, irrespective of polarity.
- *Protection Unit:* To protect the base unit from voltage surge coming through telephone line.
- *Line Filter:* To filter high frequency AC noise. It offers high impedance (proportional to frequency) to noise and blocks the same.
- *Polarity Guard:* Consists of diodes in bridge configuration to protect the unit from polarity reversal at Tip and Ring.
- *Ring Detection:* In idle mode relay (RY1), it is open and the signal goes to the ring detector. It detects the ring coming through the telephone line and gives output in terms of high and low signal to the CPU (IC 3).
- *CPU (IC3):* CPU receives the output from the ring detector and compares the received data from the pre-programmed data stored in ROM (IC 4). If the match is found, CPU calls are sub-routine.
  - *Sub-routine 1:* The power supply of the transmitter section is switched on (in idle mode transmission power is off).
  - *Sub-routine 2:* The ring digital signal is available at the data output point of CPU.
- *Frequency Modulation of Signal (Varactor diode-D1 & Crystal X3):* Crystal oscillation frequency varies according to variation in voltage across varactor diode and thus varactor diode and crystal work as a frequency modulator.
- *Multiplier Unit:* The next stage is multiplication. In this stage, three times multiplication of the crystal frequency (15.59 MHz) is done. The output of this stage is 15.59 MHz $\times$ 3 = 46.77 MHz (say $\cong$46.7).
- *Driver and Power Amplification:* The output of the multiplier stage is delivered to the power amplifier stage through a driver circuit and transmitted through antenna to the handset unit at 49.8MHz.

    Referring to Fig. 13.26 of the handset unit, the following sequence of operations takes place in the handset unit:

- *RF Amplifier:* The signals transmitted from the base is picked up by the handset antenna. The signal is tuned to a pre-determined frequency 46.7 MHz and is amplified by the RF amplifier.
- *Mixer and Filter:* The output of the RF amplifier (46.7 MHz) and local oscillator frequency, i.e. 36 MHz is fed to the mixer circuit to produce the sum (46.7 + 36 = 82.7MHz) and difference frequencies (46.7 – 36 = 10.7 MHz). Sum frequency is grounded and difference frequency, i.e. 10.7 MHz is passed by the filter circuit (CF1).
- *Second Stage Mixer and Filter (First Stage of IC 3361 & CF2):* The output of the previous

stage, i.e.10.7 MHz and second local oscillator frequency, i.e. 10.245 MHz from crystal X2 is fed to the first stage of IC 3361 which is mixed to produce sum (10.7 + 10.245 = 20.945 MHz) and difference frequencies (10.7 – 10.245 = 0.455 MHz). The sum frequency is grounded and the difference frequency, i.e. 0.455 MHz (455 kHz) is passed by filter CF2 and fed to IC 3361 again.

- In the next stage of IC 3361, audio signal and data are separated. If it is audio, it goes to the speaker amplifier and then to speaker. If it is data, it goes to data-in terminal of handset CPU (IC2).
- CPU compares received data with pre-programmed data and calls the 'sub-routines'.

  Sub-routine: The electrical signal is available to the buzzer amplifier and the buzzer rings. In this way the handset receives the ring coming through the telephone line.

- After receiving the ring, the talk key is pressed on the handset keyboard. The handset CPU (IC 2) receives the signal, compares with pre-programmed data and if match is found it calls the sub-routines.

  – *Sub-routine 1:* The power supply of the transmitter section is switched on (in idle mode the transmission power is off).
  – *Sub-routine 2:* The talk digital signal is available at data out point of the handset CPU for transmission to base and the speech path is ready in the handset.

- *Frequency Modulation of Signal (Varactor Diode-D3 and Crystal X4):* The crystal oscillation frequency varies according to variation in the voltage across the varactor diode and thus the varactor diode and crystal work as a frequency modulator.
- *Multiplier Unit:* The next stage is multiplication. In this stage, three times multiplication is done. The output of this stage is 16.6 MHz × 3 = 49.8 MHz.
- *Driver and Power Amplification:* The output of the multiplier stage is delivered to the power amplifier stage through a driver circuit and transmitted through antenna for the base unit at 49.8 MHz.

13.4.3.2 Receiving of Talk Request at Base Unit

During the time of receiving of the talk request, the circuit performs the following functions as per Fig. 13.25:

- *RF Amplifier:* The signal transmitted from the handset is picked up by the base antenna. The signal is tuned to a pre-determined frequency 49.8 MHz and is amplified by this amplifier.
- *Mixer and Filter:* The output of RF amplifier (49.8 MHz) and local oscillator frequency i.e. 39.1 MHz is fed to the mixer circuit to produce the sum (49.8 + 39.1 = 88.9 MHz) and difference frequencies (49.8 – 39.1 = 10.7 MHz). The sum frequency is grounded and the difference frequency, i.e. 10.7 MHz is passed by filter CF1.

- *Second Stage Mixer and Filter (First stage of IC 3361 and CF2):* The output of the previous stage, i.e.10.7 MHz and the second local oscillator frequency, i.e. 10.245 MHz from crystal X2 is fed to the first stage of IC 3361 which is mixed to produce the sum $(10.7 + 10.245 = 20.945$ MHz$)$ and difference frequencies $(10.7 - 10.245 = 0.455$ MHz$)$. The sum frequency is grounded and the difference frequency, i.e. 0.455 MHz (455 kHz) is passed by filter CF2 and fed to IC 3361 again.
- In next stage of IC 3361, the audio signal and data are separated. If it is audio, it goes to speaker amplifier through a band pass filter and then to the speaker and if it is data, it goes to the data-in terminal of the base CPU (IC3).

  In this case (talk request), it is data and it goes to the data-in terminal of the CPU.
- The CPU compares the received data with pre-programmed data stored in ID-ROM. If the match is found, it calls the sub-routines.

  Sub-routine: The telephone line is made available by switching on a relay.

  In this way, the telephone line is made available for communication or dialling.

13.4.3.3 Speech Path during Conversation

When the talk key is pressed from the keyboard of the handset, the CPU makes the speech path available for conversation or dialling. The microphone and speaker (4 wires) are connected to the 2-wire telephone line (tip and ring) through the hybrid transformer , which does the 2/4 and 4/2 wire conversion. In a hybrid transformer, the primary winding is connected to the telephone line (tip and ring) and secondary windings are half-split and connected to the microphone and speaker. It also facilitates isolation between the telephone line and the following circuit.

Speech Path from Telephone Line to Speaker of Handset: The signal coming on the telephone line follows the path of the protection unit, line filter, polarity guard, relay hybrid transformer, AGC, AF amplifier, frequency modulator, multiplier, driver and power amplifier, and then transmitted through the antenna of the base unit.

The signal transmitted from the base is received by the handset antenna and follows the path of the RF amplifier, mixer, ceramic filter, demodulator IC 3361, speaker amplifier and then delivered to the speaker.

Speech Path from Microphone of Handset to Telephone Line: The signal from the microphone of the handset follows the path of Automatic Gain Control (AGC), amplifier, frequency modulator, multiplier, driver, power amplifier and is then transmitted through the handset antenna.

The signal transmitted from the handset is received by the base antenna and follows the path of the RF amplifier, mixer, ceramic filter, demodulator IC, band-pass filter, variable resistor-VR2, AF amplifier, hybrid transformer, relay, polarity guard, line filter, protection unit and then delivered on the telephone line.

The heart of the controller section of the base unit is microprocessor IC3. The software is stored in an in-built ROM in the processor itself. EPROM contains ID codes and the operation codes. Any code received from the remote is compared with those stored in the EPROM for its validity by the processor. If the codes received match with those stored in the EPROM, then the microprocessor provides the supply to the transmitter section and the in-use LED of the base unit glows.

13.4.3.4 *Inter-communication Function Remote Unit*

Whenever the intercom switch of the remote is pressed, the CPU generates the data pulses that are taken from IC3 and fed to the base after modulation. At the same time, the CPU gives the output for the buzzer. The output of the buzzer is amplified and the buzzer sound will be heard.

The intercom data received by the base is demodulated and fed to IC3 for detection. This detector data is amplified by the transmitter and is heard in the loudspeaker.

13.4.3.5 *Incoming Call Procedures*

When a 20Hz ringing signal is received by the base unit from the telephone line, a 20Hz ring detector turns on the base-unit transmitter and a ring signal to the transmitter.

The receiver section of the handset unit has power, if the handset unit is in either the talk or standby mode, as long as the power is turned on and the batteries are charged. The incoming signal is demodulated by the receiver and fed to the ring signal detector. The ring signal detector is a filter that only passes a ring signal of a certain frequency. If the ring frequency transmitted by the base unit is correct (passed by the filter), the ring signal is passed to an audio amplifier. There the signal is amplified and fed to the speaker.

When someone answers the call, the handset unit is switched from the standby mode to the talk mode. That disconnects the ring signal from the amplifier and turns on the RF transmitter, pilot signal generator, and both the audio amplifier and gates. The pilot signal (guard tone) and audio (from the microphone) are fed to the transmitter where they modulate a 49-MHz band carrier. That modulated signal is then re-transmitted over the telescopic antenna to the base unit.

13.4.3.6 *Battery Charging*

When the handset is placed on the base unit, the Ni–Cd battery is charged through contact PINS. The charging of the battery is detected by the transistor and is indicated by the charge LED.

13.4.4 Troubleshooting of Cordless Telephones

Cordless telephones are fairly complex pieces of electronic equipment that, apart from containing all the circuits of a standard telephone, also include two FM transmitters and receivers. If you have ever done any troubleshooting of communication equipment, you should then have little trouble servicing cordless telephones. The radio portion is pretty straightforward, once you understand the frequency scheme and the modulation technique used.

For troubleshooting, a knowledge of the integrated circuit chips used in the cordless telephone is a must.

The datasheets of the ICs used in the instrument should be available. In case of any possible fault in a cordless telephone (one could be that there is a lot of noise in the communication channel and a humming sound is heard), conduct a visual inspection of the telephone cards to look out for any visible loose connection.

After it has been ensured that there is no loose connection, the next thing would be to check whether all integrated circuits are getting the requisite power supply.

If any one of them is not getting the supply, this could be the breeding point of the trouble.

If both these checks don't prove helpful, look for the expected outputs of different sections of the base unit and the handset and isolate the section which is not providing the expected output. The transistor or chip in the section could be faulty. Replacement of the same shall rectify the problem.

Don't try to tune the coils as the attempt at tuning could possibly create an irreparable damage, in which case an authorized dealer for the telephone will have to be contacted as these coils are not readily available in the retail market.

If it is found that the system needs repair, get everything including the base unit, handset, AC adaptor, charges unit, etc. together and follow the given procedure.

Fault: No Communication between Base and Remote Unit: Check if the installation procedure has been correctly followed and the base and remote antenna are fully extended. Check whether the power-on indication LED on the base unit is glowing. If not, then check whether the DC voltage coming from the adaptor is OK. This can localize the fault to the adaptor. If the adaptor is giving correct voltage, check if the remote handset's battery switch is set to ON and the batteries are charged. If the batteries are charged, check for transmission from the remote by tracing through the RF sending circuit. If the remote is sending, then check for 'receive by' the base. The circuit description of the remote and base will help to localize the fault to a specific component or assembly.

Fault: Battery Power of Remote is Running Out Very Fast: Check if the contacts on the charging unit and the remote unit are clean. If OK, check the charging voltage of the base unit. If the voltage is OK, then probably the battery life is nearing its end. Replace with a new battery in the handset.

Fault: Memory Dialling between Base and Remote Unit (Handset): Check if your memory is not registered or erased. Re-enter the memory.

Fault: Ring not Coming on Remote Unit: Check the ring voltage across the bridge circuit. If it is available, check the ring signal at the ring detector. If it is not available, the opto-coupler is probably damaged.

A detailed fault diagnosis tree for troubleshooting of cordless telephones is generally provided in the sesvice manual which may be consulted before attempting any major repair work.

13.4.5 Maintenance of Cordless Telephones

Certain precautions need to be taken to ensure a trouble-free service from the cordless telephone:

1. It should be ensured that while the telephone is not being used, it should be hooked up for charging the battery. This will, in turn, ensure longer battery life.
2. The switch buttons should be gently handed to avoid possible wear and tear to avoid loose contacts.
3. If even after proper charging, the operation does not last for the specified hours, it is time to replace the batteries.
4. Care is the keyword for maintenance.

In case of malfunctioning of the instrument a systematic and logical approach is necessary to localize and rectify the fault.

Periodic Maintenance: Do not let dust accumulate on cordless phones. The cleaning operation should ideally be carried out daily. If the instrument is not to be used for longer periods of time then it is better to keep it off and preferably in packed condition. In such a case before restarting regular use, the handset batteries should be charged for 12–15 hours.

Avoid using the cordless phone in a humid place. Remove fungus, grease with a dampened cloth. Remove dust and dirt from plugs or other receptacles so as to have perfect connections which is very important for trouble-free operation. It is a good practice to charge the handset batteries at regular intervals or to keep the handset on charger unit when the equipment is not in use.

The cable connecting the instrument with service line and the coil cord should be checked for visual signs of damage like twisting, cutting, etc.

14

Preventive Maintenance

14.1 PREVENTIVE MAINTENANCE

Preventive maintenance is an organized activity designed to prevent the wear and tear or sudden failure of equipment components. The mechanical, process, control or any other type of equipment failure can have adverse effects in both human and economic terms. In addition to down-time and the costs involved to repair and/or replace the equipment parts or components, there is the risk of injury to operators, and of acute exposure to chemical and/or physical agents. Preventive maintenance is, therefore, very important and ongoing accident prevention activity in workplaces should be integrated into the operations/product manufacturing process.

Preventive maintenance of equipment should therefore be carried out to:
(a) increase the system reliability by decreasing the chance for failure;
(b) provide better functional performance; and
(c) ensure a presentable looking piece of equipment at all times.

Preventive maintenance involves a policy of replacement of components of a system before the component actually fails. This is an anticipatory action and often demands the reliable prediction of wearing out components. In some cases, where the components are subjected to continuous wear, it is possible to do so, for example, in the case of rollers in a paper drive in chart recorders, servo potentiometers, motors, filament lamps and contacts on relays. By taking timely action, the reliability of a system can be greatly improved.

The frequency of maintenance is determined by the severity of the environment to which the equipment is subjected during its operation. Usually, a convenient time to perform preventive maintenance is preceding electrical adjustments of the equipment if necessary.

Preventive maintenance routines include the following steps:

(a) Inspection;

(b) Servicing;

(c) Repairing and replacing defective parts; and

(d) Validating and checking.

Basically, preventive maintenance involves the planned replacement of components designed around the following information:

- Reliability of components—Equipment failure is visually caused by its least reliable component. Check the manufacturer's information.
- Maintaining equipment service records.
- Scheduling replacement of components at the end of their useful service life.
- Acquiring and maintaining inventories of:
  - Least reliable components;
  - Critical components; and
  - Components scheduled for replacements.
- Replacing service-prone equipment with more reliable performers.

Introducing the element of planning into your maintenance function is likely to reduce the need for repair and manpower requirements.

14.2 INDICATIONS FOR PREVENTIVE MAINTENANCE ACTION

Besides the scheduled maintenance routines, many types of devices are used to provide indicators to help maintenance technicians identify or pre-empt the problems that may be developing. Some of these are:

(a) Visual indicators–, such as lights, fuses and switches;

(b) Meters and gauges;

(c) Audio indicators such as noisy motors and gears, squeaky belts, etc.; and

(d) Corrosion and dirt.

The tasks in these areas are related to both mechanical and electrical parts of the electronic equipment. The technician should, therefore, be conversant with the working of the mechanical system commonly employed in electronic equipment. It is important to know and be able to identify information regarding the removal and replacement of mechanical components from the technical literature. Such information is usually provided in the equipment service manual.

Visual inspection usually leads to the detection of such defects as broken connections, improperly seated semiconductors, damaged circuit boards and heat damaged parts. The corrective procedure for most visible defects is simple and can be immediately attempted. However, special attention should be paid to heat damaged parts. Over-heating usually indicates other trouble in the equipment. Therefore, it is important that the cause of over-heating be corrected to prevent recurrence of the damage.

Periodic checks of semiconductor devices are not recommended. The best check of performance of such devices is actual operation in the equipment.

In order to ensure accurate measurements, with some of the test and measuring electronic equipment, it is often necessary to check the electrical adjustments after each 1,000 hours of operation or every six months, if used infrequently. Also, the replacement of components may sometimes necessitate adjustment of the affected circuits. The service manuals of the equipment usually give information on performance checks and adjustments which can be helpful in localizing certain troubles in the equipment and in correcting them.

14.3 PREVENTIVE MAINTENANCE OF ELECTRONIC CIRCUITS

Dirt and dust are great enemies of electronic circuits. The accumulation of dust on printed circuit boards consisting of integrated and other solid state devices results in malfunctioning of the circuits in a number of ways.

Dirt can cause over-heating and component breakdown. It acts as an insulating blanket and results in inefficient heat dissipation. It also provides an electrical conduction path which may result in equipment failure. Therefore, the equipment must be cleaned periodically both on the exterior as well as in the interior.

Loose dust accumulated on the outside of the equipment can be removed with a soft cloth or small brush. The use of a brush should be preferred for dislodging dirt on and around the front panel controls. Dirt which remains can be removed with a soft cloth dampened in a mild detergent and water. It should be ensured that abrasive cleaners are not used. When cleaning the front panel of an oscilloscope, the CRT (cathode ray tube) face plate should be cleaned with a soft, lint-free cloth dampened with denatured alcohol.

The need for cleaning the interior of the equipment arises only occasionally. The suggested method to clean the interior is to blow off the accumulated dust with dry low velocity air (approximately 5 lb/in^2). Remove any dirt which remains with a soft brush or a cloth dampened with a mild detergent and water solution. Use a cotton-tipped applicator for cleaning in narrow spaces or for cleaning more delicate circuit components.

When cleaning the interior of an equipment, the high voltage circuits should receive special attention. Excessive dirt in this area can often cause high-voltage arcing and result in improper and unreliable equipment operation.

The use of chemical cleaning agents which might damage the plastics used in the equipment should invariably be avoided. Use a non-residue type of cleaner, preferably isopropyl alcohol or total denatured ethyl alcohol. Avoid chemicals which contain benzene, toluene, xylene, acetone or similar solvents.

The following two precautions are essential while cleaning the equipment:

1. In order to avoid electric shock, power must be disconnected from the equipment before attempting to remove the cabinet panels and operating the equipment.
2. After completing the cleaning operation, ensure that circuit boards and components must be dry before applying power to the equipment to prevent damage from electrical arcing.

14.4 PREVENTIVE MAINTENANCE OF MECHANICAL SYSTEMS

Preventive maintenance in mechanical systems is much more important than those for electronic systems. This is because the operation of mechanical systems often comes to a standstill due to failure of components requiring regular inspection and replacement. For example, a cracked or worn out belt on a motor drive will cause equipment failure and results in unscheduled downtime. Therefore, it is essential to periodically inspect mechanical system visually, clean them properly and lubricate as recommended.

The visual appearance of the equipment often indicates what type of maintenance is performed. If the equipment is clean and spotless, it is probably reasonable to assume that it has been well-maintained and its operational capability is good. On the other hand, if the equipment looks dirty, it is evident that maintenance work has been crudely performed, and the system failure is inevitable.

The reliability of potentiometers, switches and other moving parts can be maintained if they are kept properly lubricated. However, over-lubrication is as detrimental as too little lubrication.

However, in many cases, it is either difficult to accurately predict the time at which a particular component would enter the wearing out stage or it may be uneconomical to carry out preventive maintenance.

The following activities are involved in preventive maintenance of mechanical components/systems:

- Cleaning external surfaces to remove corrosion, dust, dirt and other deposits;
- Cleaning internal components such as blowers, filters, fans, coils, heat exchangers and debris, etc.;
- Replacing batteries prophylactically;
- Lubricating motors, gears, bearings and casters;
- Replacing motor brushes;
- Aligning and tightening external control knobs, switches and indicators;
- Tightening fasteners and mounts; and
- Replacing cracked/damaged tubing and flushing fluid lines and reservoirs.

14.5 GENERAL GUIDELINES FOR CLEANING AND LUBRICATING

Any equipment would quickly lose its factory freshness when put into use, particularly in hostile environments in an industry or hospital as it gathers dust, dirt or other materials. Besides looking ugly, it may be a contamination hazard and may adversely affect the device performance and safety, and may even shorten device life. Properly planned and instituted preventive cleaning can improve performance and safety, prolong equipment life and reduce the probability of sudden failure.

Cleaning schedules depend on the type of equipment and its use, as well as the working or storage environment. In most cases regular cleaning intervals cannot be rigidly fixed. The equipment should be cleaned when it appears dirty on periodic inspection. Cleaning becomes necessary to remedy a malfunction such as a noisy potentiometer, sticky casters or faulty switch contacts. Cleaning methods and materials vary according to the type of dirt, its location and the affected surfaces. Table 14.1 gives an equipment cleaning guide for different types of applications along with the suggested materials.

Table 14.1 ■ *Equipment Cleaning Guide*

Applications	*Suggested Materials*
Exterior Surfaces	
Housing	Vacuum cleaner, general cleaners, disinfectant
Displays	Vacuum cleaner, general cleaners, anti-static compound
CRT	Vacuum cleaner, general cleaners, anti-static compound
Magnetic tape head	Alcohol, proprietary head cleaner
Interior	
General	Vacuum cleaner, proprietary solvent
Mechanical	
Wheels, casters	Vacuum cleaner, general cleaner, alcohol, lubricant
Rubber rollers	Alcohol
Hinges, latches	General cleaner, alcohol, lubricant
Drive system	Proprietary solvent, lubricant
Electrical	
PC boards	Vacuum cleaner, general cleaner, alcohol, proprietary solvent, eraser, de-ionized water, proprietary solvent
Potentiometers	Proprietary solvent, contact cleaner
Heat Sinks	Vacuum cleaner, general cleaner, alcohol
Connectors	Alcohol, contact cleaner
Switch and relay contacts	Contact cleaner, lubricant, proprietary solvent

Most cleaning solvents emit irritating, toxic and/or flammable vapours. Adequate ventilation should thus be provided in the work area and manufacturers' cautions should strictly be observed while using these solvents.

14.6 TYPICAL EXAMPLES—PREVENTIVE MAINTENANCE OF PERSONAL COMPUTERS

Preventive maintenance has to be treated as an organized activity and accordingly preventive maintenance schedules based on manufacturers recommendations and users' experience need to be developed for each equipment on the inventory of the organization. The following example illustrates the procedure for preventive maintenance of today's most commonly used equipment, i.e. the personal computer (PC). The various physical parts of a computer are shown in Fig. 14.1.

Fig. 14.1 *Various parts of a Personal Computer*

14.6.1 Tools and Chemicals Required

- *Portable Computer Tools Kit* : Carrying case with assorted small hand tools: screw drivers (Philips, standard, torque drivers), pliers, cutters, wire stripper, IC inserter and extracting tools;
- *Vacuum Cleaner:* One that sucks and blows for dust/dirt removal;
- *Can of Compressed Air:* Can be used instead of a vacuum cleaner;
- *Brushes:* To sweep away dust that cannot be blown away;
- *Diskette Drive Cleaning Kit:* Used to clean dirt from floppy disk drive heads;
- *Static Grounding Strap:* Used to protect the PC from electro-static discharge (ESD);
- *Flashlight:* To look for dust in inconspicuous places;
- *Chemicals:* All-purpose cleaner, 3-in-1 oil and alcohol; and
- *Miscellaneous:* Paper towels, rags.

14.6.2 Maintenance Schedule

Daily (before switching on the PC)

- The computer should always be covered, when not in use. This will prevent dust build-up.
- Use a keyboard 'skin'. This prevents dust, liquids and other undesirable objects from entering the keyboard.
- Always use a surge protector for power connection. One strong spike of high voltage electricity can destroy your PC.
- Check all the connections to ensure that everything is properly secured.
- Try to have back-up for all data files created each day. This provides protection if original files are lost or damaged.

Weekly

- Run Utilities Programmes (e.g. Norton Disk Doctor and HD Optimizer) to ensure that the hard disk is running at peak performance.
- Check PC for viruses. They can hide in the hard disk drive and in the PC's memory. The following utilities and diagnostic software are recommended:
  - *Norton Utilities:* Very useful for keeping the hard disk drive operating optimally;
  - *Virus Checking Programmes* or DOS 6.0; and
  - *Checkit* : System diagnostics used to run tests, troubleshoot suspected problems with PC.

Monthly

- Clean the outside of the PC using an all-purpose cleaner or alcohol;
- Clean floppy drive heads using a disk cleaning kit to avoid build-up of metal oxide coating on the heads, otherwise, this can cause data to be corrupted on the floppy diskette, causing read/write errors; and
- Keep a back-up of the entire hard drive on a tape back-up.

Semi-annually

Inside the PC:

(i) Disconnect power cord and all other cables hooked up to the PC case. Use an appropriate screw driver to remove the screws holding the cover on the PC.
(ii) Inspect for anything out of the ordinary such as corrosion, loose connections and burnt parts.
(iii) Clean the inside of the PC, particularly the following parts:

- Blow out all dust using a vacuum cleaner or with a can of compressed air. Start with the power supply and vents, then disk drives, and finish off with the printed circuit board (PCB).

Outside the PC

- Use a general cleaner and paper towels to remove dust and any scruff marks on the cover and monitor.
- Clean the mouse:
  - Use a small Philips screw driver to remove the screws holding the mouse together.
  - Turn the mouse up and down and remove the ball.
  - Inspect the rollers inside for any dirt build-up. Use standard screw drivers to scrape a dirty roller clean.
  - Use a can of compressed air or vacuum cleaner to blow out any remaining dust.
  - Re-assemble the mouse.
- Cleaning the keyboard:
  - Use a can of compressed air or vacuum cleaner to clean the internal part of the keyboard. Blow out all dust and other foreign objects.
  - Use a rag and general cleaner to wipe off the keyboard and caps.
- Floppy Drives:
  - Cleaning of the floppy disk drive's read/write heads should be performed after the outside cleaning of the PC is completed.
  - Turn on your PC. Complete 'Boot up' sequence.
  - Insert disk drive cleaning kit diskette into the disk drive.
  - At A> or B> prompt, type *DIR A* :
    Press F12
- Blow the outside vents of the Monitor. Do not open the Monitor as extremely high voltage inside the Monitor can cause injury.

Cleaning the PC regularly is a very important procedure. It prevents the PC from over-heating. If the PC is kept in a dusty environment, comprehensive cleaning should be done every three months.

Bi-annually

The lithium 6V DC battery holds the set-up time and date configuration. The PC goes into the sleep mode when turned OFF. The battery provides enough voltage on the ROM BIOS (Read Only Memory Basic Input Output System), so that the memory won't be lost.

The battery will lose power over time. Replace the battery regularly to avoid this problem. Battery replacement is recommended every two years or earlier if the date, time and set-up configuration is lost.

To replace the battery, disconnect the leads from the motherboard. If the battery is soldered on, use a pair of cutters to cut it out. Then replace it with a plug-in type battery. Most of the PCs use the plug-in type 6V DC lithium battery.

14.6.3   Care of Software and Storing Diskettes

- Keep the original disks in their original package.
- Store back-up copies of original software disks in a proper diskette storage box. If security is important, use diskette storage boxes with a built-in lock.
- Store all diskettes in a cool dry place. Heat can melt and warp disks. On the other hand, damp air can cause disks to oxidize and corrode.
- Do not pack diskettes into tight bunches. This can cause physical distortion of disks and damage to data.

Keep diskettes away from magnets (e.g. stereo speakers). Magnets can destroy data on floppy diskettes.

15

Maintenance Management

15.1 OBJECTIVES OF MAINTENANCE MANAGEMENT

Physical wear and tear with the passage of time and the action of environmental elements is inevitable, in the case of both plants as well as equipment. The aim of maintenance is to prolong the life of equipment/systems and increase Mean Time Between Failures (MTBF). Some of the general objectives of a maintenance management programme are to:

- Ensure maximum operating time of equipment, at minimum maintenance cost;
- Provide a means of collecting data including costs, etc. for analysis and improving maintenance;
- Establish methods of evaluating work performance of technicians, useful as feedback to the maintenance manager;
- Aid in establishing safe working conditions for both the operating department and maintenance personnel; and
- Improve the skills of supervisors and technicians through training and continuing education programmes.

Good maintenance practices aim at minimizing downtime. Therefore, the need for better maintenance organization, adequate controls and effective planning and scheduling must be supported by the greater use of new technologies and a systematic approach.

15.2 MAINTENANCE POLICY

With the introduction of electronic equipment in almost all fields of activity, it is considered essential for organizations to evolve a sound policy for the maintenance of equipment to ensure continuity of service from the equipment. The goals for equipment maintenance management are to provide:

- A manageable and economical maintenance system by minimizing the amount of time required for maintenance; and
- Essential documentation required for all equipment.

Usually, reputed manufacturers of electronic equipment offer efficient and effective after-sales service, which could be classified as follows:

- *Breakdown service* which is provided on call when the equipment breaks down and fails to function satisfactorily;
- *Contract service* under which contract terms are agreed upon by the supplier of the equipment and the user for preventive as well as for corrective maintenance services.

Large establishments like defence services, telecommunication departments and hospitals cannot depend solely on the services offered by the manufacturers. More often, such services tend to be expensive and may not be available when needed during an emergency breakdown of the system. Therefore, it is necessary to set up in-house service facilities and it is only in the event of a highly complex fault that the services of the manufacturers are requested for.

In some situations, wherein the equipment is heavy and cannot be transported to the repair station, it is essential to arrange for some mobile servicing facilities. While in some cases, it is adequate to carry a tool box and a good range of electronic components to attend to corrective and preventive maintenance requirements, in other cases, one may need to transport test equipment like an oscilloscope, power supplies, pulse generator and digital multimeter, etc. Specially designed mobile servicing vans can be fitted with suitable electronic test equipment and mechanical workshop facilities to cater to such specialized requirements.

The maintenance policy applicable to a particular situation will obviously depend upon several factors. Some of these factors are:

(a) Type and complexity of the equipment;
(b) Location of the system;
(c) Operating conditions;
(d) Environmental conditions;
(e) Expected level of availability;
(f) Background of maintenance staff;
(g) Availability of spare parts;
(h) Expected calibration frequency; and
(i) Preventive maintenance requirements.

For troubleshooting electronic equipment, qualified persons must be employed. Unlike mechanical equipment, in which sometimes merely oiling and greasing are sufficient to restore the functioning of the faulty system, electronic equipment needs a thorough understanding of the theory of operation of the system and the knowledge of testing for both active and passive components before it can be handled. It must also be remembered that any attempt by unqualified persons to repair an electronic equipment can lead to greater troubles due to mishandling and would make the job of the qualified technician more difficult if it is later on referred to him.

15.3 EQUIPMENT SERVICE OPTIONS

The organizations normally have the option of establishing in-house facilities for maintenance of equipment or depend upon the suppliers/manufacturers for scheduled and breakdown maintenance.

- *In-house service* provides the most cost-effective and timely service for equipment maintenance. However, it is essential for the staff to have undergone training with the manufacturers or some independent organization. Also, service manuals technical support and availability of parts always remain as prime concerns. The frequency of repairs and the quantity of similar equipment will have a direct impact on the proficiency of the technical staff maintaining the equipment.
- *Manufacturer/supplier service* is sometimes the only option available in the case of many types of equipment, especially of the complex type. Typically, manufacturers are in the best position to provide a full range of services, preventive as well as corrective maintenance, to their equipment. Additionally, field modifications and some system upgrades can be made as a part of the manufacturer's contract.
- *Third party service* can be an alternative to the service provided by the manufacturer. This facility is usually availed of for common type of equipment, particularly consumer electronic products, for which the companies have set up large number of maintenance centres. For professional equipment, the availability of service parts, software and specialized test equipment should be addressed before handing over any equipment for third party service, because manufacturer support may be an important issue.

15.3.1 Types of Contract

- *Full Service Contract:* It includes both scheduled (preventive maintenance) and unscheduled (repair) service. This type of contract requires the least amount of involvement by in-house service personnel or operators.
- *Repair Only Contract:* This type of contract is viable when the capability exists for in-house preventive maintenance. In some critical fields like hospital equipment, a minimum of one annual preventive maintenance is recommended.

- *Preventive Maintenance Only Contract*: For highly technically complex equipment in which outside service is needed for preventive maintenance, it is preferable to enter into a preventive maintenance only contract. Such type of contracts are justified where the frequency of repair and/or costs do not justify the need for a full service contract.

15.3.2   General Contract Provisions

The following essential provisions should be incorporated in the contract with the service providers for maintenance of equipment:

   (i) *Scope of Service :* Preventive maintenance, equipment repair, engineering improvements, routine parts replacement, etc.;
  (ii) *Response Time :* Travel time, *per diem* expenses;
 (iii) *Payment terms :* Instalments;
 (iv) *Contract Term :* Effective date of commencement and termination of contract;
  (v) *Service Limitations :* Service/repair/replacement items not included in the contract;
 (vi) *Termination of Contract :* Conditions to define situations calling for termination of contract such as non-payment of agreed fees, inadequacy of maintenance services, etc.;
(vii) *Liability Insurance :* Coverage for contracted liability, products and operations;
(viii) *Warranties :* Company to provide warranty for all products or services to comply with all applicable local, state and federal laws, regulation and standards;
 (ix) *Indemnification :* The responsibility of the service provider to indemnify and hold the facility and its employees for any and all claims, loss, damage, liability or expense, that facility may suffer as a result of the acts or omissions of the contracting company or its employees;
  (x) *Documentation :* All activities covered under the contract with the periodic action taken to be well-documented in the service report;
 (xi) *Uptime Guarantee:* To be well-defined and the penalty in case of non-performance to be specified;
(xii) *Disputes :* Competent authority to decide the cases of dispute; and
(xiii) *Authorized Signatories :* To be specified by both sides.

15.4   MAINTENANCE ORGANIZATION

Any plant or facility which depends for its operations on particular equipment needs to have a proper organization for the maintenance of its assets. The maintenance departments in most of the organizations look after a variety of activities to ensure the smooth running of the facility. The following groups are generally constituted to streamline the working of the maintenance department:

(i) *Central Services Department: For maintaining*
  - Electrical sub-station and generating plants;
  - Air-conditioning;
  - Water lines, steam lines, air/gas lines;
  - Pollution monitoring and control; and
  - Fire fighting, vehicle maintenance, etc.

(ii) *Civil Engineering Department:*
  - Building maintenance;
  - New Civil Works and the modifications;
  - Sanitary maintenance; and
  - Horticulture.

(iii) *Engineering Department:*
  - Equipment inspection;
  - Equipment installation;
  - Maintenance, stores and purchase; and
  - Estimating and costing.

(iv) *Equipment Repair and Maintenance Cell:*
  - Breakdown maintenance of electrical/electronics/mechanical equipment;
  - Preventive maintenance; and
  - Meter and instrument calibration unit.

The organization of the maintenance department varies from company to company depending upon the size of the company, the nature of its activities and the availability of trained staff. It has been observed that equipment specialists combine in them all the functions like repairs, preventive maintenance and calibration. In many organizations, the functions of the engineering department and equipment repair and maintenance cell are combined in one unit, under the same supervisor.

15.4.1 Training Maintenance Personnel

The increasingly complex hardware used in industry today requires competent technical personnel to keep it running. The need for well-trained engineers and technicians has never been greater than at present. A comprehensive training programme can help to prevent equipment failures that impact productivity, worker morale and income. It should be remembered that good maintenance is good business.

Maintenance personnel today must think in a 'systems mode' to troubleshoot much of the hardware now in the field. New technologies and changing economic conditions have reshaped the way maintenance professionals view their jobs. As technology drives equipment design forward, maintenance difficulties will continue to increase. Such problems can be met only through the use of improved and sometimes dedicated test equipment and adequate technician training.

15.4.2   Planning of Spare Parts Inventory

During the course of preventive or corrective maintenance, there is often a need for procuring some components or parts for the replacement of defective ones. General purpose components are usually available from a trading agency or a shop, but special components specific to manufacturers of equipment need to be obtained from them. For this reason, it is necessary to properly specify the parts required. The following information is generally needed by the equipment manufacturers:

- Equipment Details   : Name, type, serial no., model no.; and
- Component Details   : Assembly reference and component reference, name and description.
  Some components may not have assembly parts like CRT, fuse, front panel control, etc.

When using components purchased from the local market, the physical size and shape of the components should be kept in mind so that the replacement components do not disturb the other parts of the assembly.

Information pertaining to the ordering of the parts is usually available in the equipment manual. Many manufacturers now offer supply of parts through e-mail. It makes communication easier and faster to and from anywhere in the world. The service manuals of some of the common equipment have been put on the Internet by the manufacturers. Make use of this facility if you have lost the manual.

Timely repair and maintenance of equipment depends, to a large extent, on the availability of the parts which may be required to take corrective action. It is therefore necessary to understand the importance of the various types of parts which need to be stocked as spares. The following are the broad categories of parts:

- *Wear Parts:* These are the parts which are exposed to wear and tear. They are designed to be changed and the exchange can be pre-planned. These parts generally protect other more critical parts. Examples: ball bearings, gaskets.
- *Safety Breaking Parts:* These protect other parts from excessive (mechanical) stress by disintegrating at a pre-defined force. Example: thermo-fuse.
- *General Spare Parts:* These can be universally utilized and are not designed for specific equipment. Examples: circuit breakers, switches, castors.
- *Specific Spare Parts:* They are designed for specific equipment and cannot be used independently from this equipment. They are preferably procured from the suppliers of equipment. Examples: photo-cells, photo-detectors, various types of transducers and sensors.
- *Consumable Parts:* They are designed to be consumed during utilization of the equipment and cannot be repaired. Their replacement can also be pre-planned. Examples: Brake and clutch lining, carbon brushes in DC motors.

- *Minor Parts*: They can be utilized universally and are of minor value. Examples: Nuts and bolts, fuses.

15.4.3 Assessing Spare Parts Requirement

For implementing an effective preventive maintenance programme and to achieve maximum uptime of equipment, the availability of spare parts is a pre-requisite. Complex mathematical methods have been developed to assess and set the priorities for the spare parts requirements. However, the most practical method is to study the typical breakdown behaviour of equipment and decide on the frequently required parts for its maintenance. Also, if possible, parts specific to an equipment should be procured together with the equipment itself.

A routine to help define priorities for the purchase of parts as spares is given in Fig. 15.1.

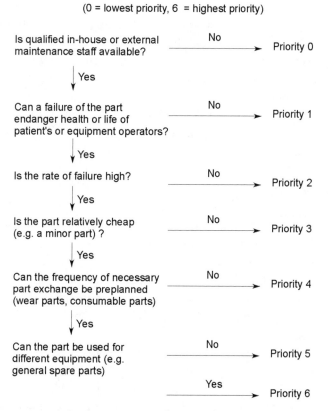

Fig. 15.1   *Prioritization procedure for purchase of spare parts*

15.5   ESSENTIALS OF A GOOD EQUIPMENT MANAGEMENT PROGRAMME

A well-organized equipment management services department has a great responsibility towards assuring that the production in a factory does not suffer or that patients in a hospital always get proper treatment. This is possible only if proper care is taken from the time of procurement of equipment till the time it is ready for condemnation. It is therefore, necessary for the maintenance personnel to be directly involved in the procurement process. This can help to ensure that the equipment purchased matches the proposed applications and in addition, can lead to a well-planned maintenance programme. The following steps are considered vital in the equipment management programme:

15.5.1   Planning for New Equipment

Depending on the need and priorities, the planning for new equipment should start when:
- Existing equipment becomes obsolete and does not meet Standards in force for improved service or new activity, etc.
- Reliability/Regulatory compliance, i.e. age, performance unreliable, parts and service are no longer available for existing equipment, or are not as per requirement of regulatory agencies; and
- Economy (excessive repair costs of existing equipment, increased load, reduced recurring cost, potential to generate more income with new equipment etc.) becomes necessary.

15.5.2   Acquisition Process

- Prepare generic specifications in consultation with the end-users; these should include both functional and technical specifications.
- Prepare a list of potential suppliers.
- Issue a notice inviting tenders (NIT) specifying provision of technical service manuals, training (operator and maintenance personnel), accessories and spare parts warranty and some earnest money to bind the bidder for honouring the bids if they are accepted.
- Preferably get bids in two parts: technical and financial/commercial. Evaluate the technical bids first followed by the financial bids. Consider only those financial bids which satisfy the laid-down technical specifications.
- Develop clear recommendations for the equipment to be procured along with recommended accessories and spare parts. Do not forget to mention the service manual in the purchase order.

- Ask for a list of utilities which need to be established and are necessary for installation of the equipment.

15.5.3 Planning of Utilities

The equipment may require utilities such as a three-phase power supply, compressed gas, water line at a certain pressure, fluid disposal arrangement, special enclosures, etc. These utilities should be well-planned and implemented in consultation with the equipment supplier well before the equipment is received, otherwise it would remain lying in the corridors packed in wooden crates, locking precious space and capital.

15.5.4 Acceptance Testing (Incoming Inspection)

A formal acceptance procedure is needed to ensure that the entry of all equipment into service is properly controlled. Acceptance testing and commissioning of equipment involve the initial electrical and mechanical tests, appropriate radiation safety tests and calibration. Checks are then carried out to verify compliance with the technical and functional performance specifications. Checks also need to be carried out to ensure compliance with appropriate standards and regulations, particularly for medical equipment and the equipment to be used in hazardous conditions. The technical manuals, spares and accessories should be checked to ensure that they are complete and functional. A formal acceptance test certificate should be prepared and signed.

15.5.5 Inventory Control

Once the equipment acceptance certificate has been signed, it should be taken on the inventory maintained in the form of an 'Asset Register'. The following information should be included in the inventory:
- Equipment name and category;
- Model no.;
- Serial no.;
- Purchase date;
- Cost;
- Maintenance arrangement: in-house, maintenance contract; and
- Location of equipment: department, section, person (user) in-charge.

It is likely that the requirements of a comprehensive inventory could be met only through the use of a computer-based system.

15.5.6   User Training

This should be carried out *in situ* in the form of live demonstrations based on the user manual. The potential errors and faults commonly found should be explained, together with the details of the appropriate corrective action. Those unfamiliar with equipment should be forbidden to operate it unless supervised or until they are considered competent in its use. There are many occasions on which equipment malfunction can be attributed to incorrect operation, with an attempt to blame the machine or non-existent equipment operational problems.

15.5.7   Technical Training

It is essential for those employed in the servicing and repair of equipment to undergo a course or training. Servicing and repair technicians should, on no account, be deployed or permitted to work on any apparatus on which they have not had adequate training.

15.5.8   Management of Service Manuals and Reference Library

The success of any maintenance programme depends, to a large extent, on the availability of service manuals of equipment, reference books, data books and other associated documents. A well-maintained library of all these would prove to be a good asset in the overall equipment management effort.

15.5.9   Maintenance Arrangement

Different types of maintenance arrangements, with respect to each equipment, should be considered. The selected arrangement should be most cost-effective and capable of providing reliable availability of equipment. In some cases, it may be possible to carry out full in-house maintenance. Alternatively, it may be more cost-effective to take out a manufacturer's comprehensive maintenance contact. This type of contract is to be monitored by local maintenance staff so as to ensure that the appropriate work is carried out by the contractor within the agreed time scale.

15.5.10   Calibration Check

For some kinds of equipment, calibration is particularly important, either periodically or after repairs. It is thus essential for the procedures and standards to be confirmed by the maintenance staff and carried out as recommended.

15.5.11 Preventive Maintenance

Preventive maintenance involves inspection and timely replacement of vulnerable components. It increases equipment reliability and reduces the likelihood of major faults. It normally results in a longer equipment life. Proper arrangement for preventive maintenance, either in-house or through companies, should be made, documented and implemented. Preventive maintenance should cover both performance and safety testing.

15.5.12 'ALERT' Issue

Maintenance departments should document all sorts of accidents, incidents and potentially harmful products, even if suspected. Such information should be circulated to all concerned in the form of 'Alert Issues' to prevent a recurrence of the incident or problem.

In addition, all such incidents should be investigated to build up a broad database for future correlation.

15.5.13 Quality Assurance

A procedure for conducting the equipment management services should be established, especially with reference to some approved standard such as ISO 9000. In general, quality assurance manuals need to define policy procedures and working instructions. All test equipment must be calibrated, with such calibration traceable to national standards. The basic concept is that the level of service to be provided should be defined and properly audited.

15.6 INSTALLATION PROCEDURES

Several factors play an important role in ensuring good uptime of equipment and less requirements for maintenance. Therefore, it is necessary to ensure that adequate attention is paid to these factors at the time of planning, procurement and installation of equipment.

15.6.1 Environmental Considerations

Most of the electronic equipment is best operated at the air-conditioning temperature (say 22–24°C). This is neither always possible nor practical to do so, especially for consumer electronic products, communication equipment, shopfloor equipment, etc. This is particularly true in tropical countries like India.

At higher temperatures, the expansion of materials and acceleration of chemical changes take place. It is estimated that a failure rate doubles with every 10°C rise in temperature. At very low temperatures, some materials may harden and transistor gains may drop. If there is rapid temperature cycling, continuous expansion and contraction may lead to accelerated failure rate.

So, carefully read manufacturer's manual to ensure that the equipment is maintained and operated at the recommended temperature.

15.6.2   Humidity

High levels of relative humidity (RH) may result in lower insulation resistance leading to insulation breakdown and corrosion of contacts, etc. On the other hand, low levels of RH may cause damage to CMOS devices due to static problems. The worst condition is that of high temperature with high humidity. It is therefore, necessary to keep the humidity levels above 50% or as per the recommendations of the manufacturer. Usually, it is possible to maintain required levels of humidity in air-conditioned rooms.

15.6.3   Altitude

At high altitudes, the density of the air is less which may result in less cooling effect, even though the temperature may be low. Consult the manufacturers manual for any appropriate information relating to the altitude or its effect on the working of the equipment.

15.6.4   Shock and Vibrations

Equipment in shopfloor may get subjected to shock and vibrations due to the working of heavy machines, presses, etc. in the vicinity. Therefore, special foundations, fixtures, damping and shock absorbers need to be arranged at the time of installation of the equipment.

Shock and vibrations experienced by the equipment during transportation may result in a loosening of the brackets, bolts, studs, etc. of heavier components such as transformers or long, thin electrolytic capacitors. Shake the equipment a little bit before installing to ensure that no loose component is rolling around inside the equipment. The other sources of shock and vibration which may result in some damage are mainly due to unbalanced rotating mechanical parts and vibration due to excessive sound pressure.

15.6.5 Protection from Electro-magnetic Interference (EMI)

Highly sensitive equipment gets affected due to electro-magnetic radiation/interference resulting from nearby electrical or electronic equipment/conductors/transmitters, etc. Equipment specifications normally include information on the level of interference which can be tolerated by the equipment. EMI specifications have also been issued by various professional agencies which must be kept in view while planning and implementing installation procedures.

15.6.6 Safety

Adopt safe working practices. Working with high voltages, machines and motors are all potentially hazardous. Great care needs to be taken in handling electrical power and hand tools during the installation procedure.

Inflammable materials, particularly liquids, should not be left exposed. Sparks often occur which may cause fire hazards. You must know the position of fire extinguishers and fire alarms that exist in the premises.

15.6.7 Electrical Installation Requirements

The mains supply in many countries including India is 230V, 50 Hz for single phase and 415 volts in three-phase systems. In US (United States)-made equipment, it is 110 volts, 60Hz with the possibility of working at 220–240V with the help of a selector switch. Check the setting of the selector switch before installing the equipment, European equipment usually has an operating voltage of 380V. In single-phase systems, the colour code followed is:

Live Wire	:	Brown (red in some cases)
Neutral Wire	:	Blue (black in some cases)
Earth Wire	:	Green (yellow green in some cases)

The planning of installation is an important step for ensuring the reliable functioning of equipment and for reducing maintenance and its associated costs. Therefore, well-installed equipment would give higher productivity and assure safe operation.

15.7 SERVICE AND MAINTENANCE LABORATORY

15.7.1 Workbench

A repair and maintenance workshop needs to be equipped with suitable working space, component storage space, record maintenance facility and requisite test equipment and tools. The most important of all these requirements is the workbench where the actual work is carried out and the technician spends a substantial amount of his time. It is, therefore, imperative that the workbench (Fig. 15.2) be designed for maximum comfort, safety and minimum fatigue.

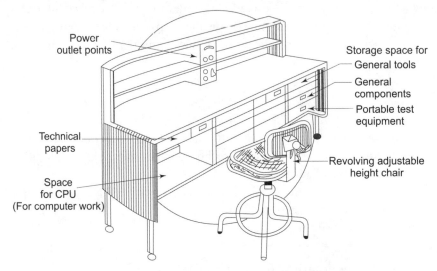

Power outlet points

Storage space for
General tools

General
components

Portable test
equipment

Technical papers

Revolving adjustable
height chair

Space
for CPU
(For computer work)

Fig. 15.2   *Design of work bench in a maintenance laboratory*

The basic requirement for an electronics workbench is that it should provide a flat, rugged and non-metallic surface on which one can work. The minimum work area should be at least 30 inches (76 cm) deep and 60 inches (152 cm) wide. A work area of that size will allow you to spread out the circuit diagrams, conveniently open the instrument and place the cover and chassis, accommodate tools and test equipment and still have some elbow room. A smaller work area is likely to lead to crowding, fatigue, impatience and hazardous situations.

The workbench should have one or more shelves at the rear side above the main work area. Such shelves allow the technicians to place frequently used test equipment like multimeter, oscilloscope, power supplies, etc. on them. In this way, these equipment will be within easy reach without permanently occupying substantial portions of limited space on the workbench. The work area surface should be covered with a ribbed rubber runner. The ribbed surface will prevent hand tools, hardware, and small components from rolling off the workbench and falling onto the floor.

The selection of a chair is equally important in ensuring the technician's comfort. It should be high enough to put one's elbows at the same level as the work area. The preferred chair is a drafting chair with adjustable seat height and back support.

15.7.2 Power for the Workbench

An AC electric power source should be available near the workbench. There should be at least two electrical wall outlets. The workbench should have bench-mounted distribution arrangement with a minimum number of six sockets and an ON/OFF master switch. This would be used to channel power to the test equipment, table lamp, soldering iron, etc. The current rating for this distributing board should be at least 15 amps which should be fitted with three conductor sockets. Some boards include either a master fuse or even individual ones for each outlet for safety.

15.7.3 Lighting

The work area should be well-illuminated and if possible, the workbench should be directly below a ceiling light fixture. Sometimes, it is desirable to supplement the light from the fixture with a lamp on the workbench. A drafting lamp with an articulated arm is quite convenient for repair and maintenance work. It has a clamp so that it can be mounted on the edge of the workbench. For working on densely packed circuit boards, a lamp with an illuminated magnifier is an invaluable aid.

15.7.4 Storage

In order to ensure efficiency in work, tools, hardware, electrical components and similar items should be stored in a systematic manner that ensures quick accessibility. Certain tools can either be kept in the workbench drawers or hung on a piece of keyboard mounted on a nearby wall.

The best method to store hardware and small electrical components is in suitably sized storage bins. The bins are made of transparent polystyrene to permit quick visual inspection of their contents. Modular storage cabinets can be stacked either vertically or horizontally on shelves above the main work area.

15.8 DOCUMENTATION

Documentation is an essential requirement in a service and maintenance laboratory. It helps in programme monitoring and obtaining data on which to base equipment maintenance decisions. In addition, documentation allows the generation of reports for various user departments and administration and can be used to demonstrate compliance with the requirements of the government and other regulatory bodies and to meet risk management requirements. It is necessary to have adequate and acceptable evidence of repair, inspections and calibrations in the event that the plant becomes involved in a liability case.

Although extensive documentation is often considered safe, yet the cost of this is high which includes the time spent on recording and filing the information as well as physical storage space for the records. Computerized records can greatly reduce the time and space required for documentation and are therefore preferred. However, it is necessary to ensure the accuracy and security of the information stored in the computer. It is also a good practice to have regular back-ups of all data on a suitable storage media.

The laboratory needs to keep a record of all incoming equipment for repairs, the details about the manpower deployed on repair and maintenance jobs, spare parts used, maintenance of spare parts inventory, test equipment and their regular calibration, costing of jobs undertaken, inspection and preventive maintenance schedules and their implementation, etc.

15.8.1   Maintenance System Overview

The service laboratory has to perform multifarious activities including inspection of incoming equipment, installation and taking it over in the inventory; carrying out repair on the defective equipment received in the laboratory and undertaking preventive maintenance activities. A proper computerized record of all these activities is required in the interest of efficient working and management of assets of an organization. Figure 15.3 gives an overview of the maintenance system.

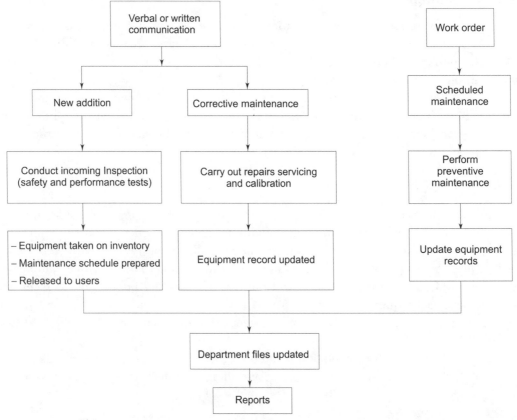

Fig. 15.3   *Maintenance system overview*

15.8.2 Sample of a Work Order for Repairs

It is a good practice to prepare a work order for any request received for repair or maintenance of a piece of equipment. This helps to understand the history of the equipment before it became non-operational, the details of the manpower deployed, the time spent on the repair work and the parts used in restoring the equipment to its functional state. Figure 15.4 gives a sample of a work order incorporating the above-mentioned requirements.

WORK ORDER

Sr. No. _____   Date _____

Request received from _____   Deptt. _____

Tele No. _____   Fax No. _____   e-mail _____

Name of the Equipment _____

Manufacturer _____

Model No. _____ Sr. No._____

Accessories Received _____

Defects Reported   _____

                         _____

                         _____

Work allotted to _____   Signature of Supervisor

Progress of Work

Name of the Engineer/ Technician	Dates		Cost Estimates
	From	To	

Parts Used

Sr. No.	Name of the Parts used	Cost

Total Cost _____

Work completed on _____

Fig. 15.4   *Sample work order for maintenance jobs*

15.8.3 Information Tags

For an efficient control and management of the equipment inventory and functional status, it is necessary to take such steps as may help in identification of the equipment, its location in the organization, information about its periodic inspection and preventive maintenance, etc. This is achieved by using various tagging devices. Some of the important tags which are used in practice are as follows:

Control Number Tags: Each device that passes an acceptance test after receipt in the organization should be permanently tagged with a control number. A sample of the Control Number is given below:

CONTROL NUMBER
00678

Inspection Tags: Inspection tags are placed on equipment following each inspection to indicate that it has been inspected and is acceptable for use. These tags are helpful in identifying equipment requiring inspection, including the due date of the next inspection.

INSPECTED BY:
DATE

Caution Tag: This tag is placed on equipment with controls that might not be checked or re-set before each use. This will warn the operator to check control settings that might have been altered during inspection. The following wording is suggested for the Control Tag (as per recommendations of the ECRI, USA):

CAUTION
This device recently underwent inspection and preventive maintenance. Please check the position and setting of controls, valves, alarms and indicators since they may have been changed from their previous positions. Please remove this tag upon first post-inspection use. By ....Date........................

'Do-Not-Use' Tag: A DEFECTIVE or DO NOT USE TAG is placed over the controls or in a prominent position on the equipment which is unsafe or ineffective. All deficiencies must be corrected and the equipment re-inspected before it is put to use.

DANGER

DEFECTIVE

DO NOT USE

DO NOT REMOVE
THIS LABEL

Date...

Name ....

15.9 PROFESSIONAL QUALITIES AND WORK HABITS

15.9.1 General Skills

Effecting repairs to an electronic equipment is a challenge to your professional experience, technical knowledge and mental discipline. The situation in which the equipment is sometimes to be serviced could be quite demanding. It is only with experience on the job that you may begin to appreciate the state of mind needed to undertake a repair and servicing job. An accomplished technician is a cool and rational worker and follows a systematic routine to attack a problem.

It may, however, be mentioned that repair work can sometimes be very frustrating, as not everything can be repaired. Sometimes, the basic design is flawed or someone before you has messed up the whole thing. Also, after repairs, the solution could be elusive to the extent that it drives one mad. It is also possible that you may find a solution to a particular problem, only to have the problem appearing again after some time. A sustained effort must decide that you can repair the equipment to a perfect level. Experience will be your most useful companion. Therefore, you need to develop a general troubleshooting approach—a logical, systematic method of narrowing down the problem to the ultimate solution.

15.9.2 Work Habits

Work habits decide, to a great extent, your efficiency and quality of work in repair and servicing. Some of the useful tips are:

(a) Always start with a clean area. Remove all pieces of paper, solder, wire bits, tools from the place where the equipment for repairs is to be placed.

(b) The equipment must be turned off when removing or replacing parts, plugging and unplugging connectors. The equipment should be turned on only when measurements are to be made.

(c) Use minimum force while pulling out circuit boards from the connectors. Use special circuit board pullers if available for this purpose.

(d) While opening an equipment, be careful in collecting all screws and washers in small plastic boxes. You could easily lose them if they are not kept properly. Any screw which may accidentally fall in the equipment during opening must be retrieved otherwise it is likely to cause a short circuit.

(e) Learn from your mistakes. A simple problem can turn into an expensive one due to a slip of the probe or of being over-eager to try something before thinking it through. Make it a point not to make the same mistake again, for that is what experience means.

15.9.3 Personal Safety

It is important to observe some essential precautions for your personal safety. These do not cost anything, but can save you from injury or even a fatal situation, if properly observed. Depending on the type of equipment, you will be working on, there can be variety of dangers, some of them potentially lethal. Ensure that you follow these safety precautions:

• The main danger in servicing electronic equipment is the hazard of electric shock. Even though most of the modern equipment is solid state and works on low voltage, there are pieces of equipment which have solenoids and motors operating at higher voltages. Also, the real danger is from the power supply part of the equipment where the mains supply of 220 volts will usually be present. It can be injurious or even fatal if your body becomes a part of the electric circuit. The amount of current flow through the body depends upon the body's electrical resistance which is unfortunately quite low. Therefore, you must be careful about where you sit, stand and what you wear when attempting repair and servicing of electronic equipment.

Make it a habit to use a pair of dry sneakers with rubber soles, keep your feet separated from ground by a thick rubber insulating mat and work with insulated hand tools. Always keep one hand in your pocket when anywhere around a powered line–connected or high voltage system. Even if you take all these precautions you cannot predict the effects of a shock when you touch a hot AC line. So, never take a chance when working with mains-operated equipment.

- While working on the equipment, always keep your fingers away from the moving parts and mechanisms. If it is absolutely necessary to work upon a mechanism while it is operating, use a tool rather than your fingers.

  Don't wear any jewellery or other articles that could accidentally contact some mechanism and get caught in moving parts or circuitry and conduct current.
- The risk of CRT implosion also exists from equipment using large CRTs.
- Be careful of vision hazards from the lasers in CD players and CD-ROM drives, DVD players and DVD-ROM drives, optical data storage devices and laser disk players.
- Do not work alone. In the event of an emergency, another person's presence may be helpful and essential.
- Set up your work area away from possible points that you may accidentally contact.
- Electric shock hazard is particularly present when servicing TV, computer and other video monitors, microwave ovens, switch mode power supplies, electronic flash units and power tools.
- It is advisable that you know your equipment thoroughly as TVs and Monitors may use parts of the metal chassis as ground return yet the chassis may be electrically live with respect to the earth ground of the AC line. Microwave ovens use the chassis as ground return for the high voltage. In addition, do not assume that the chassis is a suitable ground for the test equipment.
- Perform as many tests as possible with power off and equipment unplugged. For example, the power supply section of a TV or Monitor can be tested for short circuits with an ohmmeter.
- Never assume anything without checking it out for yourself. Don't take short-cuts. It is, therefore, imperative that safety precautions are observed while working inside any equipment.
- Don't attempt repair work when you are tired. Not only will you end up being careless but your primary diagnostic tool–deductive reasoning–will not be functioning at full capacity.

15.9.4 Smoking in the Repair Workshop

Precision electronic equipment gets affected by the chemical compounds present in the tobacco smoke. The effects of smoke coating have been observed on the following equipment:

- Tape path of VCRs and audio decks including the audio, video and control heads and cassettes and tapes inside;
- Read/write heads of floppy disk drives, zip drives and the media they use;

- Precision optics of CD and DVD players, CD-ROM and DVD-ROM drives and other optical disk equipment and the media they use;
- Mechanical parts; and
- Screens of TVs and monitors, display windows of VCRs and other devices and the outside and inside of everything resulting in an ugly look.

In addition, tobacco smoke promotes loss of lubrication in equipment and may also contribute to a deterioration of plastic and rubber parts. Contamination will often find its way into critical places that are not accessible and to the media, which is irreplaceable. Therefore, smoking in the repair workshop should be strongly discouraged.

Bibliography

1. Bosshart, W.C. (1983), *Printed Circuit Boards: Design and Technology,* Tata McGraw-Hill Publishing Company, New Delhi. ISBN 0-07-451549-7.
2. Clark, R.H. (1987), *Handbook of Printed Circuit Manufacturing,* CBS Publishers & Distributors, New Delhi.
3. Coombs (Jr), C.F. (1988) *Printed Circuits Handbook,* McGraw-Hill Book Company, U.S.A., ISBN 0-07-012609-7.
4. Cooper, W.D. (1984), *Electronic Instrumentation and Measurement Techniques,* Prentice-Hall of India Private Ltd., New Delhi. ISBN 0-87692-060-1.
5. Datta-Barua, L. (1988), *Digital Computer: Maintenance and Repair,* Tata McGraw-Hill Publishing Company Ltd., New Delhi.
6. Floyd, Thomas (1986), *Digital Fundamentals,* 3rd ed., Charles-E Merril Publishing Co., Columbus. ISBN–0675-20517-4.
7. Gangopadhyay (1980), *Production Technology-HMT,* Tata McGraw-Hill Publishing Co. Ltd., New Delhi. ISBN 0-07-096443-2.
8. Gottlieb, I.M. (1985), *Power Supplies; Switching Regulators, Invertors and Convertors,* PBP Publications, Delhi.
9. Govindarajalu, B. (1991), *IBM PC and Clones: Hardware, Troubleshooting and Maintenance,* Tata McGraw-Hill Publishing Co. Ltd., New Delhi.
10. Gumhalter, H. (1985), *Power Supply Systems in Communication Engg.,* Wiley Eastern Limited, Delhi. ISBN 81-224-0099-X.
11. Harper, C.A. (1977), *Handbook of Components for Electronics,* McGraw-Hill Co., USA, ISBN 0-07-026682-4.
12. Haskard, M.R. (1997), *Electronic Circuit Cards and Surface Mount Technology,* Technical Reference Publications Ltd., U.K. ISBN 1-872422-05-5.
13. Herrmann, G. and Egerer, K. (1992), *Handbook of Printed Circuit Technology,* Electrochemical Publications Limited, U.K. ISBN 3-87480-056-3.
14. Khandpur, R.S. (1996), *Handbook of Analytical Instruments,* Tata McGraw-Hill Publishing Co. Ltd., New Delhi. ISBN 0-07-460186-5.
15. Khandpur, R.S. (1987), *Handbook of Bio-medical Instrumentation,* Tata McGraw-Hill Publishing Co. Ltd., New Delhi.
16. Loveday, G (1989), *Electronic Testing and Fault Diagnosis,* Longman (Scientific and Technical), England. ISBN –582-03865-0
17. Lenk, J.D. (1982), *Handbook of Electronic Test Procedures,* Prentice-Hall, England. ISBN 0-13-377457-0.
18. Leonida, G. (1981), *Handbook of Printed Circuit Design, Manufacture,* Components & Assembly, Electrochemical Publications Ltd. U.K. ISBN 0-901150-09-6.

19. Malvino, A.P. (1989), *Electronic Principles,* 4th ed. McGraw-Hill, U.S.A. ISBN 0-07-099479-X.

20. Patton, J.B. (1983), *Preventive Maintenance, Instrument Society of America,* Prentice-Hall, Inc., USA. ISBN 0-13-699325-3.

21. Robinson, V (1974), *Handbook of Electronic Instrumentation Testing and Troubleshooting,* Reston Publishing Co., U.S.A. ISBN 0-87909-327-7.

22. Scarlett, J.A. (1984), *An Introduction to Printed Circuit Board Technology,* Electrochemical Publications Limited. U.K. ISBN 0-901150-16-9.

23. Sharma, D.D. (2000), *Total Quality Management,* Sultan Chand & Sons, New Delhi. ISBN 81-7014-623-2.

24. Wassink, R.J.K. (1984), *Soldering in Electronics,* Electrochemical Publishers Limited, Britain. ISBN 0-901150-142.

Index